Agnes Mainka
Smart World Cities in the 21st Century

Knowledge & Information

―――
Studies in Information Science

Editor-in-chief
Wolfgang G. Stock (Düsseldorf, Germany)

Editorial Board
Ronald E. Day (Bloomington, Indiana, U.S.A.)
Richard J. Hartley (Manchester, U.K.)
Robert M. Hayes (Los Angeles, California, U.S.A.)
Peter Ingwersen (Copenhagen, Denmark)
Michel J. Menou (Les Rosiers sur Loire, France)
Stefano Mizzaro (Udine, Italy)
Christian Schlögl (Graz, Austria)
Sirje Virkus (Tallinn, Estonia)

Agnes Mainka

Smart World Cities in the 21st Century

—

DE GRUYTER
SAUR

ISBN 978-3-11-070960-5
e-ISBN (PDF) 978-3-11-057766-2
e-ISBN (EPUB) 978-3-11-057532-3
ISSN 1868-842X

Library of Congress Cataloging-in-Publication Data
Names: Mainka, Agnes.
Title: Smart world cities in the 21st century / Agnes Mainka.
Description: Boston : De Gruyter, [2018] | Series: Knowledge and information
  | Includes bibliographical references and index.
Identifiers: LCCN 2018025230 (print) | LCCN 2018026375 (ebook) | ISBN
  9783110577662 (electronic Portable Document Format (pdf)) | ISBN
  9783110575255 (hardback) | ISBN 9783110575323 (e-book epub) | ISBN
  9783110577662 (e-book pdf)
Subjects: LCSH: Smart cities. | Information society. | City planning. |
  BISAC: LANGUAGE ARTS & DISCIPLINES / Library & Information Science /
  General. | SCIENCE / General.
Classification: LCC TD159.4 (ebook) | LCC TD159.4 .M35 2018 (print) | DDC
  307.760985–dc23
LC record available at https://lccn.loc.gov/2018025230

Bibliographic information published by the Deutsche Nationalbibliothek
The Deutsche Nationalbibliothek lists this publication in the Deutsche Nationalbibliografie;
detailed bibliographic data are available on the Internet at http://dnb.dnb.de.

© 2020 Walter de Gruyter GmbH, Berlin/Boston
This volume is text- and page-identical with the hardback published in 2018.
Typesetting: Integra Software Services Pvt. Ltd.
Printing and binding: CPI books GmbH, Leck

www.degruyter.com

Contents

Acknowledgement —— VII
List of Figures —— VIII
List of Tables —— XII

1	**Informational world cities** —— 1	
	References —— 5	
2	**The rise of the knowledge society** —— 8	
	References —— 13	
3	**Measuring cities of the knowledge society** —— 16	
3.1	Infrastructures —— 16	
3.1.1	ICT infrastructure —— 17	
3.1.2	Cognitive infrastructure —— 23	
3.2	Political will —— 29	
3.3	World city —— 34	
3.4	Hypothesis overview —— 43	
	References —— 44	
4	**Methods** —— 53	
4.1	Grounded theory: Combination of research methods to investigate new fields —— 54	
4.2	Literature review: Identifying prototypical cities of the knowledge society —— 55	
4.3	Interviews: A qualitative approach —— 56	
4.4	Gaining first-hand experience through field studies —— 59	
4.5	Measuring cities and nations on secondary data —— 60	
4.6	Retrieval of patents and scientific publications —— 61	
4.7	Limitations of the results —— 62	
	References —— 63	
5	**Identifying prototypical cities of the knowledge society** —— 67	
5.1	Digital and smart cities —— 70	
5.1.1	ICT infrastructure —— 71	
5.1.2	ICT networks —— 71	
5.1.3	Strategic master plan —— 73	
5.1.4	Economy and labor —— 76	
5.1.5	Smart and sustainable city —— 78	

5.1.6	Smart city applications —— 80	
5.2	Creative and knowledge cities —— 81	
5.2.1	Historic development —— 81	
5.2.2	Economic transformation —— 82	
5.2.3	Strategic master plan —— 84	
5.2.4	Face-to-face facilities —— 87	
5.2.5	Knowledge output —— 88	
5.2.6	Knowledge economy and labor market —— 89	
5.2.7	Creative milieu —— 90	
5.2.8	Knowledge city benchmarks —— 92	
5.3	Conclusion of the identification process —— 94	
	References —— 94	
6	**Case study investigation of 31 informational world cities —— 105**	
6.1	Infrastructures —— 106	
6.1.1	ICT infrastructure —— 106	
6.1.2	Cognitive infrastructure —— 131	
6.2	Political will —— 189	
6.3	World City —— 224	
	References —— 246	
7	**Conclusion —— 259**	
7.1	Infrastructures —— 259	
7.2	Political will —— 262	
7.3	World city —— 264	
7.4	The typical informational world city —— 265	
	References —— 266	

Appendix I: List of all interview partners —— 267

Appendix II: Literature Review —— 274

Appendix III: Bike sharing —— 281

Appendix IV: Best ranked university in city —— 283

Appendix V: Patents granted 2000–2012 —— 284

Appendix VI: Coworking spaces —— 286

Acknowledgement

I wish to express my sincere appreciation to those who have contributed to this book and supported me in one way or the other during this amazing journey. First of all, I am extremely grateful to my main supervisor, Professor Wolfgang G. Stock, for his guidance and all the useful discussions and brainstorming sessions, especially during the difficult conceptual development and writing stages. His guidance helped me in all the time of research and writing of this book.

My sincere gratitude is reserved for Professor Isabella Peters for her invaluable insights and suggestions. I really appreciate her willingness to discuss a draft version of my thesis. This has helped me a lot to reflect my own work. She is an inspiration. I thank her, the Sonder Forschungs Fond of the Heinrich-Heine University Düsseldorf, and the Department of Information Science for giving me the opportunity to visit all the cities and meet all the experts around the world.

Special thanks go as well to the experts who took the time for giving me an interview and inspired the work at hand. The interviews further have profited from colleagues and friends that have accompanied me during the field studies. I like to thank Christine Meschede, Sarah Hartmann, Janina Nicolic, Duwaraka Murugadas, Stefanie Vieten, Sviatlana Khveshchanka, Anika Stallmann, Lisa Orszullok, Adriana Kosior, Evelyn Dröge, and further, Katrin Weller for conducting the interviews in Melbourne. Especially I like to thank Isabella Peters and Wolfgang G. Stock who have advised me during my first field studies and interviews.

Further, I like to thank all colleagues and friends that have helped me to finalize this work by reading several draft versions or transcribing the interviews: Lisa Beutelspacher, Tobias Siebenlist, Margarethe Rütten, Christine Meschede, Julia Göretz, Tamara Heck, Kathleen Haking, Anika Stallmann, Lisa Orszullok, and Alexander Richter. And I like to thank Fabian Koglin for the final proofreading.

In particular, I like to thank Carsten Brinker for spending his holidays to accompanying me during the field studies and interviews in the US and Brazil and making these beautiful photos that I am allowed to use in this work. He is always motivating me and helping whenever it is possible for him. Thank you! Finally, I like to thank my parents and my sister who have supported me during my study and my research.

List of Figures

Figure 1.1 Infrastructures of an informational world city —— 2
Figure 1.2 The development of an informational city from the perspective of network economies —— 4
Figure 2.1 Economic cycles and the knowledge society (Own interpretation based on discussed literature) —— 12
Figure 4.1 Grounded theory cycle representing the investigation process of the work at hand —— 55
Figure 5.1 A Literature review of city research which uses at least one of the 31 informational cities as a case study or example (n=138) —— 68
Figure 5.2 Informational world cities on the world map. Cf. numbers in Table 5.1 —— 70
Figure 6.1 Quantitative interview results according to SERVQUAL (quality value = perception – expectation) for H1 (Informational world cities are hubs for companies with information market activities, e.g. telecommunication companies) —— 107
Figure 6.2 The relative size of information industries in the OECD. Percentage points of total value added and employment calculated —— 109
Figure 6.3 Amount of firms listed as Global Fortune 500 by industry "Technology"* and "Telecommunication." —— 109
Figure 6.4 Telecommunication and technology hub according to headquarters locations of companies listed in the Global Fortune 500 in 2016 within the sector telecommunication or technology —— 110
Figure 6.5 Quantitative interview results according to SERVQUAL (quality value = perception – expectation) for H2 (The ICT infrastructure in an informational world city is more important than automotive traffic infrastructure) —— 114
Figure 6.6 Correlation of the Human Development Index (HDI) and Networked Society City Index (NSCI) for the year 2014 with inverted ranking scores —— 116
Figure 6.7 Media Poles in Seoul. A u-city street project. Left: One media screen in use by Prof. Wolfgang G. Stock. Middle: Street in Gangnam Districts with 22 media poles. Right: One multimedia information totem in Milan —— 121
Figure 6.8 Public transportation modes. Examples of Dubai Metro (upper pictures) and trams in Hong Kong (bottom-left) and Barcelona (bottom-right) —— 123
Figure 6.9 From highway to greenway —— 125
Figure 6.10 Sharing services. Left: Car-sharing in Berlin. Right: Bike-sharing in Milan —— 126
Figure 6.11 Quantitative interview results according to SERVQUAL (quality value = perception – expectation) for H3 (Science parks or university clusters that cooperate with knowledge intensive companies are important in an informational world city) —— 133
Figure 6.12 Number of students as a percentage of total population per city (left axis) and number of international students as a percentage of total students (right axis). Cities are ordered in descending order according to their rank as "QS Best Student City." *Not available in "QS Best Student City" —— 137

List of Figures — IX

Figure 6.13 Correlation of university scores in the rankings CWUR, SRC, THE, and QS for the fiscal year 2015 for 90 universities —— 139

Figure 6.14 Correlation of the ranking positions within the three rankings: CWUR (x-axis), SRC (y-axis), and QS (bubble size). Universities = 29 —— 140

Figure 6.15 Quantitative interview results according to SERVQUAL (quality value = perception – expectation) for H4 (An informational world city needs to be a creative city) —— 143

Figure 6.16 Cultural amenities of informational world cities in numbers and shares. Cities ordered from left to right in ascending order according to the total sum of cultural amenities —— 147

Figure 6.17 Cultural amenities as a percentage per 100 citizens (right axis) and the total sum of cultural amenities for each city (left axis) —— 148

Figure 6.18 Top-left: Walt Disney Concert Hall in Los Angeles. Bottom-left: Sydney Opera House. Top-right: Shanghai Grand Theater. Bottom-right: Louvre in Paris. Photos: Carsten Brinker (top and bottom-left), Wolfgang G. Stock (top-right), Agnes Mainka (bottom-right) —— 149

Figure 6.19 Left: Total numbers of granted patents in 2000 and 2012. Top ten informational world cities listed in descending order according to the number of patents granted in 2012 from left to right. Data source: Derwent World Patent Index (retrieved in 2013). Right: Number of patents granted as percentage per 1,000 citizens for 31 informational world cities. San Francisco, Munich and Boston are the top three of the list —— 151

Figure 6.20 Gay neighborhoods. Left: Los Angeles – West Hollywood. Middle and Right: Toronto Church Street —— 153

Figure 6.21 Quantitative interview results according to SERVQUAL (quality value = perception – expectation) for H5 (Physical space for face-to-face interaction is important for an Informational world city) —— 157

Figure 6.22 Boston – Cambridge office buildings with glass front —— 158

Figure 6.23 Total number of coworking spaces per city and the number of spaces in relation to the population —— 161

Figure 6.24 Quantitative interview results according to SERVQUAL (quality value = perception – expectation) for H6 (A fully developed content infrastructure, e.g. supported by digital libraries, is a characteristic feature of an informational world city) —— 166

Figure 6.25 Digital library service provided by Informational world cities' public libraries —— 173

Figure 6.26 Ranking of informational world cities' public library service represented in descending order according to total scores reached in digital and physical library service —— 175

Figure 6.27 Quantitative interview results according to SERVQUAL (quality value = perception – expectation) for H7 (Libraries are important in an informational world city as a physical place for face-to-face communication and interaction) —— 177

Figure 6.28 Physical space of public libraries. Left: Stockholm Stadsbibliotek. Right: National Library of China in Beijing —— 179

Figure 6.29 Maker spaces. Top both and right: Maker space at the Chicago Public library. Photo: Carsten Brinker. Bottom left both: Toronto public library "Espresso Book Machine." —— 183

Figure 6.30 Spaces in physical libraries of 31 informational world cities —— 184

Figure 6.31 Spaces in the Amsterdam public library —— 185

Figure 6.32 Public libraries as "place makers." Top left to right: Amsterdam Openbare, Vancouver Public Library, Stockholm Stadsbibliotek. Bottom left to right: São Paulo Public Library, Paris Public Library, New York Public Library —— 186

Figure 6.33 Dubai physical public library —— 187

Figure 6.34 Quantitative interview results according to SERVQUAL (quality value = perception − expectation) for H8 (Political willingness is important to establish an informational world city, especially with regard to knowledge economy activities) —— 190

Figure 6.35 Quantitative interview results according to SERVQUAL (quality value = perception − expectation) for H9 (An informational world city is characterized by e-governance (including e-government, e-participation, e-democracy)) —— 199

Figure 6.36 The maturity of municipal e-government for the 31 informational world cities —— 202

Figure 6.37 Social media platforms used for governmental purposes in informational world cities —— 205

Figure 6.38 Quantitative interview results according to SERVQUAL (quality value = perception − expectation) for H10 (A free flow of all kinds of information (incl. mass media information) is an important characteristic of an informational world city) —— 214

Figure 6.39 Freedom of information based on the internet freedom and the world press freedom index. Nations represent cities investigated in the work at hand and listed in ascending order according to average scores of both indices, with higher scores indicating less freedom —— 219

Figure 6.40 Access to information on the national level according to the "Global Right to Information Rating" for the year 2016 and the "Open Data Barometer" for the year 2015. Nations are listed in descending order according to the scores reached in the Open Data Barometer. The higher the score, the more open data/broader right to information is available —— 221

Figure 6.41 Quantitative interview results according to SERVQUAL (quality value = perception − expectation) for H11 (An informational world city has to be a financial hub with a large number of banks and insurance companies) —— 226

Figure 6.42 Cities' performance within the Global Financial Center Index (GFCI) and the best performing university in each city according to the scores of the Center for World University Rankings. Cities presented in descending order of their GFCI score. GFCI scores are adjusted (n/10) —— 227

Figure 6.43 Combined change in employment and GDP per capita per metropolitan region for the time period 2013–2014. Metros are listed in descending order according to their GDP per capita in US$ for the fiscal year 2014 —— 230

Figure 6.44 Quantitative interview results according to SERVQUAL (quality value = perception − expectation) for H12 (An informational city has to be a global city ('world city') —— 236

Figure 6.45 Hubs of power, knowledge, and information according to the number of company headquarters, STN publications, and patents granted in a city —— **239**

Figure 6.46 Share of foreign-born population of informational world cities (Year of statistics **2011, 2015, *2016) —— **242**

Figure 6.47 Global Talent Competitiveness index 2015 on national data in comparison to the share of foreign born population based on city data —— **243**

List of Tables

Table 3.1 World city categories by King (2012, p. 32) —— 42
Table 5.1 A literature review of city research which uses at least one of the 31 informational cities as a case study or example (n = 138) —— 69
Table 5.2 Finalists of MAKCi 2007–2012 —— 93
Table 6.1 Startup rankings according to Global Startup Ecosystem Ranking 2015 (B. L. Herrmann et al., 2015) on the left and City Initiatives for Technology, Innovation and Entrepreneurship (Gibson et al., 2015) on the right —— 113
Table 6.2 Pearson correlation of the cities ranking position within the QS Best Student City and the number of students as % of total population and with the number of international students as % of total students —— 137
Table 6.3 Pearson correlation of university scores in the rankings CWUR, SRC, THE, and QS for the fiscal year 2015 for 90 universities —— 138
Table 6.4 Position of University College London in three global university rankings for the fiscal year 2015 —— 139
Table 6.5 Top ten metro areas of the Tolerance Index —— 152
Table 6.6 Indicators of the digital library —— 171
Table 6.7 Indicators of the physical library —— 184
Table 6.8 Five pillars of e-government based on the five-stage model according to Hiller and Bélanger (2001) —— 201
Table 6.9 Overview of rank, score, and category of nations ranked in the "World Press Freedom Index 2016" (Reporters without Borders, 2016) with reference to the 31 cities investigated —— 217
Table 6.10 The ten biggest stock exchanges according to the highest market capitalization value of shares in 2015 —— 229
Table 6.11 Overview of metro areas according to their venture capital investment in million US$. Only metro areas that are investigated in the present work as informational world cities and are mentioned within the top 20 venture capital cities within their geographical region are listed —— 232
Table 6.12 Top three cities and countries according to their number of headquarters of FinTech companies based on the 100 most successful global providers of financial technology —— 233
Table 6.13 Connectivity of the top five busiest international airports for the year 2008/2009,*2007 —— 240

1 Informational world cities

World cities are centers of human interaction. Urbanized areas and cities are home to more than half of the world's population, and by 2030 almost 5 billion people will live in cities or towns (UNFPA, 2007). The fact that only 135 metropolises contribute 37% of the world's total GDP clearly demonstrates that cities are important hubs of the global economy (Clark, Moonen, & Couturier, 2015). The increasing amount of people living in a dense area confronts city governments, citizens, and local business with new problems but also opportunities. What does this mean for the development of world cities and why is it so important that they are informational?

The origins of city research can be traced to Weber's (1921) sociological analysis *The City*, in which he discussed the anonymity of cities in contrast to rural areas, and the city's role as a market center in a capitalized economy. World cities in particular have been the focus of further urban research like in Hall's (1966) *The World Cities*. World cities are agglomerated areas which are not defined by their political boundaries, such as the conurbation area Randstad Holland. These cities are centers of governmental, economic, and human activity overall. According to Jacobs (1969, 1984), economic growth as well as the development of the modern world depends on the interaction between cities. As a simplification, city interaction is defined in terms of economic networks of firms. Thus, city researchers have emphasized the importance of world cities in a global network (Friedmann, 1986; Hall, 1966; Sassen, 2001; Taylor, 2004). How these networks are interlinked has increasingly been changed by the advent of information and communication technology (ICT), which has ushered in an entirely new era of economic trade, transaction, and communication. Following Castells (1989), this is the rise of the "networked society" in which spaces of flows (information, capital, and power) determine the spaces of places. He introduces "informational cities" as prototypical cities of this development.

The term "informational city" has not yet been firmly established as a common term to describe the development of future or modern cities (Stock, 2011). In the literature terms like "digital city" (Ishida, Ishiguro, & Nakanishi, 2002), "network city" (Craven & Wellman, 1973), "ubiquitous city" (Shin, 2009), or "smart city" (Hollands, 2008) refer to cities with an emerging and growing digital infrastructure based on ICT as well as the quality of life and green infrastructure (Mainka et al., 2013; Stock, 2011). "Smart city", however remains a somewhat fuzzy term (Albino, Berardi, & Dangelico, 2015; Nam & Pardo, 2011), sometimes used synonymously to informational city and sometimes as a reference to the digital networks and enhanced sustainability dominant in such cities. Furthermore, informational cities cannot be observed as an isolated phenomenon based on a highly developed digital infrastructure. Yigitcanlar (2010) and Stock (2011) define informational cities as prototypical cities of the knowledge society. The essential factor here are human beings who are able to use information adequately and transform it into knowledge and vice versa (Kuhlen, 1995; Linde & Stock, 2011; Stehr, 1994; Yigitcanlar, 2010). Thus, we can establish a connection to the importance of the cognitive infrastructure (Stock, 2011) that is observed in "knowledge city" (Ergazakis, Ergazakis, Metaxiotis, & Charalabidis, 2009) and "creative city" (Florida, 2002; Landry, 2008) research. Additionally, cities are complex constructions which today are equally based on digital and physical infrastructures. So far, physical infrastructures and networks have always been the focus of world or global city research (Friedmann, 1986; Hall, 1966; Sassen, 2001; Taylor, 2004). Therefore, I have adopted the term "informational world city" (Mainka et al., 2013) which combines the different types of cities and its infrastructures as illustrated in the following figure (Figure 1.1).

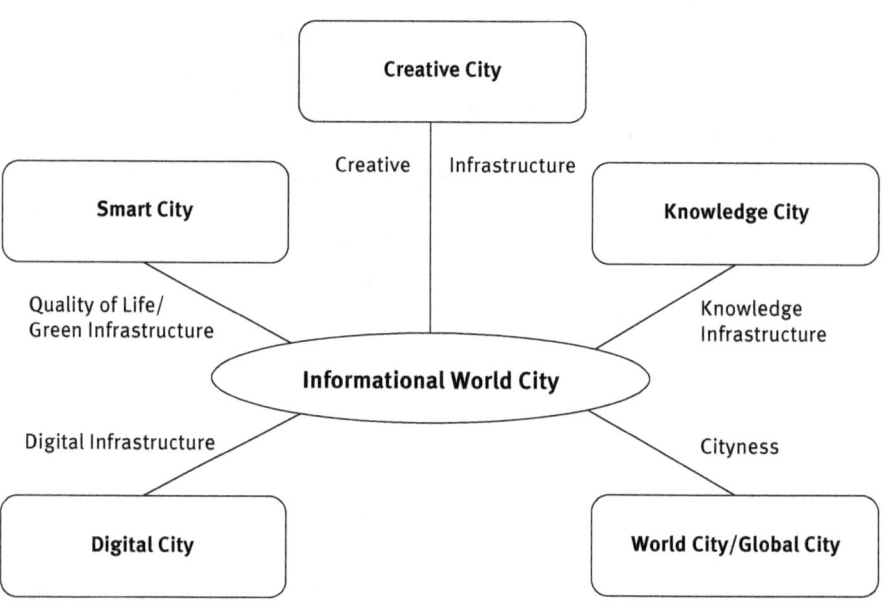

Figure 1.1: Infrastructures of an informational world city. Source: Mainka et al. (2013, p. 296).

The term "informational city" was recently defined by the researchers Manuel Castells, Tan Yigitcanlar and Wolfgang G. Stock. Yigitcanlar (2010) uses the terms "informational city" and "knowledge city" synonymously, defining knowledge as the crucial factor of the human and economic development. Most prominent is the definition by Manuel Castells (1989), who has defined cities according to their space of flows (information, capital, and power). Accordingly, cities are the space of places that are determined by the space of flows. For Castells the emergence of the digital infrastructure has incited a dramatical change of the economy and society in general. Global connectivity has been revolutionized through this new exchange of flows. Since information and knowledge have always been important, it is now possible to transform and enhance knowledge production, sharing, and consumption through these networks. This is a global trend evolving in different cultures and places and resulting in a new understanding of distance. Distances do not only shrink in relation to geographic locations, but also in the form of communication between different stakeholder such as civil society, business, and government. *Online* is a new space for communication, participation, and creativity (Lor & Britz, 2007). Mostly choosing social media or Web 2.0 platforms as their digital venues, different parties meet to share content and to communicate (de Vries, Gensler, & Leeflang, 2012; Mainka, Hartmann, Stock, & Peters, 2014). The result is a tremendous amount of digital information and data. Finally, through digitization and online communication we have arrived in the age of big data and ubiquitous computing (Bryant, Katz, & Lazowska, 2008).

Beyond this exclusively technical aspect, knowledge has also become an important factor of economy in relation to land, labor, and capital (Machlup, 1962). According to Yigitcanlar (2010), an informational or knowledge city is characterized by its knowledge infrastructure, including universities as well as research and development institutes. Further basic characteristics are a well-educated population, digital infrastructure, a globally acting economy, spending on research and development, and the creation of high value-added products. Of course, ICT networks and global connections play an equally important role as knowledge cluster agglomerations within the city which lead to personal contact and share of tacit knowledge.

The agglomeration of talented persons is also acknowledged as key driver of innovation and economic growth (Florida, 2003). The quality of life and place according to public services (e.g. health and education) are important for attracting the "creative class." A wide range of cultural activities and amenities, as well as cultural heritage, is attractive to these talents. To build a successful knowledge city, Yigitcanlar (2010, p. 395) argues that policies and visions are crucial:

> The common strategies include political and societal will; strategic vision and development plans; financial support and strong investments; agencies to promote knowledge-based urban development; an international, multicultural character of the city; metropolitan Web portals; value creation for citizens; creation of urban innovative engines; assurance of knowledge society rights; low-cost access to advanced communication networks; research excellence; and robust public library networks.

Stock (2011) defines informational cities similar to Yigitcanlar (2010) and Castells (1989). For him, informational cities are complex constructions. Different factors have to be considered when analyzing this development. In his work *Informational Cities: Analysis and Construction of Cities in the Knowledge Society*, he defines an indicator catalogue which consists of six main aspects that influence the "informativeness" of a city: infrastructures, labor market, corporate structure, soft locational factors, political willingness, and world city (cityness). All aspects are highly correlated (Figure 1.2). For example, the political willingness to become an informational city positively influences the infrastructure. This may positively impact the labor market and so on. He calls this the positive feedback loop.

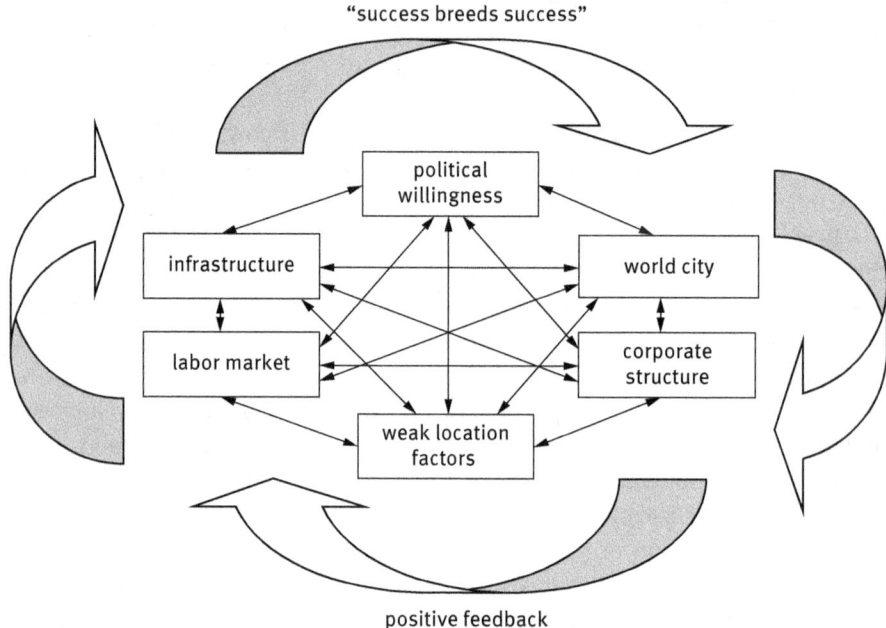

Figure 1.2: The development of an informational city from the perspective of network economies. Source: Stock (2011, p. 980).

Stock (2011, p. 966) argues that

> [w]hen Castells published The Informational City in 1989, he could not have known how existing informational cities would look like (since the Internet had not yet happened), but the theoretical foundation for dealing scientifically with informational cities had been laid. Today, we have informational cities in front of our eyes: Singapore, Seoul, and Dubai set themselves the explicit goal of creating such cities (and are very far along the way); London, New York, San Francisco (and environs), Shanghai, and Hong Kong predominantly bank on hightech industry and services and are modifying their regions into informational cities. Today – at the beginning of the 21st century – we can fill Castells' theory with life.

According to the indicators identified by Stock (Figure 1.2), global comparisons of cities are lacking. There exist some rankings and indices on isolated aspects, like the political willingness of a city, as he mentioned Singapore, Seoul, and Dubai. But whether political willingness, in general, has a positive effect on the transition towards an informational city is not demonstrated in his paper. Furthermore, he refers to findings relating to the knowledge society that are based on isolated regional data, like the comparison of the labor markets of the San Francisco Bay Area and the Los Angeles region, and on state-level data, like the Human Development Index and ICT Development Index. Detailed data on the state and development of concrete cities on their path into the knowledge society all over the world as case studies, however, are nowhere to be found. Drawing on Castell's, Yigitcanlar's and Stock's theory, this thesis is a first attempt to investigate the influence of political willingness, infrastructures, and the status as a world city on the state and development of prototypical cities of the knowledge society on a global scale.

The work at hand will provide an introduction to the development of the city in the knowledge society, in particular the economic development and innovations that have led to the emergence of the knowledge society (chapter 2). Many current publications deal with the future development of our cities and in this work, I will connect them and investigate the main indicators of the world, knowledge, creative, digital and smart city. The focus will be to identify indicators that are relevant for this development by investigating the infrastructure (ICT and cognitive), political willingness, and, finally, cityness. Provoking hypotheses will be developed with the intent of animating experts to argue for or against a characteristic feature of informational city research (chapter 3). The main method employed is grounded theory, allowing the combination of literature review, online research, field study, and expert interviews (chapter 4). The investigation in this work is based on a case study of 31 cities which have been identified through their prominence in the existing literature (chapter 5). The results will then emphasize best practice examples and will be discussed within the context of informational world city development (chapter 6). Finally, a summary and an outlook on future work will conclude the investigation (chapter 7).

References

Albino, V., Berardi, U., & Dangelico, R. M. (2015). Smart cities: Definitions, dimensions, performance, and initiatives. *Journal of Urban Technology, 22*(1), 3–21. https://doi.org/10.1080/10630732.2014.942092

Bryant, R., Katz, R. H., & Lazowska, E. D. (2008). Big-data computing: Creating revolutionary breakthroughs in commerce, science and society. *Academic Press*. Retrieved from http://www.cra.org/ccc/initiatives

Castells, M. (1989). *The informational city: Information technology, economic restructuring, and the urban-regional process*. Oxford, UK: Blackwell.

Clark, G., Moonen, T., & Couturier, J. (2015). *The business of cities 2015*. Retrieved from http://www.jll.com/Research/jll-business-of-cities-report.pdf

Craven, P., & Wellman, B. (1973). The network city. *Sociological Inquiry, 43*(3–4), 57–88.

de Vries, L., Gensler, S., & Leeflang, P. S. H. (2012). Popularity of brand posts on brand fan pages: An investigation of the effects of social media marketing. *Journal of Interactive Marketing, 26*(2), 83–91. https://doi.org/10.1016/j.intmar.2012.01.003

Ergazakis, E., Ergazakis, K., Metaxiotis, K., & Charalabidis, Y. (2009). Rethinking the development of successful knowledge cities: An advanced framework. *Journal of Knowledge Management, 13*(5), 214–227. https://doi.org/10.1108/13673270910988060

Florida, R. L. (2002). *The rise of the creative class: And how it's transforming work, leisure, community and everyday life*. New York, NY: Basic Books.

Florida, R. L. (2003). Cities and the creative class. *City and Community, 2*(1), 3–19. https://doi.org/10.1111/1540-6040.00034

Friedmann, J. (1986). The world city hypothesis. *Development and Change, 17*(1), 69–83. https://doi.org/10.1111/j.1467-7660.1986.tb00231.x

Hall, P. (1966). *The world cities*. London, UK: Weidenfeld and Nicolson.

Hollands, R. G. (2008). Will the real smart city please stand up? *City, 12*(3), 303–320. https://doi.org/10.1080/13604810802479126

Ishida, T., Ishiguro, H., & Nakanishi, H. (2002). Connecting Digital and Physical Cities. In M. Tanabe, P. van den Besselaar, & T. Ishida (Eds.), *Digital cities II: Computational and sociological approaches* (pp. 246–256). Berlin, Heidelberg, DE: Springer. https://doi.org/10.1007/3-540-45636-8_19

Jacobs, J. (1969). *The economy of cities*. New York, NY: Random House.

Jacobs, J. (1984). *Cities and the wealth of nations*. New York, NY: Random House.

Kuhlen, R. (1995). *Informationsmarkt. Chancen und Risiken der Kommerzialisierung von Wissen [Information market. Opportunities and risks of the commercialization of knowledge]*. Konstanz, DE: UVK-Universitätsverlag.

Landry, C. (2008). *The creative city: A toolkit for urban innovators* (2nd ed.). New York, NY: Earthscan Publications.

Linde, F., & Stock, W. G. (2011). *Information markets: A strategic guideline for the i-commerce*. Berlin, De; New York, NY: Walter de Gruyter.

Lor, P. J., & Britz, J. J. (2007). Is a knowledge society possible without freedom of access to information? *Journal of Information Science, 33*(4), 387–397. https://doi.org/10.1177/0165551506075327

Machlup, F. (1962). *The production and distribution of knowledge in the United States*. Princetown, NJ: Princetown University Press.

Mainka, A., Hartmann, S., Orszullok, L., Peters, I., Stallmann, A., & Stock, W. G. (2013). Public libraries in the knowledge society: Core services of libraries in informational world cities. *Libri, 63*(4), 295–319. https://doi.org/10.1515/libri-2013-0024

Mainka, A., Hartmann, S., Stock, W. G., & Peters, I. (2014). Government and Social Media: A Case Study of 31 Informational World Cities. In *47th Hawaii International Conference on System Sciences* (pp. 1715–1724). Wahington, DC: IEEE Computer Society. https://doi.org/10.1109/HICSS.2014.219

Nam, T., & Pardo, T. A. (2011). Conceptualizing smart city with dimensions of technology, people, and institutions. In *Proceedings of the 12th Annual International Digital*

Government Research Conference on Digital Government Innovation in Challenging Times (pp. 282–291). New York, NY: ACM. https://doi.org/10.1145/2037556.2037602

Sassen, S. (2001). *The global city: New York, London, Tokyo* (2nd Ed.). Princeton, NJ: Princeton University Press.

Shin, D.-H. (2009). Ubiquitous city: Urban technologies, urban infrastructure and urban informatics. *Journal of Information Science, 35*(5), 515–526. https://doi.org/10.1177/0165551509100832

Stehr, N. (1994). *Knowledge societies*. London, UK: Sage.

Stock, W. G. (2011). Informational cities: Analysis and construction of cities in the knowledge society. *Journal of the American Society for Information Science and Technology, 62*(5), 963–986. https://doi.org/10.1002/asi

Taylor, P. J. (2004). *World city network: A global urban analysis*. London, UK: Routledge. https://doi.org/10.4324/9780203634059

UNFPA. (2007). *The state of the world population 2007: Unleashing the potential of urban growth*. New York, NY. https://doi.org/ISBN 978-0-89714-807-8

Weber, M. (1921). Die Stadt. Eine soziologische Untersuchung [The city. A sociological study]. *Archiv Für Sozialwissenschaft Und Sozialpolitik, 47*, 621–772.

Yigitcanlar, T. (2010). Informational city. In R. Hutchison (Ed.), *Encyclopedia of urban studies* (pp. 392–395). New York, NY: Sage.

2 The rise of the knowledge society

In endeavoring to define the knowledge society, we first need to define the term knowledge in our society. According to Hack (2006), knowledge can be of different dimensions: He distinguishes between scientific knowledge and common knowledge and between lay and expert knowledge. Knowledge may take different forms, such as the foundation of education and training, a repository of patents and licenses, a form of organizational knowledge like management strategies, investment decisions, or the algorithm of search engines. Finally, knowledge may also mean a binding truth or a tentative interpretation. The crucial and unifying factor of all these forms is that knowledge is firmly interconnected with the society (Drucker, 1993). Knowledge without human interpretation is merely information (Kuhlen, 1995). In combination with the economic resources land, capital, and labor, the resource knowledge is vital for the efficiency of the other resources (Choo, 2002). Thus, the turning point from the industrial to the post-industrial capitalism is the acknowledgment of knowledge as a significant factor of economic growth (Bell, 1973; Drucker, 1993; Hepworth, 1987; Machlup, 1962; Porat, 1977), along with the change of status of education and learning (Stock, 2011). For Heidenreich (2002), education and life-long learning are essential drivers of our society, as knowledge institutions and an adequate educational system are needed to educate potential knowledge workers (Lim, 1999).

Hence, we cannot define the knowledge society without understanding the economic development of our capitalist world. Fortunately, economic or societal change has been the subject of detailed analysis for decades (Bornschier & Suter, 1992). According to Kondratieff (1926), capitalism has thus far experienced four cycles of economic situation[1]. For Schumpeter (1939), each economic cycle is driven by a basic innovation. Two types of innovation are acknowledged: (1) product innovation (a new consumer good e.g. the car in the nineteenth century) and (2) process innovation (a novel production technique e.g. assembly line production in the early twentieth century) (Bornschier & Suter, 1992; Freeman, Clark, & Soete, 1982; Van Duijn, 1983). Each kind of innovation has led to a transition which also embeds the emergence of a network (Linde & Stock, 2011). The first cycle began in the eighteenth century. This was the shift from a largely agricultural to an industrialized society. The basic innovation was the steam engine, heavily influencing textile manufacturing. The second cycle was based on steel, advancing the railroad network and allowing mass transportation. The third

---

[1] Economic growth and decline were measured going back to the eighteenth century (prior data was not available (Stiller, 2005)), with the findings based on economic data (prices and products).

https://doi.org/10.1515/9783110577662-002

cycle was fueled by stable electricity grids, based on the electrotechnology and chemical industry. Finally, the fourth cycle was based on the automobile and petrochemical industry, resulting in road networks that allowed independent mobility and individual transportation. As the fifth cycle emerges, so does a new type of society (Nefiodow, 1991), with information as its basic innovation, leading to global telecommunication networks, particularly the internet. For Bell (1973) this marks the shift from industrial to post-industrial society. A similar approach can be found in Toffler's (1980) definition of *The Third Wave*. The first wave refers to agriculture, the second to industry, and the third to post-industrial society. Accordingly, the social shift is not only defined as post-industrial society but also as *informational* or *knowledge society* (Stehr, 1994; Stock, 2011; Webster, 1995, 2002). Its main aspect is the economic transformation of knowledge, which now can be stored and shared through digital networks (Drucker, 1993; Machlup, 1962; Porat, 1977). Therefore, Castells (1996) uses the term *network society*. He refers to the information and telecommunication technology network based on electronic devices such as phones, fax machines, printers, computers, and, nowadays, tablets, smartphones, wearables (smart watches or glasses) and other networked devices. They form the network and allow a new kind of information sharing and communication in real time (Castells, 1996; Melzi, 2009; van Dijk, 2012). Thus, information can be shared and stored anytime and anywhere, which Bonitz (1986) takes to mean that we have arrived at a new holographic and time principle: "The entirety of human knowledge is one gigantic hologram, which consists of all storage units, databases etc. available to mankind" (Bonitz, 1986, p. 192). Information can now be consumed simultaneous to its production (Stock, 2011).

Nefiodow and Nefiodow (2014) claim that we arrived at the end of the fifth Kondratieff cycle with the financial crisis in 2008/2009. Thus, researchers and economists started to debate about which innovation or technology could drive our future from here on out. It is assumed that future innovations will enhance the productivity of our society as prior Kondratieff cycles have, e.g. in the manufacturing of clothes, mobility on railroads or streets, and digitization of routine work (Nefiodow, 2006). Nefiodow (2006) speculates that the sixth Kondratieff cycle could be based on biotechnology and health. His idea is rooted in the growth rates of the health sector in the US and in Europe. Additionally, he suspects that sustainability, environment protection, and alternative energy supply will be growing sectors as well.

For the economist Rifkin (2014) energy supply and sustainability are the twin drivers of our future economy. But he goes one step further and declares that the Internet of Things (IoT) will have a tremendous impact on our society and the market, predicting the emergence of the "new smart society" (Rifkin, 2014, p. 15). His ideas are futuristic, but corporations are already exploring the possibilities of networks that do not only connect information but also things in our world. Examples are General Electric's "Industrial Internet," CISCO's "Internet of Everything," IBM's "Smarter Planet," and Siemens' "Sustainable Cities" programs. Sensors on roads, in cities, or in households are connected via the internet and can be managed remotely through electronic devices like smartphones. For Rifkin, this is the *Third Industrial Revolution*, which consists of three internets: the communication internet, the energy internet, and the logistics internet. He claims that in the future, shared networked commons will change our lives, grounded on a network of sharing (sharing of information, energy, and transportation). This vision may be referred to Castells' (1996) definition of the network society in which spaces of places are determined by spaces of flows (flows of information, capital, and power). Both ideas are rooted in information and communication technology.

Other researchers have not followed Rifkin's idea of the sharing society since. They prefer to conceive the current state as a new form of capitalism, one which is based on information and the knowledge labor force (Fuchs, 2012; Vercellone, 2007). Furthermore, they argue that we have not yet reached the end of the fifth Kondratieff cycle's economic wave and that, therefore, ICT is still the driving innovation.

A further idea developed by Hall (2010) is that we have arrived in *The Age of The City*. He has identified three main urban innovations: (1) the cultural/intellectual, (2) the technological-productive, and finally (3) the joining of the cultural and the technological. The third innovation is a new phenomenon which cannot be found in past developments. This is the new knowledge economy which brings together technological and cultural innovations, especially in cities like Los Angeles, San Francisco, New York City, and London. Accordingly, Batty (2015) argues in his essay *Creative Destruction, Long Waves and the Age of the Smart City* that the next cycle could be the cycle of the smart city, marrying the development of the IoT with the innovation density of cities. For him, the next step is based on the kind of communication which develops from "anywhere at any time" in the fifth cycle to "through anything, at anytime and anywhere" in the sixth. Not everybody agrees with this assumption, e.g. Hollands (2008) claiming that the smart city is merely a new marketing label intended to promote the products of IBM, CISCO, Siemens and other companies which have established many IoT projects in various cities. ICT networks are therefore not in the spotlight, but simply the basic infrastructure in cities of the knowledge society.

If we refer to the new knowledge economy or to the IoT, then information, knowledge, and ICT are still the drivers of our economy. Hence, on a global scale we are not able to proclaim the end of the fifth cycle. Looking at the development of world cities, traditional paragons like New York and London have suffered heavily under the effects of the financial crisis that began in 2008 (Bassens,

2012). But there are also winners, for example emerging cities such as Dubai, Doha and Abu Dhabi in the Gulf region (Kosior, Barth, Gremm, Mainka, & Stock, 2015) or Beijing, Shanghai, and Hong Kong in China. Thus, the hypothesis that we are currently in a heavy fifth-cycle downturn is not borne out by global evidence. Figure 2.1 illustrates the past and present economic cycles according to Kondratieff, as well as the emergence of the knowledge society. The shift from the fifth to the sixth cycle is interpreted as a much smoother transition than has been observed in previous cycles. However, conclusive evidence for this will only become available in hindsight.

In this, we are able to see a development which is not exclusively driven by commercial interests. In many cities and regions of the world (most pronounced in Western countries) communities are starting to engage with the future of their city or region to develop a more open, innovative, and democratic world (Open Knowledge Labs, n.d.). This is evident through the growing number of open data activists like the labs associated with the Open Knowledge Foundation and the growing importance of the Open Data Day (Akbaba, 2016). These activities are about helping citizens, governments, and businesses in a city with creating valuable engines and becoming an empowered partner rather than just a city user. This development is based on digital networks and the personal engagement of communities (physical networks). The EU, amongst other organizations, has begun to speculate about the economic profit that could be gained through open data in the future (European Commission, 2015).

Summing up, the basis of our current societal development consists of information, knowledge, and networks. Stock (2011, p. 965) has defined the factors of the information and knowledge society as follows:

Information society:
- "basic innovations are carried by the resource information"
- "computers are of great importance"

Knowledge society:
- "displays all the characteristics of an information society,
    o at which digital information and
    o at which computer networks play important roles"
- "information contents of every kind are available in any place and at any time (holographic and speed principles) and are intensively taken advantage of"
- "lifelong learning (including learning how to learn) is necessary"

According to Nefiodow (2006), Hall (2010), Rifkin (2014), and Batty (2015), the productivity of the knowledge society will increase through further developments. Thus, we have to define the indicators of the knowledge society 2.0 or smart society.

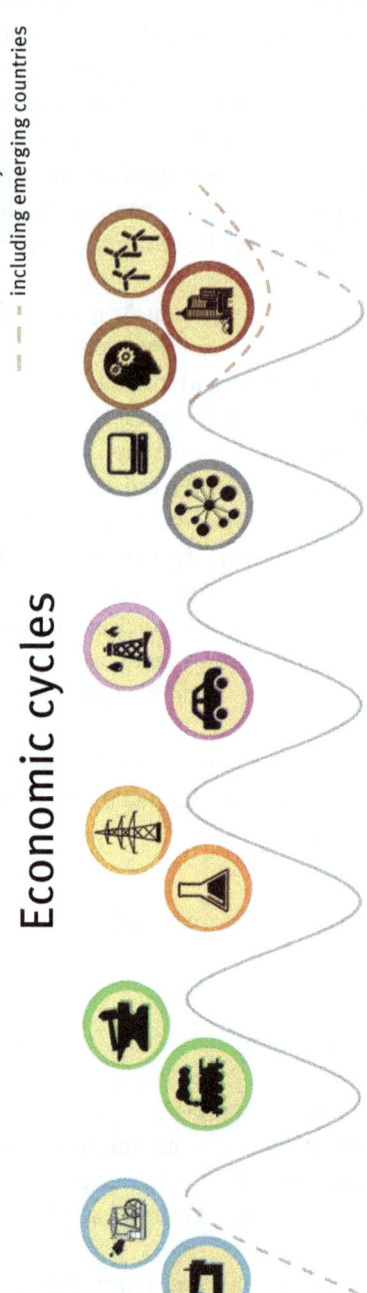

Figure 2.1: Economic cycles and the knowledge society (Own interpretation based on discussed literature).

Smart society:
- displays all characteristics of an information and knowledge society
  - in which the networks of information are increasingly growing (including the Internet of Things and Open Data)
  - with advanced holographic and speed principles which grow from "at any time and anywhere" to "through anything, at any time and anywhere"
- sustainability and health become important factors for the society and economy
- empowered citizens that engage in a more creative, innovative, and democratic future (open innovation on city level)

Finally, the concepts of the smart society are still based on the innovation "information and communication", which shows its assets through the combination of ICT, physical networks, and human capital. Because knowledge still remains the crucial factor of the society – humans transform information into knowledge – I will adopt the term knowledge society to describe this "era" of human development in the twenty-first century (Stock, 2011).

References

Akbaba, C. (2016). Open Data Day 2016 in Karlsruhe: Wie digitales Ehrenamt unsere Gesellschaft verändert? [Open data day 2016 in Karlsruhe: How digital volunteering influences our society]. *Techtag*. Retrieved from http://www.techtag.de/netzkultur/open-data-day-2016-karlsruhe/

Bassens, D. (2012). The world city concept travels East: On excessive imagination and limited urban sustainability in UAE world cities. In B. Derudder, M. Hoyler, P. J. Taylor, & F. Witlox (Eds.), *International handbook of globalization and world cities* (pp. 530–537). Cheltenham, UK; Northampton, MA: Edward Elgar Publishing.

Batty, M. (2015). *Creative destruction, long waves and the age of the smart city* (No. 200). UCL Working Paper Series. London, UK. https://doi.org/10.1103/PhysRevE.78.016110

Bell, D. (1973). *The coming of the post-industrial society*. New York, NY: Basic Books.

Bonitz, M. (1986). Holographie-und Tempoprinzip: Verhaltensprinzipen im System der wissenschaftlichen Kommunikation [Principle of holography and principle of speed. Characteristics of the systems of scientific communication]. *Informatik*, *33*(5), 191–193.

Bornschier, V., & Suter, C. (1992). Long waves in the world system. In V. Bornschier & P. Lengyel (Eds.), *Waves, formations and values in the world system* (2nd ed., pp. 15–50). New Brunswick, NJ; London, UK: Transaction Publishers.

Castells, M. (1996). *The rise of the network society*. Malden, MA: Blackwell.

Choo, C. W. (2002). *Information management for the intelligent organization: The art of scanning the environment*. Medforfd, NJ: Information Today.

Drucker, P. F. (1993). *Post-capitalist society*. New York, NY: HaperCollins.

European Commission. (2015). *Creating value through open data: Study on the impact of re-use of public data resources*. Luxembourg, LUX. https://doi.org/10.2759/328101
Freeman, C., Clark, J., & Soete, L. (1982). *Unemployment and technical innovation: A study of long waves and economic development*. London, UK: Frances Printer.
Fuchs, C. (2012). Capitalism or information society? The fundamental question of the present structure of society. *European Journal of Social Theory*, 16(4), 413–434. https://doi.org/10.1177/1368431012461432
Hack, L. (2006). Wissensformen zum Anfassen und zum Abgreifen. Konstruktive Formationen der „Wissensgesellschaft" respektive des transnationalen Wissenssystems [Knowledge forms for touching and tapping. Constructive formations of the "knowledge society" ...]. In U. Bittlingmayer & U. Bauer (Eds.), *„Die Wissensgesellschaft" Mythos oder Ideologie* (pp. 109–172). Wiesbaden, DE.
Hall, P. (2010). The age of the city: The challenge for creative cities. In O. G. Ling & B. Yuen (Eds.), *World Cities. Achieving liveability and vibrance* (pp. 47–70). Singapore, SG: World Scientific Publishing.
Heidenreich, M. (2002). *Merkmale der Wissensgesellschaft. Lernen in der Wissensgesellschaft [Characteristics of the knowledge society. Learning in the knowledge society]*. Innsbruck, AUT: StudienVerlag.
Hepworth, M. E. (1987). The information city. *Cities*, 4(3), 253–262. https://doi.org/10.1016/0264-2751(87)90033-3
Hollands, R. G. (2008). Will the real smart city please stand up? *City*, 12(3), 303–320. https://doi.org/10.1080/13604810802479126
Kondratieff, N. D. (1926). Die langen Wellen der Konjunktur [Long waves of economic cycles]. *Archiv Für Sozialwissenschaft Und Sozialpolitik*, 56, 573–609.
Kosior, A., Barth, J., Gremm, J., Mainka, A., & Stock, W. G. (2015). Imported expertise in world-class knowledge infrastructures: The problematic development of knowledge cities in the Gulf Region. *Journal of Information Science Theory and Practice*, 3(3), 17–44.
Kuhlen, R. (1995). *Informationsmarkt. Chancen und Risiken der Kommerzialisierung von Wissen [Information market. Opportunities and risks of the commercialization of knowledge]*. Konstanz, DE: UVK-Universitätsverlag.
Lim, E. (1999). Human resource development for the information society. *Asian Libraries*, 8(5), 143–161. https://doi.org/10.1108/10176749910275975
Linde, F., & Stock, W. G. (2011). *Information markets: A strategic guideline for the i-commerce*. Berlin, De; New York, NY: Walter de Gruyter.
Machlup, F. (1962). *The production and distribution of knowledge in the United States*. Princetown, NJ: Princetown University Press.
Melzi, C. (2009). Mobility and consumption of "informational" city. New perspectives and fields of study. The location-based services. In *City Futures in a Globalising World. An International Conference on Globalism and Urban Change*. Madrid, ESP.
Nefiodow, L. A. (1991). *Der fünfte Kondratieff: Strategien zum Strukturwandel in Wirtschaft und Gesellschaf [The fifth Kondratieff: Strategies for structural change in the economy and society]* (Vol. 2). Frankfurt, Wiesbaden, DE: FAZ, Gabler.
Nefiodow, L. A. (2006). *Der sechste Kondratieff: Wege zur Produktivität und Vollbeschäftigung im Zeitalter der Information. Die langen Wellen der Konjunktur und ihre Basisinnovationen [The Sixth Kondratieff: ways of productivity and full employment in the age of information...]* (6th ed.). Sankt Augustin, DE: Rhein-Sieg-Verlag.

Nefiodow, L. A., & Nefiodow, S. (2014). *The Sixth Kondratieff: A New Long Wave in the Global Economy*. North Charleston, SC: Createspace Independent Pub.

Open Knowledge Labs. (n.d.). Members – Open Knowledge Labs. Retrieved from http://okfnlabs.org/members/

Porat, M. U. (1977). *Information economy (Vol.9), Office of Telecommunications Special Publication 77-12[1]–77-12[9]*. Washington, DC.

Rifkin, J. (2014). *The zero marginal cost society: The internet of things, the collaborative commons, and the eclipse of capitalism*. New York, NY: Palgrave Macmillan.

Schumpeter, J. A. (1939). *Business cycles: A theoretical, historical, and statistical analysis of the capitalist process*. New York, NY: McGraw-Hill Book Company, Inc.

Stehr, N. (1994). *Knowledge societies*. London, UK: Sage.

Stiller, O. (2005). *Innovationsdynamik in der zweiten industriellen Revolution: Die Basisinnovation Nanotechnologie [Innovation dynamics in the second industrial revolution: The basic innovation nanotechnology]*. (Marburger Förderzentrum für Existenzgründer aus der Universität, Ed.). Norderstedt, DE: Books on Demand GmbH. Retrieved from http://archiv.ub.uni-marburg.de/diss/z2006/0002/pdf/dos.pdf

Stock, W. G. (2011). Informational cities: Analysis and construction of cities in the knowledge society. *Journal of the American Society for Information Science and Technology*, *62*(5), 963–986. https://doi.org/10.1002/asi

Toffler, A. (1980). *The third wave*. New York, NY: Morrow.

van Dijk, J. A. (2012). *The network society* (3rd ed.). London, UK: Sage.

Van Duijn, J. J. (1983). *The long wave in economic life*. London, UK: Allen & Unwin.

Vercellone, C. (2007). From formal subsumption to general intellect: Elements for a Marxist reading of the thesis of cognitive capitalism. *Historical Materialism*, *15*(1), 13–36.

Webster, F. (1995). *Theories of the information society*. London, UK: Routledge.

Webster, F. (2002). The information society revisited. In S. Livingstone & L. Lievrouw (Eds.), *Handbook of new media* (pp. 255–266). London, UK: Sage.

3 Measuring cities of the knowledge society

The development of the knowledge society is intertwined with many factors that impact each other. One aspect is economic growth, as increased economic productivity results in market growth, which can be statistically measured by the gross domestic product (GDP), i.e. "the monetary value of all goods and services produced in a nation during a given time period, usually one year" (Brezina, 2011, p. 4). Thus, all goods and services that are bought by ordinary consumers (e.g. in a supermarket), corporations (e.g. all computers for their employees), or the government (e.g. construction of new metro lines) increase the GDP. The United Nations Development Program uses statistical data such as the GDP, together with the life expectancy of the population at birth, the rate of alphabetization of adults, and the ratio of students in their respective age groups to analyze the human development status on a national level. Their aim is to discover and observe development trends in diverse areas such as mobility, climate change, and gender and ethnic inequality (UNDP, 2014). Depending on the topic of investigation, different indicators are included in studies and finally, in most cases, compared to the GDP growth rate.

On the city level, different terms and approaches exist to describe the measurement of the knowledge society. Depending on the type of city (digital, smart, knowledge, or creative), different indicators are used. As explained in the first chapter, I will bring different approaches in line with the definition of an informational world city. Figure 1.2 represents an overview of the main characteristics which are identified by Stock (2011) and are used in several case studies as well as in development reports. Some indicators like the GDP are firmly established as an indicator of growth but others are not and used experimentally. For those that are not established, I will formulate hypotheses that will be investigated later on.

3.1 Infrastructures

Referring to the development of the knowledge society, two main infrastructures have evolved, namely the "ICT infrastructure" and the "cognitive infrastructure." The former is allocated to the technical development of the fifth Kondratieff cycle. It is the basic innovation of this economic cycle. The latter is not new, but has grown in importance and refers to the knowledge and creative institutions and their output.

3.1.1 ICT infrastructure

ICT is the infrastructure which has become central in everyday life and essential for many businesses. The approach of being connected everywhere and anytime is mostly investigated under the heading of ubiquitous computing or ubiquitous cities (Schumann & Stock, 2015). Internet connectivity does not only include private individuals and businesses. We have by now arrived in the age of the Internet of Things and have to include sensors and things with Wi-Fi connections in this ubiquitous environment (Hollands, 2008; Rifkin, 2014). However, cities are still the center of human interactions (Sassen, 2001) and physical mobility must therefore not be ignored.

With the advent of ICT, investigations related to human development have referred to the information society. Bruno, Esposito, Genovese, and Gwebu (2011) have examined the progress of different indices that measure the ICT access, use, and skills on a national level. *Access* is related to telephone landlines per 100 inhabitants, mobile phone subscriptions per 100 inhabitants, internet bandwidth (bit/s) per internet user, and proportion of households with a computer and with internet access at home. ICT use is calculated by the number of internet users per 100 inhabitants, landline broadband internet subscribers per 100 inhabitants, and mobile broadband subscribers per 100 inhabitants. Finally, ICT skills[1] are estimated with the help of the adult literacy rate and to secondary and tertiary education gross enrollment rate.

In the early stage of the fifth Kondratieff cycle in 2003, the ITU published their "Digital Access Index" (ITU, 2003) and Orbicom the "Infostate" index, which was established to explore the digital divide on a national level (Sciadas, 2003). For future publication, Orbicom and ITU merged their indices and published the "ICT Opportunity Index" (ICT-OI) from 2001 to 2005 (ITU, 2007a). Other indices that analyze the ICT development on national level are the "ICT Diffusion Index" (ICTDI), published by the United Nations Conference on Trade and Development (UNCTAD) for the year 2005 (United Nations, 2006) , the "Digital Opportunity Index" (DOI) published by ITU for the years 2005 and 2006 (ITU, 2007b), the "ICT Development Index" (IDI), which has been continuously published by ITU since 2007 (its most recent publication is ITU (2010)), the "E-Readiness Index" published annually by the Economist Intelligence Unit and IBM from 2000 to 2009 (Economist Intelligence Unit, 2009), the "Digital Economy Rankings" published

---

[1] The indicator ICT skills is thus somewhat underdeveloped, since a direct connection between the education obtained by individuals and their ability to use ICT is not evident. Media or information literacy is a far superior indicator, but almost impossible to be evaluated on a global scale (Beutelspacher, 2014; Henkel. & Stock, 2016).

by the Economist Intelligence Unit and IBM in 2010 (Economist Intelligence Unit, 2010), and finally, the "Networked Readiness Index" published annually by the World Economic Forum (World Economic Forum, 2015).

According to a study by Paparwekorn (2015, p. 385), ICT indices can generally be grouped into five categories:
1. infrastructure and accessibility,
2. national regulatory agency (NAR) and ICT policy,
3. business environment,
4. education and innovation,
5. usage and security.

Paparwekorn (2015) argues that indices are important for gauging the degree of development of a certain region or nation. Especially when it comes to development plans, a region or city is able to benchmark its own development against other regions and then able to set realistic goals for further ICT strategies. However, a comparison of a certain set of variables always comes with a number of limitations, such as missing data or methodical assumptions that are not applicable for every region on a global scale. Thus, indices have to be read and interpreted very carefully.

Nevertheless, Bruno et al. (2011) find that ICT indices are highly correlated with the GDP of a nation. Thus, the ICT-OI has a correlation of +0.946 and the IDI a correlation of +0.916 with the GDP (Bruno et al., 2011, p. 27). This implies that national growth and wealth impacts ICT access, use, and skills, and vice versa. A similar trend is determined by Stock (2011), who identifies a high correlation of +0.90 (Pearson) for the Human Development Index (HDI) and the ICT Development Index (IDI) for 112 countries.

In a further study, Jin and Cho (2015) use different variables (e.g. PC adoption rate, internet use rate, profit rates of ICT supplier, ICT trade, the amount of ICT workers, policies and governmental investments) to evaluate the impact of ICT on national growth rates for a total of 128 countries. According to their findings, the adoption rate of PCs has no significant correlation. This is not surprising since, nowadays, many other devices also allow internet access. Thus, they argue that the adoption rate of multiple devices with internet access has a positive effect on economic growth. Further, a statistically significant correlation was found for the variables internet use, profit of telecommunication suppliers, and governmental investments. No correlation could be found for the variables amount of ICT workers and ICT trade. In addition, they investigated the effect of education and literacy rates on economic growth and conclude that both variables do not have as strong an impact as, for example, ICT use. They do not, however, discuss the limited significance of the characteristics used (school enrollment rates and literacy rates), which by themselves do not tell us anything about the level of information literacy (Spitzer, Eisenberg, & Lowe, 1998) of the population or the quality of the educational system. Further variables with some influence are national corruption and the consumer price index, as "both are shown to affect national economic growth positively, with significance levels of 95% and 99%, respectively" (Jin & Cho, 2015, p. 258). It is argued that "[t]he positive effect of CPI (corruption perception index) on economic development is consistent with the assertions that emphasize good governance as a critical factor for economic development" (Jin & Cho, 2015, p. 258).

Overall, we therefore see a complex mix of indicators used to measure the state and progress of the knowledge society on the national level. Summarizing, the knowledge society has been investigated on national level since the early twenty-first century. Its development is highly intertwined with economic growth and information and communication technology. Rapid technological innovations have influenced which kind of technology infrastructure is needed for such a society. One example is the indicator "wireless broadband subscription," which has replaced "fixed internet subscription" since slow internet access via modem is no longer relevant. Education has increased in importance, as well as freedom of information and access to public information. In addition, the gross enrolment and literacy rate of adults are used as indicators of ICT skills and HDI. In the following, I will use these indicators and bring them into line with investigations on the city level, since cities can be defined as the hubs that are the cause of the development of the modern world (Jacobs, 1969, 1984). According to Hall (1985) cities are part of the "geography of the fifth Kondratieff." Furthermore, Stock (2011) emphasizes that typical cities for a societal era have always existed, as for example Manchester was a typical city of the industrial society or Manhattan of the service society. Accordingly, he is convinced that the prototypical cities of the knowledge society in the twenty-first century are informational cities.

One index, which focuses on ICT and the development of smart cities, is the Networked Society City Index 2014, which was assembled by the Swedish IT Corporation Ericsson. Because Ericsson is a private corporation, their index has to be read with some degree of caution. Further indices by independent agencies are under development, for example by the ITU. A report on "Smart Sustainable Cities" will be developed by the ITU-Study Group 20. According to the "Smart cities – Preliminary Report 2014" (ISO/IEC JTC 1. InformationTechnology, 2015), city indices like "Global City Indicators" or the "Green Cities Index" use variables that do not reflect or relate to the ICT infrastructure in cities. Accordingly, smart cities have different demands which can be defined as technical, market, and social needs. Those require standardization to facilitate interoperability. The established indicators of ICT developments on a national level, as well as the discussion on ICT indicators to compare city developments, are never isolated ICT infrastructure measurements. For example, the investigation of European medium-sized cities as smart cities uses the ICT infrastructure as a sub-indicator of "smart mobility" (Giffinger et al., 2007). Economic and social factors have to be considered as well and have been included, for example, by Nesta (2015). They have established a "European Digital City Index" with the aim of comparing entrepreneurship in the digital economy. The lack of ICT city indices on a global scale and the approach to define standards of smart cities indicates the need for further research in this field.

For example, one indicator that is valuable only on the city level is accessibility of WiFi hotspots. Those are especially important for tourists, business travelers, but also for people who cannot afford (mobile) broadband contracts. Ergazakis, Ergazakis, Metaxiotis, and Charalabidis (2009) define low-cost access to Wi-Fi networks as one of their criteria of a knowledge city. Furthermore, low-cost access to broadband networks is assumed to be present in most global cities (Ergazakis et al., 2009). The amount of Wi-Fi hotspots is useful for a comparison of dense areas like a city or touristic hubs where a lot of people will use this access. This was acknowledged within the E-Readiness Ranking, a national index, in which Wi-Fi hotspots are not compared due to concerns about the data comparability on a national level (Economist Intelligence Unit, 2009).

The ICT infrastructure is subject to different quality and availability options depending on the market conditions. Quality and availability of internet access are influenced by the interplay of government development (plans) and economic interests. Taking a look at the Unite Arab Emirates (UAE), with the two big cities Dubai and Abu Dhabi, it could be observed that the telecommunication market used to be monopolistic with Emirates Telecommunications Corporation (Etisalat) as the only provider (Kosior, 2013). Since 2007, however, a second provider named Emirates Integrated Telecommunications Company (du) is able to offer universal ICT support like Etisalat. Further corporations have been entering the market since 2009, but they are merely allowed to support niches. The main providers are under governmental control. Due to this oligopolistic or governmentally-owned market, the prices are relatively high. They range from 1% (starting with 1 Mbps) up to 8% (for 100 Mbps) of the GDP on a per-capita basis in fiscal year 2013. In contrast, the ICT providers on a polypolistic market such as can be found in Scandinavian cities – Oslo, Helsinki, and Stockholm – offer comparably cheap contracts which make up only between 0.5% and 0.7% of the GDP per-capita in 2011 (Stallmann, 2014). In both cases, the majority is using broadband connections if available. On the one hand, the government as monopolist at the market can keep prices at an artificially high level. On the other hand, governments can also help to provide broadband access to all areas if they are working as a provider, as in the case of UAE, or offer incentives for providers of the private economy, e.g. through subsidization or other reward systems.

The increasing importance of ICT with regard to city development and planning is already a common topic for IT firms, science, city mayors, and citizens. Cities have become data-driven hubs as envisioned in many science fiction films (Hollands, 2014). Recent studies on future city development refer to the term "Smart City" (Hollands, 2008). The IT companies Cisco and Gale International are involved in building the Smart City Songdo in South Korea and share a common vision: "... in cities of the future urban services will be delivered more innovatively, and cities will be managed more efficiently using technology that enables newer models of managed and hosted services within public-private partnerships" (Songdo IBD, 2009). Furthermore, IBM is solving urban problems through technology, as well (IBM, 2010). The goal is not to build a whole new city, but to develop smarter city projects for different urban problems, such as health care or traffic. Hollands (2014) criticizes the development of "Smart Cities," as they are advanced by IT firms and the huge amount of data produced by citizens will be owned by corporations instead of the cities themselves.

However, IT firms located in a region may bring further advantages as they require an excellent ICT infrastructure and may therefore support its development simply as a sensible business decision. One example is the IT Corporation Nokia. Nokia is located in Espoo, which is located near the Finnish capital Helsinki, and has been one of the leading global providers of cellular phones of the first generation. Today, Helsinki is a city with an enhanced and highly developed ICT infrastructure. A further example is Google. Google has implemented a large number of Wi-Fi hotspots around Palo Alto, the location of Google headquarters. This is not merely limited to IT firms, as other businesses also depend on a highly developed ICT infrastructure and therefore may potentially fund it out of economic self-interest.

The connection of physical space, digital data, and citizens enables new means to solve urban problems created by increasing urbanization. To mention one example, we could envision traffic jam warnings calculated through the traveling speed of car drivers using a smartphone with GPS connection (Herrera et al., 2010). These kind of services are possible with an enhanced ICT infrastructure in which users generate and retrieve real-time data through mobile IT devices.

A further example of how ICT has changed and continues to change our lives is the way we communicate, work collaboratively, and retrieve information. In many cases emails, messengers (e.g. WhatsApp or Facebook Messenger), group discussion apps (e.g. Slack), video calls (e.g. Skype or Hangout), and social media platforms (e.g. Facebook or Twitter) have opened up significantly more opportunities than the classic telephone call or posted letter. If we are looking for information, we are probably going to use Google (Rowlands et al., 2008). For more profound information, deep web databases are at our disposal through digital libraries (Mainka et al., 2013). These possibilities are also based on ICT. The innovation is that we access these sources of information anytime and anywhere through our mobile devices. Most importantly, the ICT infrastructure is the basis for future economic development and represents the fifth Kondratieff cycle, which has evoked the network of ICT, namely the internet, and outpaced the network of the fourth Kondratieff cycle, namely automotive traffic in cities. Smart city developments are always concerned about efficiency and sustainability (Hollands, 2008). Climate change and new modes of mobility are in focus (Nam & Pardo, 2011), and car and bike sharing has entered the mobility market in metropolises next to green buses and (underground) trains in enhancing sustainable public transport.

Summing up, the ICT infrastructure of a city or region is measurable through hard factors like broadband speed, accessibility, and quality as well as the amount of internet users as determined by the number of broadband subscribers and Wi-Fi hotspots. Political willingness to improve the ICT infrastructure or support businesses of the ICT market are understood as advantage. The ICT infrastructure is conquering the city and automotive traffic is increasingly reduced. Finally, information literacy needs more attention and adequate investigation.

Which technical factors are crucial in an investigation of informational world cities? The fact that ICT access, use, and skills are considered important is already evident through the high number of indices on national level which use them as part of their calculation. From the economic perspective, it is still controversial whether ICT companies have to be located in a city to improve the city's ability to become an informational world city. Thus the following hypothesis will be proven:

H1   Informational world cities are hubs for companies with information market activities, e.g. telecommunication companies.

The importance of the ICT infrastructure with regard to informational world city development may usurp old traditions. According to Stock (2011), highways have to be abandoned in favor of more space for recreation, as in the case of Seoul, where the highway in the city center has been demolished to open up the space for walking and leisure. Seoul is a leading city with regard to its ICT industry. Contrasting these two development paths the following hypothesis will be investigated:

H2   The ICT infrastructure in an informational world city is more important than automotive traffic infrastructure.

3.1.2 Cognitive infrastructure

Humans are in the focus in the investigation of informational world cities. In an economy in which knowledge is understood as crucial growth factor (Machlup, 1962), humans are important to transform information into knowledge and vice versa (Kuhlen, 1995). As already described in the first chapter, the topic knowledge city or creative city is often referred to when it comes to the investigation of the knowledge society. The infrastructure of these city types can be combined as "cognitive infrastructure" (Stock, 2011). It is a soft infrastructure, as it is not measurable according to hard facts like the ICT infrastructure (Setunge & Kumar, 2010). According to Stock (2011, p. 970), the cognitive infrastructure of informational cities can be investigated by two types of activity:
– Scientific–technical–medical activities and the results thereof ('knowledge city'),
– Creative–artistic activities and the results thereof ('creative city').

As activities need a place of origin, I will add institutions of the knowledge and creative city as further part within the cognitive infrastructure. In particular, libraries inhabit a special role in the twenty-first century as meeting, learning, and creative spaces as well as knowledge providers. In the following I will discuss the key factors that define a knowledge city as well as a creative city and the role of libraries within this context.

Knowledge city
The knowledge city infrastructure includes institutions like universities, enterprises of the knowledge economy, and research and development units. Kunzmann (2004), for example, defines the knowledge city as a home for knowledge institutions like public and private universities, publicly funded research institutions (e.g. the Frauenhofer Institute or Max Planck Institute in Germany), and research institutes or centers operated by private corporations. The agglomeration of these institutes is also called the "triple helix" (Kunzmann, 2004). Universities and enterprises in such cities benefit from each other. Corporations have the money to invest in research or pay for professors, and universities educate knowledge workers, which are needed in corporations, and help knowledge workers to open innovative startups (Van Winden, 2009). According to Carrillo (2004, p. 44) "[knowledge cities] are becoming the scenario for probably the most substantial development in human evolution."

The basis of knowledge is education. Education has an overwhelmingly positive impact and is related to the status and income of a person in society (Heidenreich, 2002). Thus, enrolment in secondary and tertiary education is part of the HDI, and the rate of adult alphabetization is included in several ICT indices as well. In addition, life-long learning as well as an adequate educational system are seen as part of the continuous development of the knowledge society (Lim, 1999). Therefore, institutions like universities and libraries are needed (Stock, 2011), as libraries are able to offer life-long learning.

Referring to the education level, scientific and technical competencies could be used as an indicator of knowledge flows. Stock (2011, p. 970) emphasizes that "[t]he knowledge society may be concerned with all sorts of knowledge," among them scientific knowledge of STM – science, technology, and medicine – , because economic development is based on scientific innovations. In the biotechnology sector, for example, scientists are currently working on new biofuels that could end the global dependency on limited fossil fuels such as oil and gas (Department of Energy, n.d.), a development that would have a substantial impact on future economic growth. In the field of medicine, scientific innovation can, for example, lead to an increase of the population's life expectancy (Nefiodow, 2006), which is why the HDI includes the indicator "life expectancy at birth" (UNDP, 2014).

Stock (2011) proposes to use informetric investigations to analyze scientific output by the number of patents and scientific articles and its impact by the number of citations of these publications. These are the information flows from a scientific point of view. Powell and Snellman (2004, p. 203) show that a patent investigation reflects the development of the economy, as proven by the evaluation of patents granted in the US: "The upsurge in overall patenting activity is driven by the emergence of new industries, highly fertile in terms of generation of novel ideas and new products. In tandem, there is a decline in traditional sectors."

In summary, on the city level the investigated infrastructure related to the knowledge city is based on places, education, and information flows. Places refer to institutions of higher education, science parks, libraries, and enterprises of the knowledge economy (Carrillo, Yigitcanlar, García, & Lönnqvist, 2014; Kunzmann, 2004). Education plays a significant role, as secondary and tertiary enrolment are both part of the HDI. Finally, information flows refer to scientific articles, patents, and citations (Powell & Snellman, 2004; Stock, 2011).

Thus, the question arises how to measure the knowledge infrastructure? As knowledge institutions are the origin of knowledge output and are physically visible in the city they could be understood as the foundation of the knowledge output. This should not be taken to mean that all knowledge institutions produce the same output. However, the trend in informational world cities is to agglomerate science and development at one place, e.g. at science parks or university clusters (Kunzmann, 2004). Thus, the following hypothesis will be investigated:

H3 Science parks or university clusters that cooperate with knowledge-intensive companies are important in an informational world city.

The investigation of information flows as relating to informational cities should also be considered an important pursuit, but would exceed the scope of this thesis and therefore will not be included. However, quality rankings of knowledge institutions include indices like the number of patents and scientific publications as indicator to which I will refer in the following.

Creative city
According to Florida (2003b, p. 40) three types of creativity exist: "(1) technological creativity or innovation, (2) economic creativity or entrepreneurship, and (3) artistic and cultural creativity." Creativity is a driver of economic growth. Thus, Florida (2003a) refers to the "creative capital," which is a type of human capital, that, if it can be found in a place, e.g. a city, will be a magnet for further creative people, as they prefer open, tolerant and diverse environments. Further, he defines the "3 T's" of the creative class that are crucial factors of economic growth: technology, talent, and tolerance (Florida, 2002). He investigates this by correlating high-tech places with places with the highest concentration of gay people according to the "Gay Index" and his "Bohemian Index," which counts the penetration rate of artists, writers, and performers in a region (Florida, 2014). Florida (2003a) also defines the *creative class* on a broader spectrum not only limited to artists, writers, and performers, but also including occupations like scientists and engineers, university professors, non-fiction writers, editors, think-tank researchers, and analysts.

From an economic perspective the creative class is often referred to as the "creative industry," "cultural industry," or "creative economy" (Flew, 2013; Howkins, 2001; Pratt, 2008). Thus far, no consensus has been established on which sectors belong to the creative industries or even on which term should be used, creative or cultural industry (Hölzl, 2006). Singapore, for example, uses a classification framework for the creative industries which consists of three parts: arts and culture, design, and media. The British Council (Collis, Felton, & Graham, 2010) advocates dividing the creative sector into four domains: (1) books and press, (2) audio-visual, (3) performance, and (4) visual arts and design. For Stock (2011), the creative class overlaps with the definition of knowledge workers. Accordingly, Florida (2003a, p. 8) argues that *"the creative class also includes 'creative professionals' who work in a wide range of knowledge-intensive industries."* Thus, a clear separation of the knowledge and creative city is not feasible, and both parts are best jointly considered as the cognitive infrastructure of a city.

Just as there are differences in the terminology and classification of cultural and creative industries, they also have a diverse impact if we compare this development on a global scale. Murugadas, Vieten, Nikolic, Fietkiewicz, and Stock (2015) analyze the impact of creative industries (including culture and arts, and innovation in the sense of publications and patents) on the GDP on city level. Their results show that the creative industries in the American cities investigated have a higher correlation with the GDP than in European or Asian cities. Thus, they argue that Florida's results on the impact of creativity and economic growth is true for American cities, but cannot be generalized as a global phenomenon. Further research on the relevance of Florida's creative class theory was undertaken by Hansen, Asheim, and Vang (2009). Accordingly, not all hypotheses of the creative class are true for Europe. Nevertheless, tolerance and attractive environments attract the creative class and have a positive impact on the regional development. They argue for simplifying the theory to 2T's (talent and tolerance), and that a more diverse investigation of occupations would result in a better understanding of the creative class. Identifying patents in the fields of ICT, fashion design, music and other creative sectors could also help to identify locations of creative/competitive cities (Boulton, Brunn, & Devriendt, 2012).

The creative city, in its physical representation, is dedicated to creative industries. Most evident are bohemian neighborhoods in which creative talents and industries settle down (Collis et al., 2010), often established in creative quarters near the city center and with the proximity of talents and industries of the creative class within a city as an important factor. Interlinking is not based on ICT but on face-to-face communication (Storper & Venables, 2004). Thus, places in the city where creatives can meet each other become crucial (Stock, 2011). Local creatives can function as a bridge between the space of flows and the space of places (Castells, 1996). Therefore, local creative clusters which bring together local creatives and authorities are important for the development of a "milieu of innovation" (Camagni, 1995; Hitters & Richards, 2002).

Open Social Learning is a new form of the knowledge market of the twenty-first century which can be dedicated to a new form of innovative milieu:

> OPEN SOCIAL LEARNING ... [includes] spaces for action and entities (Public Private Partnerships, PPP), that use digital technology as a resource to offer LEARNING (focusing on the learner) that is SOCIAL (co-creation and collective use of knowledge) and OPEN (based on the reusability of the learning object). These new organizations "labs" co-create, collaborate, and build alternative roads to bring about real change in the way we learn, innovate, work, enjoy, live ... (OSL crowdmap, n.d., para. 1).

Carrillo et al. (2014) mention, for example, Citilabs as spaces for collaborative learning. Learning initiatives are initiated by citizens and are often related to technology. Furthermore, FabLabs, HackLabs, living lab, Tech Shops or Innovation Boot camps are events or spaces for an open community that is likely to participate and to create something. Cities that are home to creatives need to offer space that can be used accordingly.

In summary, the creative infrastructure includes creative talents, creative industries, proximity and face-to-face interaction. The correlation of the creative class and GDP growth rates is evident for American cities, but cannot be observed on a global scale. Thus, it is questionable whether Florida is right with his assumption of the high impact of creative city indicators on the development of cities in the twenty-first century. Accordingly, the following hypothesis will be investigated:

H4   An informational world city needs to be a creative city.

Further, the agglomeration of creative industries and talents are important to foster innovation. The creative class (including creative talents and knowledge workers) need space within the city to meet and interact. Therefore, the following hypothesis will be investigated in addition:

H5   Physical space for face-to-face interaction is important for an informational world city.

Libraries in informational world cities
According to Hall (1997), access to information either via ICT or face-to-face is one of the main indicators of cities of the twenty-first century. And this is exactly the role of libraries in informational world cities, as library services are dedicated to two spaces: digital and physical (Mainka et al., 2013). Following Ergazakis et al. (2009), libraries offer a tremendous amount of collections through digital access. Thus, a main characteristic of the knowledge city is the digital library that serves the needs of citizens and businesses. Access is, for the most part, cheap or even free of charge. Furthermore, access is not only limited to reference information, e.g. through the library catalog, but also extends to a great deal of full text material (Stock, 2011). This access to information via library networks is only possible through the enhanced ICT infrastructure. A precondition of this service is that citizens and companies located in the city are able to get access through the internet and their own devices. Stock (2011) emphasizes the idea of library networks as service providers independent of their location. One example mentioned is the digital library in Reykjavik, which offers access for the whole population of Iceland to all scientific publications published in specialized journals on Elsevier's Science Direct

(van de Stadt & Thorsteinsdóttir, 2007). Another example is the nation of Singapore: Its library network offers access to a wide range of selected resources of general interest through private and library devices for registered users, and access to specialized resources and databases through devices located at any library branch, e.g. the National Library in the city center (Chellapandi, Wun Han, & Chiew Boon, 2010; Sharma, Lim, & Boon, 2009). Both Iceland's and Singapore's models are very cost-intensive. In comparison, the model of Singapore is less expensive since specialized materials are not provided through the digital network but rather only accessible through devices at the library building (Stock, 2011). The available media are not only dedicated to scientific or specialized materials – fiction is part of the library collection, as well. According to Stehr (2003) research on the knowledge society should not only be interested the knowledge output, but also in the consumption of knowledge.

Libraries both public and specialized are also considered in terms of their economic value – following Mainka et al. (2013) this is evident through the studies of Aabø (2005), Ferreira dos Santos (2009), Ko, Shim, Pyo, Chang, & Chung (2012), Koontz (1992), McCallum & Quinn (2004), Missingham (2005), Morris, Sumsion, & Hawkins (2002), Poll (2003), Saxena & McDougall (2012). In the business sector public libraries are an important service for small and medium-sized companies that do not maintain any libraries or information centers of their own (Ferreira dos Santos, 2009). Libraries serve as document suppliers and as providers of current information, e.g. on mergers and acquisitions in the business sector (Yimin & Zhong, 2008). Public libraries belong to the "soft" location factors, as they cannot be investigated through "hard" facts like infrastructures or wage structure (Mainka et al., 2013). Soft location factors are important in attracting and binding companies as well as the creative class (Florida, 2003a; Umlauf, 2008). Mainka et al. (2013) argue that public libraries, even if they are just a soft location factor or a cultural institution, are important as a physical space within the city, as proven by investigations of cultural amenities and libraries (Evans, 2009).

The economic value of a public library can be investigated through the willingness of costumers to pay for its services, but can also come to light if a public library is to be shut down and citizens start to protest (Aabø, 2005; Hummel, 1990; Ko et al., 2012). As physical place the library can also act as a "place maker" (Skot-Hansen et al., 2013). New library buildings with an attractive architecture are further used as urban planning instruments to revitalize city space. Public libraries in general invigorate their communities by inviting citizens to interact with their library or by publishing content produced by the community. Public libraries offer space for working, meeting, and learning (Mainka et al., 2013). As public library buildings are public space, they offer room for the community and are understood as part of the basic infrastructure of the creative and knowledge city (Florida, 2003a; Landry, 2008; Stock, 2011).

In summary, infrastructure role of libraries comprises their physical and digital services. The digital library refers to information support including access to full text content. According to Dehua and Beijun (2012), knowledge in an informational world city is understood as a service that is ubiquitously available through digital networks and devices. Therefore, the following hypothesis will be investigated in connection with the support of digital content:

H6 A fully developed content infrastructure, e.g. supported by digital libraries, is a characteristic feature of an informational world city.

With regard to the physical space, Mainka et al. (2013) conclude their study with the question of how creative talents and knowledge workers could be attracted to library buildings. Space within the library is changing from being a large repository of printed material towards more space for the community. Thus, I will investigate this changing role of the library as formulated in the following hypothesis:

H7 Libraries are important in an informational world city as a physical place for face-to-face communication and interaction.

3.2 Political will

The political willingness of a city can be observed by its achieved goals and future master plans, which are related to a city's development e.g. in economy, society, and infrastructure. According to Yigitcanlar (2010, p. 395) the following activities are crucial with regard to the political and societal will of an informational city:
- strategic vision and development plans,
- financial support and strong investments,
- agencies to promote knowledge-based urban development,
- an international, multicultural character of the city,
- metropolitan Web portals,
- value creation for citizens,
- creation of urban innovative engines,
- assurance of knowledge society rights,
- low-cost access to advanced communication networks,

- research excellence,
- and robust public library networks.

Political willingness goes hand in hand with societal will. Thus, for example, a multinational character within a city can only be reached by an open-minded society. All of the factors required for an informational city as previously mentioned by Yigitcanlar (2010) can be tracked with the help of the "Knowledge-Based Urban Development" (KBUD) framework, which covers four pillars of development, namely economical, societal, spatial, and institutional (Carrillo et al., 2014). This framework can aid in detecting good governance in line with a knowledge society.

City mayors have recognized the importance of information and communication as a driver of economy and wealth and have started to push their city with governmental future plans. One of the early adopters is Singapore, which developed a master plan in 2005 in which they describe their goal to become "An Intelligent Nation, A Global City, Powered by Infocomm" by 2015 (IDA, n.d.-a). Today they follow a "Smart Nation Vision", which equally refers to the further development of the ICT sector (IDA, n.d.-b). Another example is Barcelona: Their vision is "to become: self-sufficient, with productive neighborhoods, living at a human speed and producing zero emissions. A productive, open, inclusive and innovative city; a living city with enterprising people and organized communities" (Dameri & Rosenthal-Sabroux, 2014, p. 14). For citizens, living in a "Smart" or "Informational City" presents advantages in everyday life. Services become gadgets in their pockets, e.g. in smartphone applications based on city data or services (Mainka, Hartmann, Meschede, & Stock, 2015a). Furthermore, citizens are able to add value to the tremendous amount of data that smart cities are generating, as "the role of sensors/information providers is the only role citizens can play that is specific (not to say exclusive) to smart cities" (Castelnovo, 2015, para. 4).

However, merely having the technological equipment is not a guarantee for having access to information. Thus, Lor and Britz (2007) ask whether a knowledge society is possible without freedom of access to information. Access to information should be affordable for all of society, implying a balance between regulation by the government and the market: "The development of a knowledge society requires freedom. It therefore requires a social system that allows critical thinking, encourages access to the ideas of others, and promotes the freedom to communicate and participate in the sharing of the global body of knowledge" (Lor & Britz, 2007, p. 395). They conclude that it is possible to be an information society but not a knowledge society – if there is no freedom of information, content, and human capacity. Freedom of information is understood as a human right and stated by the United Nations to be essential for a transparent, open, and democratic government, in addition possibly helping to avoid corruption (La

Rue, 2011). Freedom of information is measured for example by Reporters without Borders (n.d., para. 5), who compile the World Press Freedom Index using the following indicators:
- Pluralism
  Measures the degree to which opinions are represented in the media
- Media independence
  Measures the degree to which the media are able to function independently of sources of political, governmental, business and religious power and influence
- Environment and self-censorship
  Analyses the environment in which journalists and other news and information providers operate
- Legislative framework
  Analyses the impact of the legislative framework governing news and information activities
- Transparency
  Measures the transparency of the institutions and procedures that affect the production of news and information
- Infrastructure
  Measures the quality of the infrastructure that supports the production of news and information
- Abuses
  Measures the level of violence and harassment during the period assessed

In addition to freedom of press, the freedom of the internet is a basic indicator of the free flow of information. Accordingly, this is measured by Freedom House (2016, para. 10) with the following indicators on national level:
- **Obstacles to Access** details infrastructural and economic barriers to access, legal and ownership control over internet service providers, and independence of regulatory bodies;
- **Limits on Content** analyzes legal regulations on content, technical filtering and blocking of websites, self-censorship, the vibrancy/diversity of online news media, and the use of digital tools for civic mobilization;
- **Violations of User Rights** tackles surveillance, privacy, and repercussions for online speech and activities, such as imprisonment, extralegal harassment, or cyberattacks.

However, mass media information as well as government information should be open for the public, whether online or offline (Yu & Robinson, 2012). Next to government services and online initiatives, governments have started to make public data available through open data portals in the last few years (Mainka et al., 2015a; Mainka, Hartmann, Meschede, & Stock, 2015b). Those portals may be on the national level (e.g. data.gov in the U.S., or data.gov.uk in the United Kingdom), the regional level (e.g. data.qld.gov.au for the federal state Queensland in Australia or open.nrw/de/dat_kat for the federal state North Rhine-Westphalia in Germany) or the city level (e.g. nycopendata.socrata.com for New York in the US, or data.gov.hk for Hong Kong). This kind of development is often part of an open government strategy which ideally involves different stake holders participating in policy-making and decision-making processes (Bingham, Nabatchi, & O'Leary, 2005; Harrison, Burke, Cook, Cresswell, & Hrdinová, 2011). Also, the understanding of the role of government has changed. With the advent of information and communication technology came issues like e-governance and open government, which reflect new modes of government-business, government-citizen, and government-government interaction (Harrison et al., 2011; Palvia & Sharma, 2007). This is more than just the digitization processes within or between governmental agencies, but rather describes an open discourse between all stakeholders in the city.

The importance of adequate e-government in a knowledge society is stressed by the United Nations e-government survey. Since 2003, the UN has conducted a global survey of countries. Their most recent publication is the ninth edition from 2016 (United Nations, 2016). For their measurement of the "E-Government Readiness Index" they combine the "Telecommunication Infrastructure Index" (TII), "Human Capital Index" (HCI), and "Online Service Index" (OSI). Similar to the previously presented ICT indices, the TII is based on the amount of mobile-cellular subscriptions, fixed-telephone subscriptions, and fixed and wireless-broadband subscriptions, each per 100 inhabitants. In addition, the amount of individuals using the internet as a percentage of the whole population is calculated. For the HCI the adult literacy ratio, the gross educational enrolment ratio, and the expected and mean years of schooling are considered, reflecting the importance of an adequate educational system. The last part, the OSI, is based on the e-government survey undertaken by the UN. The survey focuses on the following topics (United Nations, 2014, p. 191):
- the rising importance of a whole-of government approach and integrated online service delivery;
- the use of e-government to provide information and services to citizens on environment related issues;
- e-infrastructure and its increasing role in bridging the digital divide, with a particular emphasis on the provision of effective online services for the inclusion of disadvantaged and vulnerable groups, such as the poor, the disabled, women, children and youth, the elderly, minorities, etc.;

- the increasing emphasis on service usage, multichannel service delivery, 'open government data', e-procurement;
- the expansion of e-participation and mobile government.

Following Nam and Pardo (2011), political willingness in the sense of smart or informational city initiatives includes citizen engagement. Citizens are not only passive users of the smart city infrastructure, but able to interact smoothly, reshaping the local authority's awareness of the citizens' needs (Pimbert & Wakeford, 2001). It is crucial that governmental development plans should consider the people and their lives (Orofino, 2014). An open government or e-governance strategy should include new ways of e-democracy (Palvia & Sharma, 2007). E-democracy and e-participation, in particular through social media channels, help to enhance the transparency and flow of information, but is still in its infancy (Bonsón, Torres, Royo, & Flores, 2012).

Overall, different governmental issues are of importance when characterizing an informational city. First, from the political point of view it is not clear whether an informational city can establish itself within the global network merely by way of economic market activity – through the invisible hand of the market –, or whether political willingness and more intensive governmental support are needed to establish such a city (Chandler, 2002). Silicon Valley is an example of a city that became an informational city without a master plan, whereas Singapore has made a tremendous transition based largely on governmental future vision plans (Stock, 2011). Accordingly, I will investigate the following hypothesis:

H8 Political willingness is important to establish an informational world city, especially with regard to knowledge economy activities.

Second, governmental services have changed due to the enhanced ICT infrastructure. Information and services have become available ubiquitously. Through online participation and an open governmental culture, we can arrive at a new democratic level within a city. Therefore, I will investigate whether these opportunities are established in informational world cities with the following hypothesis:

H9 An informational world city is characterized by e-governance (incl. e-government, e-participation, e-democracy).

Third, the ability to share information through different media and digital networks does not necessarily imply that there is a free flow of information. Critical thinking and a free flow of information is understood as a human right and essential for becoming a knowledge society. Thus, the following hypothesis will be investigated as part of the political will:

H10   A free flow of all kinds of information (incl. mass media information) is an important characteristic of an informational world city.

3.3 World city

The corporate structure of an informational world city is the foundation for the space of flows as defined by Castells (1989). Flows of information, capital, and power are not determined by administrative city borders but by the interplay of city networks. Thus, for some researchers the world city network is related to the economic development of cities. Taylor, Firth, Hoyler, and Smith (2010, p. 2814), for instance, argue that "world cities do not make the world city network, advanced producer service firms do." The firms located in a city build the relations to other cities that form the network of cities. World city networks are often related to the finance sector. Sassen (2001) sees finance, insurance and real estate (FIRE) as the advanced producer service sector with major impact on globalization.

With regard to the economic development, Egedy, von Streit, and Bontje (2013) state that cities are being recognized as "multi-layered." Multi-layered cities are complex constructs formed by the interplay of physical, socio-economic, and socio-culltural processes. This concept is related to the concept of historic development paths of a city according to Musterd and Murie (2010), and is also closely related to the world city and global city concepts by Taylor (2001) or Sassen (2001). Egedy, von Streit, and Bontje (2013) see Amsterdam, Milan, and Munich as examples of such cities due to the variety of businesses and industries located there, with "one-company" towns on the opposite side of the spectrum. According to Stock (2011, p. 979) the coporate structure of an informational world city covers four types of companies:
1. Capital-Intensive Service Providers
2. Knowledge-Intensive Companies (high-tech industry)
3. Companies in the Information Economy
4. Creative Companies.

Throughout history, cities have played key roles if they were located in a hegemony state (Jacobs, 1969), i.e. a state which leads in production, commerce, and finance (Taylor, Hoyler, & Smith, 2012). Finance is seen as decisive factor for cities to obtain important status within the world network. Following Sassen (2001), finance is part of the preeminent advanced producer service firms in global cities. Cities like London or New York are important world cities because of their location in hegemony states. Taylor, Hoyler, and Smith (2012, p. 29) have shown that "hegemony-making occurs in specific vibrant regions." Cities in a specific region are leading in economic growth, which is related to the plentiful amount of new work encouraged through their networking with each other. Thus, current cities recognized as world city play a key role in finance, commerce, and business (Friedmann & Wolff, 1982). For example, from the European perspective London is a financial center which covers the global market and Frankfurt the hub to serve the European market (Taylor, 2012b). This is evident, for one, through the fact that the European Central Bank has its seat in Frankfurt rather than in London.

Indicators to analyze international finance centers may vary slightly between different research approaches (Lizieri, 2012). The "Global Financial Centers" ranking by Z/Yen, for example, uses a combination of different quantitative indicators of market share, qualities (e.g. quality of the labor market and quality of the infrastructure), and openness (e.g. transparency, regulatory and tax structure), among others. Furthermore, Stock (2011) suggests to use the turnover of the regional stock exchange as indicator of the flow of capital with reference to the flows of spaces (capital, power, and information), as every financial center (e.g. New York, London, or Tokyo) has an own stock exchange market to enable those capital flows.

In summary, the corporate structure of an informational world city is characterized by a mix of companies that cover diverse economic industries (capital-intensive, knowledge-intensive, information market activities, and creative). They cooperate and compete in a global market. To be successful within the global network, the finance sector plays a crucial role. First, being located in a hegemony state and second, being a leader with respect to advanced producer service firms (finance and insurance companies) are both advantages in becoming a global hub within the network of cities. Therefore, I will focus on the financial sector and prove the following hypothesis:

H11   An informational world city has to be a financial hub with a large number of banks and insurance companies.

When describing the formation of a world city network, the term "cityness" is used in diverse ways. For Jacobs (1969), "city-ness" is the relation between cities that form the network of cities. In contrast, local processes that happen within a city are understood as "town-ness," i.e. related to its local economy and civic society (Taylor, Firth, et al., 2010). Cities that have a high number of inter-city relations are referred to as world or global cities (Friedmann & Wolff, 1982; Taylor, 2001) and represent the hubs within the world/global city network. Another definition of "cityness" by Sassen (2001) refers to local urban space. In her definition, "cityness" is the process in a city that happens when all instances interact productively. Thus, the core concept of cityness is "complexity, incompleteness, and making" (Sassen, 2013, p. 209). Cityness does not come about merely because of a certain amount of housing, offices, or factories at one place, and it is not only related to Western understanding and concepts of urbanity. It can also be a different kind of cityness, one which we may not be able to see immediately. She points out cultural differences between cities in different regions, and that western researchers need to be more open to all kinds of urbanity. Finally, both aspects of cityness are of relevance when investigating informational world cities: First, the inner city processes and second, the inter-city relations – or in other words the formation of the global/world city network.

Castells (2001) combines inner and inter-city relations within his theory. For him "glocal" connectivity is a special characteristic of informational cities. Locally, enterprises connect with each other, and globally they are connected with the rest of the world through digital networks (Stock, 2011). The physical connection plays a significant role in the development of cities. Inner city processes as well as their relations and influence on other cities need to be considered. Cities are connected through "asymmetric power relations," which are determined by cities' dominance over other cities through resources like "food, raw materials, cheap industrial goods and labour" (Taylor, 2012b, p. 65). Throughout history, cities have always been part of a network of cities, and do not become world cities simply because of their size, even though it can be assumed that larger cities profit from larger inter-city flows, e.g. trade. Jacobs (1969, 1984) defines cities in a network as the hubs that are the cause of the development of the modern world. She argues that economic growth depends on the interaction between these cities. Taylor et al. (2010) argue that the development of world city networks should be referred to as "central flow theory" because the economy, namely the firms located in a city and building networks through branches in other cities, are the main actors of this process of constitution.

For the investigation of world or global cities it is important to acknowledge that cities are not static. They are understood as processes (Castells, 1996; Taylor, 2012a): "Every city is constituted of myriad urban processes represented by a particular outcome at the point of study. Thus, the world city process is very strong in London and New York and this is reflected in their network measurements and the fact that they are often applauded as 'global cities' (Sassen, 2001)" (Derudder, Hoyler, Taylor, & Witlox, 2012, p. 3). The term world is understood as the contemporary world with respect to historical systems such as the "Roman World" (Taylor, 2012c). Thus, contemporary cities are in the focus of world city or global city network analysis. In the following I will adopt the definition of "city-ness" of Jacobs and Taylor, derived from inter-city flows that are also used in Castells' (1996) theory of spaces of flows (capital, power, and information). Furthermore, the economic definition of a city is not restricted to political-administrative boundaries (Stock, 2011), e.g. Greater London or City of London according to its two authorities. Following Castells (2010), this is due to the fact that we have arrived in the network society and networks have no boundaries. Similarly, Friedmann (1995, p. 23) defines world cities as "large, urbanized regions that are defined by dense patterns of interactions rather than by political-administrative boundaries."

Popular city researchers of world cities or global cities are Peter Hall *(The World Cities*, 1966), John Friedmann (*The World City Hypothesis*, 1986), Saskia Sassen (*The Global City*, 2001), and Peter Taylor (Director of the Globalization and World Cities (GaWC) Research Network). According to Hall (1966) the term world city has by now been in use for more than 100 years and is predominately used to describe cities that act as centers for government, economy, and human activity. Taylor (2012a, p. 11) defines a world city as a city within a network of cities. For him the "world" in "world cities" does not refer to "worldwide" but to a "distinctive fragment of the myriad societies in the world." Therefore, cities can be understood as "cities in globalization" (Taylor, Derudder, Saey, & Witlox, 2007). Because of this global process and its connection, Sassen (2001) prefers to call such cities *global city*. The status of a world city is also measured according to different approaches. In the following, I will present some popular methods and highlight those which are characteristic for an informational world city.

According to the analysis of economic flows, a hierarchical order of world cities can be identified (Friedmann, 1995). In the literature there is a difference in the analysis of cities according to whether the terms "world city" or "global city" was used. World cities are traditionally analyzed through information on multinational corporations and global cities through information on producer services firms (Derudder, De Vos, & Witlox, 2012). Since both approaches relate to the emergence of a global network based on economic processes it is useful to combine them for further investigations. Derudder, De Vos, and Witlox (2012) do so and identify a correlation between the outputs of both approaches. However, their results also show us that cities with a high producer service firm value do not always have a high multinational corporation value. In the following, I will refer just to the term world city as there is no reason to distinguish between both terms. Since both methods are based on economic data, I see no reason to maintain the distinction between both terms.

The method to analyze world cities based on multinational corporations begins with existing lists or entries of firms, e.g. "Fortune's Global 500" for the largest firms in the world, or "Polk's World Bank Directory" for international bank offices (Alderson & Beckfield, 2012). The network can be built with hierarchical orders, e.g. headquarters get higher values in the coding schema, or vertical connections, e.g. the relation between headquarters and subsidiaries are interpreted as collaborative. Advanced producer service firms cannot be found in a structured list. Taylor, Ni, and Derudder (2010) use website information of those firms and code them according to their office size and respective importance, e.g. headquarters are coded with the highest score. Their investigation results in a network that is based on three layers: The net level comprises the space of flows of the world economy, the node level represents the cities, and the third level, the so-called sub-nodal level, relates to advanced producer service firms. This model is called the interlocking network model (Taylor, 2012c; Taylor, Derudder, Hoyler, & Witlox, 2012; Taylor, Ni, et al., 2010). It is an investigation of contemporary world cities, because advanced producer services firms expand after 1990 (Taylor, Derudder, et al., 2012). From this time, the study starts to build its economic world city network.

A similar investigation is undertaken by Wall and van der Knaap (2012). They analyze companies starting with the top 100 headquarters of the Fortunes 500 list in 2005, and then city networks by the company's ownership of subsidiaries in more than 2000 cities worldwide. Their results show "that 84 per cent of the network occurs between and not within cities and that 70 per cent of European and North American ties extend beyond their supra regions, it is shown that cities have become dissociated from their local geographies (Friedmann, 1986)" (Wall & van der Knaap, 2012, p. 225). According to Friedmann's (1986) definition of city hierarchies, they argue that cities link hierarchically and horizontally between different city levels.

A world city network may also be built by other economic players. For example, the companies which provide the internet infrastructure are not part of the advanced producer service sector but they are important for the construction of informational world cities (Malecki, 2012). According to Graham (1999), world cities are providers of telecommunication infrastructure due to their agglomeration of financial service firms and transnational corporations. Thus, both kind of firms are located in world cities. To investigate ICT network flows between cities no clear measurements are available, it is, for example, not possible to get a total count of emails between employees of all companies (Taylor, 2004). However, it is possible to identify global internet hubs by "Internet backbone capacity (Townsend, 2001b), number of internet domain names (Zook, 2001; Sternberg and Krymalowski, 2002), and internet exchange points (IXPs) (Malecki, 2002; Devriendt et al., 2010)" (Malecki, 2012, p. 117). Malecki (2012) argues that it is important to include content providers such as Google and cloud computing according to their colocation facilities (co-location facilities host servers, storage, and networking equipment for content providers).

Taylor (2012b) maintains that city hierarchies do not exist on a global scale, as hierarchies are limited to political processes. One example for him is the relationship between London and Paris. London is often listed higher in city rankings than Paris. This does not mean that London is able to tell Paris "what to do," but rather indicates that London for instance has more company headquarters. However, Paris also has headquarters with branches in London. Thus, there exists a horizontal relationship in that Paris headquarters instruct branches in London and vice versa. On a political level London stands higher in a hierarchical order within other cities in the United Kingdom because of decisions which are taken in London on other cities. That there is no evidence for a hierarchy on a global level is also reflected in Friedmann's (1986) world city system, in which world cities are split into primary and secondary world cities with horizontal relations. From an economic perspective, a city like London also stands higher in a hierarchy comparing to its hinterland. Taylor, Hoyer, and Verbruggen (2010) describe this phenomenon as "central place theory." Furthermore, they introduce the "central flow theory" with reference to Castells theory of the importance of the space of place as well as space of flows, which are both prevalent in a network of cities.

Cities are also defined as world cities because of their political power. Important to a political world city is that the city plays a significant role in a global network. This could be because a city is home to political actors with a global reach or because of the global reach of decisions that are taken in such a city, e.g. Washington or Brussels (van der Wusten, 2012). Another aspect can be the presence of diverse non-governmental organizations and universities that are globally oriented. They are referred to as "civil society hubs," e.g. The World Economic Forum which aims to foster public-private cooperation worldwide (van der Wusten, 2012). A third aspect is the "symbolic significance" of a political city, e.g. the White House in Washington (van der Wusten, 2012).

Networks also can be created through political decisions to join a network of nations, such as the European Union (Taylor, 2012b). The aim is to build relations of sharing "best practice" expertise and enhanced transparency on policy information within this network. Inter-city relations also play an important role when it comes to infrastructure investments such as building new airports. But Taylor (2012b) considers the economy, specifically the agents who are the important actors in inter-city relation networks, as the main users of new infrastructure. For him the inter-city network is built through intra-firm relations of service firms.

A further method of measuring world cities is based on the physical inter-city connectivity, as calculated for example through airline networks. A high level of accessibility is analyzed by the number of direct flights originating from or destined for a city or its respective metropolitan region (Derudder & Witlox, 2005). For Grubesic and Matisziw (2012), direct connections indicate an important role of a city in the world system. They argue that further calculations can be derived from airline networks, e.g. minimum cost of travelling by calculating the minimal number of steps (e.g. Shimbel distance calculation) or minimum distance (e.g. L-matrix calculation). Airline connectivity reflects the actual need of consumption and changes depending on demand (Grubesic & Matisziw, 2012). According to Budd (2012, p. 151), airports bring nations closer together in "time and space." Airports that offer direct flights every day to global airport hubs should also be considered within such calculations (Budd, 2012). It is, however, questionable how important airports will remain in the future. Budd (2012) suspects that increasing fear of terrorism and fluctuating oil prices might result in fewer flights, which could turn flying into a luxury good in the future. Still, most global cities are also harbor cities because of the historical importance of shipping goods leading to global flows of trade and commerce (King, 2012). Thus, alternative flows could be detected in more traditional physical connections.

In addition, entrepreneurship can be seen as a globalizing phenomenon. In her book *From The Other Side of The World*, Bayrasli (2015) reports on entrepreneurs from the seven countries Mexico, Nigeria, China, India, Russia, Turkey, and Pakistan – specifically, young entrepreneurs who have learned their craft (for example in Silicon Valley), go back to their home country, and develop start-ups to solve local problems (Bayrasli, 2015). Unfortunately, flows of knowledge cannot be investigated like the flows of power of producer service firms, e.g. by counting headquarters and their branches. The early stages of entrepreneurs are not countable in such a way, but in later stages, if the business idea is fruitful, branches may go global. Cities which are following the trend of entrepreneurship can be identified through their amount of newly founded enterprises and self-employment rates (Murugadas et al., 2015). Many cities like to foster entrepreneurship through diverse initiatives, but Bayrasli (2015) argues that the implementation of a "Silicon Anything" like the Silicon Alley in New York or Silicon Cape in South Africa will not lead to the next Silicon Valley. Furthermore, an investigation of the impact of entrepreneurship on the GDP in metropolitan regions has shown that there is no significant correlation which could indicate a positive effect to the economy (Murugadas et al., 2015). This may be due to the still relatively small number of newly founded enterprises or low self-employment rates in comparison with cities' population growth. Nevertheless, the globalization of entrepreneurship as described in Bayrasli's (2015) report opens the market and space for competition, which may result in fruitful ideas and economic growth in future.

Accordingly, to measure knowledge flows from an economic perspective, Faulconbridge and Hall (2012) have identified three forms of flow and interconnectivity: First, city-related knowledge production is evident through clusters and universities in a city; second it is demonstrated by flows between cities through trade fairs or conferences; and third knowledge is produced in spaces between cities such as organizational communities of practice in which company members located in different cities share their knowledge through calls, meetings, and other methods.

Further, the cultural economy has gained in importance as a source of employment and an economic actor in many places. In London, for instance, it has become the third largest sector of the economy in the twenty-first century (Pratt, 2012). As Pratt (2012) argues, the cultural economy is very complex and its actions and flows are not easy to analyze in terms of headquarters and subsidiaries. More appropriate and sensitive tools are needed to determine the impact of the cultural economy upon and within world cities and globalization. One idea is to analyze global networks of huge media firms. Watson (2012) investigates transnational media corporations (TNMCs) because they belong to the largest firms in the world and serve a global market which is borderless due to digital connection through the internet. It is not easy to identify true indices that make TNMCs global. Watson (2012, p. 284) adapts the methodology of global investigations of advanced producer service firms (see Taylor, 2001) and analyzes the "transnational office networks of 25 of the world's largest TNMCs," referring to Castells' (2009) theory that the network of this multinational companies makes them global. Thus, the investments, labor, production, and distribution interlock these organizations. Watson (2012) concludes that the two largest media fields by media production are the Pacific Rim and the US. He argues that TNMCs are not truly global but rather regionally focused, especially considering the total sales of TNMCs, which are largely domestic. Nevertheless, he argues that the combination of TNMCs and smaller media organizations located in cities make a city global and interlock the city in diverse levels of global flows and networks. He calls such cities "global media cities."

The research methods presented so far force us to pose the question: What leads to the existing network? Is it "city networks" or a "network of cities"? According to Allen (1999), Sassen is concerned with the city as place – i.e. referring to the city networks – and Castells puts the flows between cities in the focus of his theory, thus referring to the network of cities. Both have been investigated according to connectivity between different instances. As such, the space of flows is "glocal." Secondly, we have to ask: *what* flows? Following Stock (2011, p. 981), informational world cities can be measured by three types of flows: "(a) the amount of capital flows (stock exchange turnover), (b) the amount of power flows (as indicated by the sum of profits of their companies), and (c) the amount of information flows (the connectivity of the city with regard to business and STM publications/citations)." According to Boulton, Brunn, and Devriendt (2012), world city networks can consist of more than just flows and acknowledge that there is a big research gap in analyzing cities beyond flow theories. They conceptualize smart world cities in three dimensions of cyber infrastructure: physical, human, and soft. The physical infrastructure refers to airlines, visitor facilities like hotels and conference halls, and includes the high-tech fiber optic technologies which are the basis of modern telecommunication. The human infrastructure refers to the traditional human capital (labor) and, finally, the soft infrastructure refers to informational "cloud" web domains, cyberspaces, and user-generated and geo-referenced spatial data. Drawing on King (2012), further categories need to be added: (1) political and cultural criteria, (2) infrastructural criteria, and (3) visual and architectural criteria (Table 3.1).

Table 3.1: World city categories by King (2012, p. 32).

Political and cultural criteria	Infrastructural criteria	Visual and architectural criteria
– Presence of international organizations of governance and administration – A minimally large and diverse population – Media – Communications and publishing activities – Museums – Theatres – Educational facilities – Religious sites for pilgrimage and tourism – The capacity to sponsor global sporting or cultural events	– Major airport and transport hub	– Skyscrapers – Distinctive skyline

Thus, King (2012) merges many criteria to describe the constellation of a world city. As this section has demonstrated, identifying criteria for a world city is very controversial. A multitude of different factors such as multinational corporations, advanced producer service firms, large internet firms, media corporations, entrepreneurs, physical connectivity through air or water, and many others have been proposed. At least, we do not need to differentiate between global cities and world cities – as explained above, both can be used interchangeably. Combining world city and informational city theory, both have in common that they investigate the space of flows, no matter whether they are digital, economical, or knowledge flows. The essential characteristic is the resulting network which is genuinely a feature of the network society. In summary, I will investigate if being a world city is a necessary characteristic of an informational city according to following hypothesis:

H12   An informational city has to be a global city ("world city").

3.4 Hypothesis overview

Based on the literature investigated, a number hypotheses were derived above, and are given in summary below. They will be investigated through methods used from grounded theory (see chapter 4, "Methods"), mainly because measuring cities in terms of the amount of people living in a city, internet access per household, GDP or other hard statistical facts assumes that everything can be measured with the same scale – which is clearly not the case. For instance, the educational systems between Asia and Europe differ in many ways. In Asia for nearly every profession at least a Bachelor degree is required (SH 1, personal communication, July 13, 2012), while in Europe one may enter a profession through apprenticeships or on-the-job training and then receive a vocational degree from the local Chamber of Industry and Commerce. The bare numbers are unable to reflect the culture or identity of a city, which is equally important in detecting cities of the knowledge society. Following now are the hypotheses concerning the main characteristics of informational world cities in the twenty-first century:

H1   Informational world cities are hubs for companies with information market activities, e.g. telecommunication companies.

H2   The ICT infrastructure in an informational world city is more important than automotive traffic infrastructure.

H3   Science parks or university clusters that cooperate with knowledge-intensive companies are important in an informational world city.

H4   An informational world city needs to be a creative city.

H5   Physical space for face-to-face interaction is important for an informational world city.

H6 A fully developed content infrastructure, e.g. supported by digital libraries, is a characteristic feature of an informational world city.

H7 Libraries are important in an informational world city as a physical place for face-to-face communication and interaction.

H8 Political willingness is important to establish an informational world city, especially with regard to knowledge economy activities.

H9 An informational world city is characterized by e-governance (incl. e-government, e-participation, e-democracy).

H10 A free flow of all kinds of information (incl. mass media information) is an important characteristic of an informational world city.

H12 An informational world city has to be a financial hub with a large number of banks and insurance companies.

H13 An informational city has to be a global city ("world city").

References

Aabø, S. (2005). Valuing the benefits of public libraries. *Information Economics and Policy*, *17*(2), 175–198. https://doi.org/10.1016/j.infoecopol.2004.05.003

Alderson, A. S., & Beckfield, J. (2012). Corporate networsk of world cities. In B. Derudder, M. Hoyler, P. J. Taylor, & F. Witlox (Eds.), *Handbook of globalization and world cities* (pp. 126–134). Cheltenham, UK; Northampton, MA: Edward Elgar Publishing.

Allen, J. (1999). Cities of power and influence: Settled formations. In J. Allen, D. Massey, & M. Pryke (Eds.), *Unsettling cities: movement/settlement* (pp. 181–218). London, UK: Routledge.

Bayrasli, E. (2015). *From the other side of the world: Extraordinary entrepreneurs, unlikely places*. New York, NY: Perseus Books Group.

Beutelspacher, L. (2014). Assessing information literacy: Creating generic indicators and target group-specific questionnaires. In *Information literacy. Lifelong learning and digital citizenship in the 21st century* (pp. 521–530). Springer International Publishing. https://doi.org/10.1007/978-3-319-14136-7_55

Bingham, L. B., Nabatchi, T., & O'Leary, R. (2005). The new governance: Practices and processes for stakeholder and citizen participation in the work of government. *Public Administration Review*, *65*(5), 547–558. https://doi.org/10.1111/j.1540-6210.2005.00482.x

Bonsón, E., Torres, L., Royo, S., & Flores, F. (2012). Local e-government 2.0: Social media and corporate transparency in municipalities. *Government Information Quarterly*, *29*(2), 123–132. https://doi.org/10.1016/j.giq.2011.10.001

Boulton, A., Brunn, S. D., & Devriendt, L. (2012). Cyberinfrastructures and "smart" world cities: physical, human and soft infrastructures. In B. Derudder, M. Hoyler, P. J. Taylor, & F. Witlox (Eds.), *International handbook of globalization and world cities* (pp. 198–205). Cheltenham, UK; Northampton, MA: Edward Elgar Publishing.

Brezina, C. (2011). *Understanding the gross domestic product and the gross national product.* New York, NY: The Rosen Publishing Group.

Bruno, G., Esposito, E., Genovese, A., & Gwebu, K. L. (2011). A critical analysis of current indexes for digital divide measurement. *The Information Society, 27,* 16–28. https://doi.org/10.1080/01972243.2010.534364

Budd, L. C. S. (2012). Airports: From flying fields to twenty-first century aerocities. In B. Derudder, M. Hoyler, P. J. Taylor, & F. Witlox (Eds.), *International handbook of globalization and world cities* (pp. 151–161). Cheltenham, UK; Northampton, MA: Edward Elgar Publishing.

Camagni, R. P. (1995). The concept of innovative milieu and its relevance for public policies in European lagging regions. *Papers in Regional Science, 74,* 317–340. https://doi.org/10.1111/j.1435-5597.1995.tb00644.x

Carrillo, F. J. (2004). Capital cities: A taxonomy of capital accounts for knowledge cities. *Journal of Knowledge Management, 8*(5), 28–46. https://doi.org/10.1108/1367327041058738

Carrillo, F. J., Yigitcanlar, T., García, B., & Lönnqvist, A. (2014). *Knowledge and the city: Concepts, applications and trends of knowledge-based urban development.* New York, NY: Routledge.

Castells, M. (1989). *The informational city: Information technology, economic restructuring, and the urban-regional process.* Oxford, UK: Blackwell.

Castells, M. (1996). *The rise of the network society.* Malden, MA: Blackwell.

Castells, M. (2001). *The internet galaxy. Reflections on the internet, business, and society.* Oxford, UK: Oxford University Press.

Castells, M. (2009). *Communication power.* Oxford, UK: Oxford Univ. Press.

Castells, M. (2010). Globalisation, networking, urbanization: Reflections on the spatial dynamics of the information age. *Urban Studies, 47*(13), 2737–2745. https://doi.org/10.1177/0042098010377365

Castelnovo, W. (2015). Citizens as sensors/information providers in the co-production of smart city services. In *Proceedings of the 12th Conference of the Italian Chapter of AIS.* Rome, IT.

Chandler, A. D. J. (2002). *The visible hand. The managerial revolution in American business* (16th ed.). Cambridge, UK; London, UK: Harvard University Press (Original work published 1977).

Chellapandi, S., Wun Han, C., & Chiew Boon, T. (2010). The National Library of Singapore experience: Harnessing technology to deliver content and broaden access. *Interlending & Document Supply, 38*(1), 40–48. https://doi.org/10.1108/02641611011025361

Collis, C., Felton, E., & Graham, P. (2010). Beyond the inner city: Real and imagined places in creative place policy and practice. *The Information Society, 26*(2), 104–112. https://doi.org/10.1080/01972240903562738

Dameri, R. P., & Rosenthal-Sabroux, C. (Eds.). (2014). *Smart City. How to Create Public and Economic Value with High Technology in Urban Space.* Cham, Heidelberg, New York, Dordrecht, London: Springer. https://doi.org/10.1007/978-3-319-06160-3

Dehua, J., & Beijun, S. (2012). Internet of knowledge. The soft infrastructure for smart cities [in Chinese]. In *In Smart City and Library Service. The Proceedings of the Sixth Shanghai International Library Forum* (pp. 147–154). Shanghai, CHN.

Department of Energy. (n.d.). About the Bioenergy Technologies Office: Growing America's energy future. Retrieved from http://www.energy.gov/eere/bioenergy/about-bioenergy-technologies-office-growing-americas-energy-future

Derudder, B., De Vos, A., & Witlox, F. (2012). Global city/world city. In B. Derudder, M. Hoyler, P. J. Taylor, & F. Witlox (Eds.), *International handbook of globalization and world cities* (pp. 73–82). Cheltenham, UK; Northampton, MA: Edward Elgar Publishing.

Derudder, B., Hoyler, M., Taylor, P. J., & Witlox, F. (2012). Introduction: A relational urban studies. In B. Derudder, M. Hoyler, P. J. Taylor, & F. Witlox (Eds.), *International handbook of globalization and world cities* (pp. 1–4). Cheltenham, UK; Northampton, MA: Edward Elgar Publishing.

Derudder, B., & Witlox, F. (2005). An appraisal of the use of airline data in assesing the world city network: A research note on data. *Urban Studies, 42*(13), 2371–2388.

Economist Intelligence Unit. (2009). *E-readiness rankings 2009. The usage imperative.* London, UK; New York, NY; Hong Kong, HK. Retrieved from http://graphics.eiu.com/pdf/e-readiness%20rankings.pdf

Economist Intelligence Unit. (2010). *Digital economy rankings 2010: Beyond e-readiness.* London, UK; New York, NY; Hong Kong, HK. Retrieved from http://graphics.eiu.com/upload/EIU_Digital_economy_rankings_2010_FINAL_WEB.pdf

Egedy, T., von Streit, A., & Bontje, M. (2013). Policies towards multi-layered cities and cluster development. In S. Musterd & Z. Kovács (Eds.), *Place-making and policies for competitive cities* (pp. 35–58). Oxford, UK: John Wiley & Sons. https://doi.org/10.1002/9781118554579.ch4

Ergazakis, E., Ergazakis, K., Metaxiotis, K., & Charalabidis, Y. (2009). Rethinking the development of successful knowledge cities: An advanced framework. *Journal of Knowledge Management, 13*(5), 214–227. https://doi.org/10.1108/13673270910988060

Evans, G. (2009). Creative cities, creative spaces and urban policy. *Urban Studies, 46*(5–6), 1003–1040. https://doi.org/10.1177/0042098009103853

Faulconbridge, J., & Hall, S. (2012). Business knowledges within and between the world city. In B. Derudder, M. Hoyler, P. J. Taylor, & F. Witlox (Eds.), *International handbook of globalization and world cities* (pp. 230–231). Cheltenham, UK; Northampton, MA: Edward Elgar Publishing.

Ferreira dos Santos, V. (2009). Public libraries and their contribution towards economic development: A discussion. *LIBRES. Library and Information Science Research Electronic Journal, 19*(2), 1–9.

Flew, T. (2013). *Creative industries and urban development: Creative cities in the 21st century.* New York, NY: Routledge.

Florida, R. L. (2002). *The rise of the creative class: And how it's transforming work, leisure, community and everyday life.* New York, NY: Basic Books.

Florida, R. L. (2003a). Cities and the creative class. *City and Community, 2*(1), 3–19. https://doi.org/10.1111/1540-6040.00034

Florida, R. L. (2003b). Entrepreneurship, creativity, and regional economic growth. *The Emergence of Entrepreneurship Policy*, 39–58.

Florida, R. L. (2014). The creative class and economic development. *Economic Development Quarterly, 28*(3), 196–205. https://doi.org/10.1177/0891242414541693

Freedom House. (2016). About freedom on the net. Retrieved from https://freedomhouse.org/report-types/freedom-net

Friedmann, J. (1986). The world city hypothesis. *Development and Change, 17*(1), 69–83. https://doi.org/10.1111/j.1467-7660.1986.tb00231.x

Friedmann, J. (1995). Where we stand. A decade of world city research. In P. L. Knox & P. J. Taylor (Eds.), *World cities in a world-system* (pp. 21–47). Cambridge, UK: Cambridge University Press.
Friedmann, J., & Wolff, G. (1982). World city formation: An agenda for research and action. *International Journal of Urban and Regional Research, 6*(3), 309–344. https://doi.org/10.1111/j.1468-2427.1982.tb00384.x
Giffinger, R., Fertner, C., Kramar, H., Kalasek, R., Pichler-Milanovic, M., & Meijers, E., (2007). Smart cities ranking of European medium-sized cities. (Technical Report) Vienna University of Technology. Retrieved from http://www.smart-cities.eu/download/city_ranking_final.pdf
Graham, S. (1999). Global grids of glass: On global cities, telecommunications and planetary urban networks. *Urban Studies, 36*(5–6), 929–949.
Grubesic, T. H., & Matisziw, T. C. (2012). World ctities and airline networks. In B. Derudder, M. Hoyler, P. J. Taylor, & F. Witlox (Eds.), *International handbook of globalization and world cities* (pp. 97–116). Cheltenham, UK; Northampton, MA: Edward Elgar Publishing.
Hall, P. (1966). *The world cities*. London, UK: Weidenfeld and Nicolson.
Hall, P. (1985). The geography of the fifth Kondratieff. In P. Hall & A. Markusen (Eds.), *Silicon landscapes* (pp. 1–19). Boston, MA: Allen & Unwin.
Hall, P. (1997). Modelling the post-industrial city. *Futures, 29*(4–5), 311–322. https://doi.org/10.1016/S0016-3287(97)00013-X
Hansen, H. K., Asheim, B., & Vang, J. (2009). The European creative class and regional development: How relevant is Florida's theory for Europe? In L. Kong & J. O´Connor (Eds.), *Creative economies, creative cities. Asian-European perspectives* (pp. 99–120). London, UK; New York, NY; a.o.: Springer. https://doi.org/10.1007/978-1-4020-9949-6_7
Harrison, T. M., Burke, G. B., Cook, M., Cresswell, A., & Hrdinová, J. (2011). Open government and e-government: Democratic challenges from a public value perspective. In *12th Annual International Conference on Digital Government Research* (pp. 245–253). College Park, MD: ACM.
Heidenreich, M. (2002). *Merkmale der Wissensgesellschaft. Lernen in der Wissensgesellschaft [Characteristics of the knowledge society. Learning in the knowledge society]*. Innsbruck, AUT: StudienVerlag.
Henkel., M., & Stock, W. G. (2016). "We have big plans." – Information literacy instruction in academic and public libraries in the United States of America. In *Proceedings of the 2nd International Conference on Library and Information Science* (pp. 159–175). Taipeh, TWN: International Business Academics Consortium.
Herrera, J. C., Work, D. B., Herring, R., Ban, X. J., Jacobson, Q., & Bayen, A. M. (2010). Evaluation of traffic data obtained via GPS-enabled mobile phones: The mobile century field experiment. *Transportation Research Part C: Emerging Technologies, 18*(4), 568–583. https://doi.org/10.1016/j.trc.2009.10.006
Hitters, E., & Richards, G. (2002). The creation and management of cultural clusters. *Creativity and Innovation Management, 11*(4), 234–247. https://doi.org/10.1111/1467-8691.00255
Hollands, R. G. (2008). Will the real smart city please stand up? *City, 12*(3), 303–320. https://doi.org/10.1080/13604810802479126
Hollands, R. G. (2014). Critical interventions into the corporate smart city. *Cambridge Journal of Regions, Economy and Society*. https://doi.org/10.1093/cjres/rsu011
Hölzl, K. (2006). Creative industries in Europe and Austria Definition and potential. In S. Karkulehto & K. Laine (Eds.), *Call for Creative Futures Conference Proceedings* (pp. 36–51). Oulu, FI: Department of Art Studies and Anthropology.
Howkins, J. (2001). *The creative economy: How people make money from ideas*. London, UK: Penguin Books.

Hummel, M. (1990). Kultur als Standortfaktor [Culture as Location Factor]. *Ifo Schnelldienst*, *43*(10/11), 3–10.
IBM. (2010). *Smarter cities for smarter growth. How cities can optimize their systems for the talent-based economy.* Somers, NY. Retrieved from www-01.ibm.com/common/ssi/ cgi-bin/ssialias?infotype=PM&subtype=XB&appname=GBSE_GB_TI_USEN&htmlfid=GBE 03348USEN&attachment=GBE03348USEN.PDF
IDA. (n.d.-a). iN2015 Masterplan. Retrieved from www.ida.gov.sg/Tech-Scene-News/iN2015-Masterplan
IDA. (n.d.-b). Smart Nation Vision. Retrieved from www.ida.gov.sg/Tech-Scene-News/ Smart-Nation-Vision
ISO/IEC JTC 1. InformationTechnology. (2015). *Smart cities. Preliminary Report 2014.* Geneva, CHE. Retrieved from http://www.iso.org/iso/smart_cities_report-jtc1.pdf
ITU. (2003). *Measuring the information society. Annual report of International Telecommunication Union.* Geneva, CHE. Retrieved from http://www.itu.int
ITU. (2007a). *Chapter seven. The ICT opportunity index (ICT-OI).* Retrieved from https://www.itu. int/osg/spu/publications/worldinformationsociety/2007/WISR07-chapter7.pdf
ITU. (2007b). *Chapter three. The digital opportunity index (DOI). Current Science.* Retrieved from http://unctad.org/Sections/dite_dir/docs/WISR07-chapter3.pdf
ITU. (2010). *Measuring the information society.* Geneva, CHE. Retrieved from https://www.itu. int/en/ITU-D/Statistics/Documents/publications/mis2014/MIS2014_without_Annex_4.pdf
Jacobs, J. (1969). *The economy of cities.* New York, NY: Random House.
Jacobs, J. (1984). *Cities and the wealth of nations.* New York, NY: Random House.
Jin, S., & Cho, C. M. (2015). Is ICT a new essential for national economic growth in an information society? *Government Information Quarterly*, *32*(3), 253–260. https://doi. org/10.1016/j.giq.2015.04.007
King, A. D. (2012). Imperialism and world cities. In B. Derudder, M. Hoyler, P. J. Taylor, & F. Witlox (Eds.), *International handbook of globalization and world cities* (pp. 31–39). Cheltenham, UK; Northampton, MA: Edward Elgar Publishing.
Ko, Y. M., Shim, W., Pyo, S.-H., Chang, J. S., & Chung, H. K. (2012). An economic valuation study of public libraries in Korea. *Library & Information Science Research*, *34*(2), 117–124. https://doi.org/10.1016/j.lisr.2011.11.005
Koontz, C. M. (1992). Public library site evaluation and location: Past and present market-based modelling tools for the future. *Library and Information Science Research*, *14*(4), 379–409.
Kosior, A. (2013). *Aufkommende informationelle Städte in der Golf-Region – Eine empirische Studie zu Dubai, Abu Dhabi, Doha und Manama [Emerging informational cities in the Gulf region – An empirical study on Dubai, Abu Dhabi, Doha and Manama].* Unpublished master's thesis, Heinrich-Heine University Düsseldorf, DE.
Kuhlen, R. (1995). *Informationsmarkt. Chancen und Risiken der Kommerzialisierung von Wissen [Information market. Opportunities and risks of the commercialization of knowledge].* Konstanz, DE: UVK-Universitätsverlag.
Kunzmann, K. R. (2004). Wissensstädte. Neue Aufgaben für die Stadtpolitik [Knowledge cities. New tasks for city politics]. In U. Matthiesen (Ed.), *Stadtregion und Wissen* (pp. 29–41). Wiesbaden, DE: VSVerlag für Sozialwissenschaften.
La Rue, F. (2011). *Report of the Special Rapporteur on the promotion and protection of the right to freedom of opinion and expression.* New York, NY. Retrieved from http://www2.ohchr. org/english/bodies/hrcouncil/docs/17session/A.HRC.17.27_en.pdf
Landry, C. (2008). *The creative city: A toolkit for urban innovators* (2nd ed.). New York, NY: Earthscan Publications.

Lim, E. (1999). Human resource development for the information society. *Asian Libraries*, *8*(5), 143–161. https://doi.org/10.1108/10176749910275975

Lizieri, C. (2012). Global cities office markets and capital flows. In B. Derudder, M. Hoyler, P. J. Taylor, & F. Witlox (Eds.), *International handbook of globalization and world cities* (pp. 162–176). Cheltenham, UK; Northampton, MA: Edward Elgar Publishing.

Lor, P. J., & Britz, J. J. (2007). Is a knowledge society possible without freedom of access to information? *Journal of Information Science*, *33*(4), 387–397. https://doi.org/10.1177/0165551506075327

Machlup, F. (1962). *The production and distribution of knowledge in the United States*. Princetown, NJ: Princetown University Press.

Mainka, A., Hartmann, S., Meschede, C., & Stock, W. G. (2015a). Mobile application services based upon open urban government data. In Proceedings of the iConference 2015: Create, Collaborate, Celebrate. Newport Beach, CA (p. 24.-27). University of Illinois at Urbana-Champaign: iSchools, IDEALS.

Mainka, A., Hartmann, S., Meschede, C., & Stock, W. G. (2015b). Open government: Transforming data into value-added city services. In M. Foth, M. Brynskov, & T. Ojala (Eds.), *Citizen's right to the digital city: Urban interfaces, activism, and placemaking* (pp. 199–214). Singapore, SG: Springer.

Mainka, A., Hartmann, S., Orszullok, L., Peters, I., Stallmann, A., & Stock, W. G. (2013). Public libraries in the knowledge society: Core services of libraries in informational world cities. *Libri*, *63*(4), 295–319. https://doi.org/10.1515/libri-2013-0024

Malecki, E. J. (2012). Internet networks of world cities: Agglomeration and dispersion. In B. Derudder, M. Hoyler, P. J. Taylor, & F. Witlox (Eds.), *International handbook of globalization and world cities* (pp. 117–125). Cheltenham, UK; Northampton, MA: Edward Elgar Publishing.

McCallum, I., & Quinn, S. (2004). Valuing libraries. *The Australian Library Journal*, *53*(1), 55–69. https://doi.org/10.1080/00049670.2004.10721613

Missingham, R. (2005). Libraries and economic value: A review of recent studies. *Performance Measurement and Metrics*, *6*(3), 142–158. https://doi.org/10.1108/14678040510636711

Morris, A., Sumsion, J., & Hawkins, M. (2002). Economic value of public libraries in the UK. *Libri*, *52*(2), 78–87. https://doi.org/10.1515/LIBR.2002.78

Murugadas, D., Vieten, S., Nikolic, J., Fietkiewicz, K. J., & Stock, W. G. (2015). Creativity and entrepreneurship in informational metropolitan regions. *Journal of Economic and Social Development*, *2*(1), 14–24. Retrieved from http://hdl.handle.net/10125/41496

Musterd, S., & Murie, A. (eds.) (2010). *Making competitive cities. The Oxford Handbook of Urban Politics*. Oxford, UK: Wiley-Blackwell.

Nam, T., & Pardo, T. A. (2011). Conceptualizing smart city with dimensions of technology, people, and institutions. In *Proceedings of the 12th Annual International Digital Government Research Conference on Digital Government Innovation in Challenging Times* (pp. 282–291). New York, NY: ACM. https://doi.org/10.1145/2037556.2037602

Nefiodow, L. A. (2006). *Der sechste Kondratieff: Wege zur Produktivität und Vollbeschäftigung im Zeitalter der Information. Die langen Wellen der Konjunktur und ihre Basisinnovationen [The Sixth Kondratieff: ways of productivity and full employment in the age of information...]* (6th ed.). Sankt Augustin, DE: Rhein-Sieg-Verlag.

Nesta. (2015). The European digital city index. Retrieved from http://www.nesta.org.uk/blog/launching-european-digital-city-index

Orofino, A. (2014). It's our city. Let´s fix it [Video file]. Retrieved from https://www.ted.com/speakers/alessandra_orofino

OSL crowdmap. (n.d.). Retrieved from https://opensociallearning.crowdmap.com/main

Palvia, S. C. J., & Sharma, S. S. (2007). E-government and e-governance: Definitions/domain framework and status around the world. In A. Agarwal, & V. Venkata Ramana (Eds.), *Proceedings of the Fifth International Conference on e-Governance (ICEG)* (pp. 1–12). Sonning Common: Academic Conferences Ltd.

Paparwekorn, T. (2015). Comparison of globally recognized ict and telecommunications indices. *International Journal of Future Computer and Communication*, 4(6), 381–385. https://doi.org/10.18178/ijfcc.2015.4.6.420

Pimbert, M., & Wakeford, T. (2001). Deliberative democracy and citizen empowerment. *Participatory Learning and Action [PLA Notes 40]*, 23–28. Retrieved from http://www.iied.org/pla-40-deliberative-democracy-citizen-empowerment

Poll, R. (2003). Measuring impact and outcome of libraries. *Performance Measurement and Metrics*, 4(1), 5–12. https://doi.org/10.1108/14678040310471202

Powell, W. W., & Snellman, K. (2004). The knowledge economy. *Annual Review of Sociology*, 30(1), 199–220. https://doi.org/10.1146/annurev.soc.29.010202.100037

Pratt, A. C. (2008). Creative cities: The cultural industries and the creative class. *Geografiska Annaler, Series B: Human Geography*, 90(2), 107–117. https://doi.org/10.1111/j.1468-0467.2008.00281.x

Pratt, A. C. (2012). The cultural economy and the global city. In B. Derudder, M. Hoyler, P. J. Taylor, & F. Witlox (Eds.), *International handbook of globalization and world cities* (pp. 265–274). Cheltenham, UK; Northampton, MA: Edward Elgar Publishing.

Reporters without Borders. (n.d.). Detailed methodology. Retrieved from https://rsf.org/en/detailed-methodology

Rifkin, J. (2014). *The zero marginal cost society: The internet of things, the collaborative commons, and the eclipse of capitalism*. New York, NY: Palgrave Macmillan.

Rowlands, I., Nicholas, D., Williams, P., Huntington, P., Fieldhouse, M., Gunter, B., ... Tenopir, C. (2008). The Google generation: The information behaviour of the researcher of the future. *Aslib Proceedings*, 60(4), 290–310. https://doi.org/10.1108/00012530810887953

Sassen, S. (2001). *The global city: New York, London, Tokyo* (2nd Ed.). Princeton, NJ: Princeton University Press.

Sassen, S. (2013). Does the city have speech? *Public Culture*, 25(2 70), 209–221. https://doi.org/10.1215/08992363-2020557

Saxena, S., & McDougall, A. (2012). Estimating the economic value of libraries. *Prometheus*, 30(3), 367–369. https://doi.org/10.1080/08109028.2012.702056

Schumann, L., & Stock, W. G. (2015). Acceptance and use of ubiquitous cities' information services. *Information Services & Use*, 35(3), 191–206. https://doi.org/10.3233/ISU-140759

Sciadas, G. (2003). *Monitoring the digital divide ... and beyond*. Ottawa, CA. Retrieved from http://orbicom.uqam.ca/upload/files/research_projects/2003_dd_pdf_en.pdf

Setunge, S., & Kumar, A. (2010). Knowledge infrastructure: Managing the assets of creative urban regions. In T. Yigitcanlar (Ed.), *Sustainable urban and regional infrastructure development: Technologies, applications and management* (pp. 102–117). Hershey, PA: Information Science Reference.

Sharma, R. S., Lim, S., & Boon, C. Y. (2009). A vision for a knowledge society and learning nation: the role of a national library system. *The ICFAI University Journal of Knowledge Management*, 7(5/6), 91–113.

Skot-Hansen, D., Hvenegaard, C., Jochumsen, H., Skot-Hansen, D., Hvenegaard Rasmussen, C., & Jochumsen, H. (2013). The role of public libraries in culture-led urban regeneration. *New Library World*, *114*(1/2), 7–19. https://doi.org/10.1108/03074801311291929

Songdo IBD. (2009). Collaboration transforming smart+connected communities worldwide. Retrieved from www.songdo.com/songdo-international-business-district/news/press-releases.aspx/d=98

Spitzer, K. L., Eisenberg, M. B., & Lowe, C. A. (1998). *Information Literacy: Essential Skills for the Information Age*. Washington, DC: ERIC.

Stallmann, A. (2014). *Sind Helsinki, Oslo und Stockholm Vorreiter der Wissensgesellschaft? Eine empirische informationswissenschaftliche Untersuchung [Are Helsinki, Oslo and Stockholm a pioneer of the knowledge society? An empirical study]*. Unpublished master's thesis, Heinrich-Heine University Düsseldorf, DE.

Stehr, N. (2003). The social and political control of knowledge in modern societies. *International Social Science Journal*, *55*(4), 643–655. https://doi.org/10.1111/j.0020-8701.2003.05504014.x

Stock, W. G. (2011). Informational cities: Analysis and construction of cities in the knowledge society. *Journal of the American Society for Information Science and Technology*, *62*(5), 963–986. https://doi.org/10.1002/asi

Storper, M., & Venables, A. (2004). Buzz: Face-to-face contact and the urban economy. *Journal of Economic Geography*, *4*, 351–370.

Taylor, P. J. (2001). Specification of the world city network. *Geographical Analysis*, *33*(2), 181–194. Retrieved from http://doi.wiley.com/10.1111/j.1538-4632.2001.tb00443.x

Taylor, P. J. (2004). *World city network: A global urban analysis*. London, UK: Routledge. https://doi.org/10.4324/9780203634059

Taylor, P. J. (2012a). Historical world city networks. In B. Derudder, M. Hoyler, P. Taylor, & F. Witlox (Eds.), *International handbook of globalization and world cities* (pp. 9–21). Cheltenham, UK; Northampton, MA: Edward Elgar Publishing.

Taylor, P. J. (2012b). On city cooperation and city competition. In B. Derudder, M. Hoyler, P. J. Taylor, & F. Witlox (Eds.), *International handbook of globalization and world cities* (pp. 64–72). Cheltenham, UK; Northampton, MA: Edward Elgar Publishing.

Taylor, P. J. (2012c). The interlocking network model. In B. Derudder, M. Hoyler, P. J. Taylor, & F. Witlox (Eds.), *International handbook of globalization and world cities* (pp. 51–63). Cheltenham, UK; Northampton, MA: Edward Elgar Publishing.

Taylor, P. J., Derudder, B., Hoyler, M., & Witlox, F. (2012). Advanced producer servicing networks of world cities. In P. J. Taylor, B. Derudder, M. Hoyler, & F. Witlox (Eds.), *International handbook of globalization and world cities* (pp. 135–145). Cheltenham, UK; Northampton, MA: Edward Elgar Publishing.

Taylor, P. J., Derudder, B., Saey, P., & Witlox, F. (Eds.). (2007). *Cities in Globalization: Practices, Policies and Theories*. London, UK: Routledge.

Taylor, P. J., Firth, A., Hoyler, M., & Smith, D. (2010). Explosive city growth in the modern world-system: An initial inventory derived from urban demographic changes. *Urban Geography*, *31*(7), 865–884. https://doi.org/10.2747/0272-3638.31.7.865

Taylor, P. J., Hoyler, M., & Smith, D. (2012). Cities in the making of world hegemonies. In B. Derudder, M. Hoyler, P. J. Taylor, & F. Witlox (Eds.), *International handbook of globalization and world cities* (pp. 22–30). Cheltenham, UK; Northampton, MA: Edward Elgar Publishing.

Taylor, P. J., Hoyler, M., & Verbruggen, R. (2010). External urban relational process: Introducing central flow theory to complement central place theory. *Urban Studies*, *47*(13), 2803–2818. https://doi.org/10.1177/0042098010377367

Taylor, P. J., Ni, P., & Derudder, B. (2010). Introduction: The GUCP /GaWC project. In P. J. Taylor, P. Ni, B. Derudder, M. Hoyler, J. Huang, & F. Witlox (Eds.), *Global urban analysis: A survey of cities in globalization* (pp. 1–16). London, UK: Earthscan.

Umlauf, K. (2008). *Kultur als Standortfaktor. Öffentliche Bibliotheken als Frequenzbringer [Culture as location factor. Public libraries as traffic generators]. Berliner Handreichungen zur Bibliotheks- und Informationswissenschaft 245*. Berlin, DE.

UNDP. (2014). *Human development report 2014*. New York, NY. Retrieved from http://hdr.undp.org/en/content/human-development-report-2014

United Nations. (2006). *The digital divide report. ICT diffusion index 2005*. Geneva, CHE. Retrieved from http://unctad.org/en/Docs/iteipc20065_en.pdf

United Nations. (2014). *E-government survey 2014: E-government for the future that we want*. New York, NY. Retrieved from http://unpan3.un.org/egovkb/Reports/UN-E-Government-Survey-2014

United Nations. (2016). *United Nation e-governmnet survey 2016. E-government in support of sustainable development*. New York. Retrieved from http://workspace.unpan.org/sites/Internet/Documents/UNPAN96407.pdf

van de Stadt, I., & Thorsteinsdóttir, S. (2007). Going e-only: All Icelandic citizens are hooked. *Library Connect, 5*(1), 2.

van der Wusten, H. (2012). Political global cities. In B. Derudder, M. Hoyler, P. J. Taylor, & F. Witlox (Eds.), *International handbook of globalization and world cities* (pp. 40–48). Cheltenham, UK; Northampton, MA: Edward Elgar Publishing.

Van Winden, W. (2009). European cities in the knowledge-based economy: Observations and policy challenges. *disP, 178*(3), 83–88.

Wall, R., & van der Knaap, B. (2012). Centrality, hierarchy and heterarchy of worldwide corporate networks. In B. Derudder, M. Hoyler, P. J. Taylor, & F. Witlox (Eds.), *International handbook of globalization and world cities* (pp. 209–229). Cheltenham, UK; Northampton, MA, UK; Northampton, MA: Edward Elgar Publishing.

Watson, A. (2012). How global are the "global media"? Analysing the networked urban geographies of transnational media corporations. In B. Derudder, M. Hoyler, P. J. Taylor, & F. Witlox (Eds.), *International handbook of globalization and world cities* (pp. 284–294). Cheltenham, UK; Northampton, MA: Edward Elgar Publishing.

World Economic Forum. (2015). *Global information technology report 2015. ICTs for inclusive growth*. Retrieved from http://reports.weforum.org/global-information-technology-report-2015/

Yigitcanlar, T. (2010). Informational city. In R. Hutchison (Ed.), *Encyclopedia of urban studies* (pp. 392–395). New York, NY: Sage.

Yimin, G., & Zhong, C. S. (2008). From municipal Shanghai library: How an enterprise can get the most for its competitive intelligence and project development. In W. Ratzek & E. Simon (Eds.), *Wirtschaftsförderung und Standortentwicklung durch Informationsdienstleistungen. Das unterschätzte Potential von Bibliotheken [Economic and site development due to information services. The underrated potential of libraries]* (pp. 58–75). Berlin, DE: Simon.

Yu, H., & Robinson, D. G. (2012). The new ambiguity of open government. *59 UCLA L. Rev. Disc., 178*(2012), 178–208. https://doi.org/10.2139/ssrn.2012489

4 Methods

This work is part of a research project on informational cities conducted at the Department of Information Science at Heinrich-Heine University Düsseldorf (Germany) since 2010. Therefore I will refer to prior publications on informational world cities in which I have been involved, such as the case studies of Singapore and London as prototypes of informational world cities (Khveshchanka, Mainka, & Peters, 2011; Murugadas, Vieten, Nikolic, & Mainka, 2015), the case studies of emerging informational world cities in the Gulf Region (Kosior, Barth, Gremm, Mainka, & Stock, 2015), the investigation of public services such as public libraries and e-government services in informational world cities (Mainka, Hartmann, et al., 2013; Mainka, Fietkiewicz, Kosior, Pyka, & Stock, 2013), the analysis of social media activities by public authorities and citizens in these cities (Förster, Lamerz, Mainka, & Peters, 2014; Förster & Mainka, 2015; Mainka, Hartmann, Stock, & Peters, 2015), and the investigation of the use of open data in mobile applications for city services (Mainka, Hartmann, Meschede, & Stock, 2015). Furthermore, I will also refer to publications by my colleagues such as the case study of the labor market in informational world cities (Dornstädter, Finkelmeyer, & Shanmuganathan, 2011), the case study of indicators of the "space of flow" (Nowag, Perez, & Stuckmann, 2011), the investigation of entrepreneurship and creativity on a global scale (Murugadas, Vieten, Nikolic, Fietkiewicz, & Stock, 2015), and the case study of informational world cities in Japan (Fietkiewicz & Pyka, 2014; Fietkiewicz & Stock, 2015). Wolfgang G. Stock (2011) laid the foundation for all this work with his article *Informational Cities: Analysis and Construction of Cities in the Knowledge Society*, in which he first defined indicators of informational cities and raised the question of how to investigate prototypical cities of the knowledge society. The intention of this research is to evaluate prototypical cities of the knowledge society on a global scale with regard to the indicators infrastructures, world city, and political willingness.

Research on informational world cities is based on an interdisciplinary approach and uses methods from information science and social science. I will combine qualitative with quantitative approaches, additionally using secondary data to analyze cities of the knowledge society in a broader spectrum. The methods used in this research project can be subsumed under the grounded theory, which includes literature review, semi-structured interviews, field study and statistical data investigation methods. In the following, the methods used are described.

4.1 Grounded theory: Combination of research methods to investigate new fields

As a common qualitative approach, grounded theory as defined by Glaser and Strauss (1998) is used in social science studies worldwide (Strauss & Corbin, 1996). According to Strauss and Corbin (1996), the combination of interviews, field research and statistical data can be used to define a new data-driven theory: grounded theory. This research approach does not start with a theory to be proven, but rather with a field of research to be investigated with the assistance of data and analytical methods, the focus and important aspects of the object under investigation coming to light during the research process. Based on grounded theory as defined by Glaser and Strauss (1998), Oktay (2012) has formulated four core components that have mutual effects on the developing theory:

1. *Theoretical sensitivity:* This ability refers to the researcher's ability to think analytically. It implies that the researcher is able to develop an understanding of the research field or object and to identify crucial characteristics of it. Adhering to this concept helps the researcher to conceptualize and formulate a theory based on the investigated data.
2. *Constant comparative method:* This is the basic method used in grounded theory. Here the researcher may conceptualize theories through a constant comparison of empirical data. This helps the researcher to build concepts by identifying similarities and differences in the data.
3. *Theoretical sampling:* This is not one of the first steps in grounded theory since the developing theory is a process. Thus, sampling may change during the research process e.g. after an interview. Sampling may become irrelevant or help to define the final theory.
4. *Theoretical saturation:* This means that the researcher has analyzed all data which may be relevant for defining new concepts. More data would not result in further findings. Saturation is the point that the researcher has met after a process of data evaluation and data analysis.

Thus, the basic research method used in this work is grounded theory and combines the different approaches of qualitative and quantitative methods. As a first step, the literature of informational cities and related fields was analyzed to formulate first samplings. The samplings were then used to formulate first theories which were then adopted to develop questions for interviews of experts. Further questions that arose from the data gathered in the interviews were added individually and in relation to the field, e.g. librarians were posed additional questions about citizen's participation and engagement in library activities. Further theories were able to emerge through the interview evaluation, which also led to an additional literature review during the research process. Therefore, grounded theory is a helpful method to disclose further hypotheses based on emerging topics, problems, or challenges. This approach is flexible and allows researchers to combine complementary methods like literature reviews, interviews, field studies and statistical data analysis, as it was done for this work to gain complementary and fruitful results. Figure 4.1 visualizes the process of investigation.

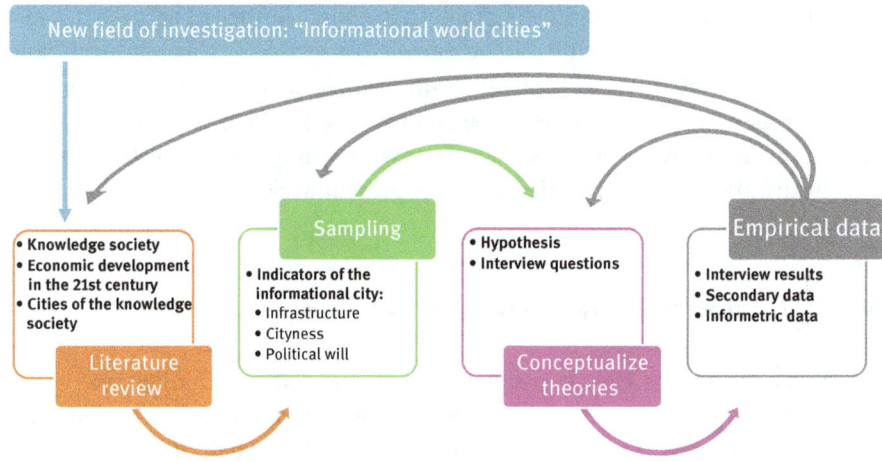

Figure 4.1: Grounded theory cycle representing the investigation process of the work at hand.

4.2 Literature review: Identifying prototypical cities of the knowledge society

To identify potential cities of the knowledge society, a literature review was undertaken on the platforms Web of Science®, Scopus, Google Scholar, and Wiley Online Library. All platforms are multidisciplinary databases which can be used for bibliometric investigations (Moed, 2009) and allow searching e.g. for topics, titles, abstracts, keywords, authors, etc. They differ in the number of journals covered, due to different authority regulations for journal integration. Thus, it is highly recommended to use more than one database to enhance the recall.

To add a city to the potential investigation corpora, two conditions had to be met: The first requirement was that the city had to be referred to as a world city in the literature, in the sense of the world city and global city definitions of the researchers John Friedmann (1986), Saskia Sassen (2001), or Peter J. Taylor (2004).

For them, world or global cities are not defined by a number of people living in a city but through the city's position in a global economy. The second precondition was that the city had to have been investigated in at least one aspect of the development of an informational city according to the research of Manuel Castells, Tan Yigitcanlar, or Wolfgang G. Stock. As those cities are complex and serve as a hub for economic, government, and human interaction, further search terms had to be included in order to represent the cities' development towards a knowledge city, digital city, creative city, or smart city.[1] Additionally, as bibliographic databases do not offer full-text research, a manual investigation of proper literature was necessary. The identified cities and their characteristics will be described in chapter 5 ("Identifying prototypical cities of the knowledge society"). This approach does not aim to assemble an exhaustive list of prototypical cities of the knowledge society, which would exceed the scope and funding of this research project. Only prominent cities have been chosen by the research group to be investigated thoroughly and are therefore investigated in the present work.

4.3 Interviews: A qualitative approach

Aside from the literature review, empirical data were gathered through interviews. We chose expert interviews as a method as such interviews are recommended when the researcher's goal is to gain orientation within a field (Flick, 2009). The investigation in this case cannot be limited to only one closely delineated field since informational world cities are highly complex structures (Stock, 2011) and require consideration of many different aspects such as culture, infrastructures, economic market, knowledge institutions, and more. Therefore, the experts interviewed are from different fields such as universities, government authorities, public libraries, urban planning, or cultural institutions. These individuals are "highly knowledgeable informants who can view the focal phenomena from diverse perspectives" (Eisenhardt & Graebner, 2007, p. 28). Appendix I provides a list of the dates and locations of the interviews. Three to five experts were interviewed in each city to gain an overview of the city and its strengths and weaknesses according to different indicators of an informational city. In total 158 experts were interviewed.

For the interviews, a prepared questionnaire was used. The questionnaire contained 18 semi-structured questions. The questions were on the one hand based on the SEVQUAL method, which is used to measure the quality of products,

---

[1] A short version of the process was published in Mainka, Hartmann, et al. (2013). The literature review has since been updated and includes literature published up to June 2016.

but on the other hand also contain qualitative elements as well, as the interviewee has the option to add his or her thoughts and knowledge to each question.

The SERVQUAL method was established by Parasumaran, Zeithaml, and Berry (1988) to analyze the quality of services and goods. Traditionally, a questionnaire that is based on SERVQUAL includes 18 questions and covers five dimensions: reliability, assurance, tangibles, empathy, and responsiveness. All questions have to be answered for the two components "expectation" and "perception." The evaluation is based each on a Likert-type scale with marks from one (strongly disagree) to seven (strongly agree). The difference is the degree of the quality value (Q) of the analyzed good or service. Thus, the quality of an item can be calculated as follows:

$$\text{Quality value} = \text{Perception} - \text{Expectation}$$

As Parasumaran, Zeithaml, and Berry (1988, p. 17) argue, "[t]he skeleton, when necessary, can be adapted or supplemented to fit the characteristics or specific research needs of a particular organization." Therefore, their approach to investigating the quality of the development of an informational city is used in this work and the prepared questionnaire for the present work consists of 18 questions with the two components expectation, to reflect the personal assumption of an interviewee, and experience, to reflect the interviewee's perceptions and observations. In line with Stock's (2011) definition of informational world cities the dimensions infrastructures, labor market, corporate structure, world city (cityness), political will, and soft locational factors are covered. The questions are based on the interview questions used by our research group to investigate Singapore as a prototypical informational world city (Khveshchanka et al., 2011) and have been adjusted to fit into the SERVQUAL method.

As mentioned before, the interviews are based on both quantitative and qualitative approaches. On the quantitative side, the data from the SERVQUAL scales was accumulated and analyzed statistically. The results were not statistically significant since the number of respondents was small and did not represent a cross-section of the society, but the results can give us a first idea about the investigated city. In the results section, the evaluation will be presented in one figure covering the expectation, perception, and the quality value. The SERVQUAL scores are scaled from one to seven and the quality value from plus seven to minus seven which needs to be considered while interpreting the results.

Furthermore, we also evaluated the interviews qualitatively. The experts were allowed to give as much information on each questioned topic as they were willing and able to. Thus, the interview was based on a semi-structured method, offering space for open answers by the interviewee (Diekmann, 2014). The evaluation of the qualitative part of the interviews was conducted according to the case- and group comparison method introduced by Mattissek, Pfaffenbach, and

Reuber (2013). Therefore, the interviews had to be transcribed. When permission was given, the interviews were recorder and subsequently transcribed. Otherwise important statements by the experts were noted during the interview.² In the following, the interview results have been categorized in types of homogeneity (statements that are very similar) and of heterogeneity (statements that differ greatly). This has been done separately for each question. The goal of this method is to identify parallels and differences between the interviewees, which could eventually lead to a generalized statement on each topic.

The interviews were conducted between June 2010 and May 2014 in German and in English, depending on the preference of the interviewee. The duration of the interviews varied between 15 minutes and two hours. The time spent for each interview was highly dependent on the contact person's willingness to provide much information, his or her schedule for the day, and how many people were in attendance for the interview. As the interviews were conducted over a period of four years, it should be noted that interviews always represent a retrospective (Diekmann, 2014). As such, the interviewees could always merely report on the state of the art at the time of the interview. Especially for the dimension of political willingness this does not present a significant limitation, since city projects towards a knowledge society often take many years from planning to completion.

However, using interviews in research has some further limitations: The researcher depends on the willingness of contact persons to take some time out of their schedule for an interview. When no personal introduction could be arranged, persons of interest had to be contacted over regular channels, i.e. via email or phone. This was not always successful and unfortunately, we were not able to arrange any interviews in Beijing. On the other side of the spectrum, we were able to conduct eight interviews in Vienna and nine in Barcelona. A further complication was caused by the standardized SERVQUAL interview, as not every expert was satisfied with this structure. Thus, alternative and more open questions were asked in those interviews, which nevertheless led to fruitful conversations and information that has enhanced the work in this research project. In addition, it was not always possible to conduct individual interviews, forcing us to conduct individual and group interviews in the investigated cities depending on the availability of the contact persons. According to Diekmann (2014), the attendance of other persons during the interview may bias the response. While this may be evident for intimate questions about individuals' private lives, my experience during the group interviews was that interviewing more people at once led to very productive conversations related to the topics under investigation. A further limitation was the personal inclinations

---

2 The transcripts need to undergo a cleaning process before they could be published. Data will be made available to all interested researchers upon request.

and experience of the interviewee. If he or she is not interested in a specific topic or service, the quality of this object might unjustifiedly be evaluated as poor. For instance, a senior male professor is, in all likelihood, not the right person to assess the quality of shopping malls as important consumer hotspots.

4.4 Gaining first-hand experience through field studies

A further method used in this research project is ethnographic field studies. To gain first-hand experience and conduct the interviews face-to-face my colleagues and I visited all cities investigated in this work.[3] Ethnographical field studies originate from ethnography and anthropology and became a popular method of social science at the end of the nineteenth century (Lichtman, 2013). The basic idea of this research method is based on the observation and participation of a given research field (Malinowski, 1922). Thus, the main aspect is the description of ethnographical and cultural characteristics of groups of people (Fischer, 2003; Meier, 2001). Ethnographical field research is further used to investigate smaller communities like a neighborhood or how employees work together in a single office (Fischer, 2003). There are many ways how ethnographical field study is applied in practice (see for further readings Melhuus, Mitchell, and Wulff (2012)), but some common main characteristics of this method can be identified.

According to Meier (2001) ethnographical field research is a process of readjusting research questions according to the field and dependent on the research task – it allows the detailed investigation of specific cases (e.g. ethnic groups or business enterprises), it enriches the investigation through the complementary nature of interviews and observations as well as unsystematic observation and attendance of situations in everyday life, and finally it induces the researcher to put the focus on understanding observations of actions rather than verifying one's own hypotheses about these actions. It is obvious that the researcher has to spend some time personally in the space of field under investigation to observe the habitat of the object or objects of research (Schmitz, n.d.), though a specific timeframe is not prescribed. Additionally, Schmitz (n.d.) adds that quantitative data collections of facts and data of the investigated field are of relevance, e.g. photo and audio documentation as well as interviews.

---

[3] Melbourne, Australia, is the only city in which I did not conduct the interviews personally. In addition, further interviews were conducted in Tokyo by another research group but with the same questionnaire, the results of which I have included in the present work.

Ethnographic field work can be used to open up many different perspectives and show the field under investigation from different points of view, e.g. those of communities and authorities (Geertz, 1983). The researcher's task is, therefore, to be open-minded, since he or she has to investigate the field from all these different perspectives (Meier, 2001). The perspective is closely related to the data that is used to investigate the field. Accordingly, Schmitz (n.d.) has prepared a list of descriptions with examples of data evaluations of possible perspectives, based on Geertz' (1983) definitions. According to Fietkjewicz (2013) and Stallmann (2014), the following are of relevance for informational city research:

- the perspective of the administration (e.g. through the investigation and analysis of statistical and demographic data),
- the perspective of local inhabitants (e.g. through interviews and documentation such as photos or videos),
- the perspective of experts in the field (e.g. through expert interviews and their cooperation in sharing relevant documents), and
- the perspective of analytical social science (e.g. according to the published literature in the field).

In my investigation of informational world cities, ethnographical field research was used as part of the larger grounded theory approach. Still, all investigated cities had to be visited and, since the research project had to be finished in a reasonable amount of time, the time spent in each city was limited to a maximum of five days. The cities were visited between June 2010 and May 2014. A total of 30 cities were visited by me and one (Melbourne, Australia) by my colleague Katrin Weller. Except for Frankfurt, a second person always attended the interviews to help with note-taking and transcription. During the stay, interviews with experts were conducted as described in the subchapter above. As ethnographic field work includes further methods of data acquisition, photographs were taken to visualize observations. Furthermore, statistical and demographic data as well as field-related literature and online publications complete the ethnographic data collection.

4.5 Measuring cities and nations on secondary data

Numerous city comparisons of various kinds can be found in reports and other publications from organizations, corporations, and academics. They investigate cities, regions, or nations in different aspects and often allow tracing the development over time if they are published regularly. They are concerned with current topics of the economy, government, society, or environment. Prominent examples of such reports are the *Human Development Index* by the United Nations (UNDP, 2014), or the *Global City Index* by A.T. Kearny. An academic research group in this field is, for example, GaWC (Globalization and World Cities). They publish many case studies and approaches to analyze world cities. They also offer a large number of open datasets which are available online (www.lboro.ac.uk/gawc/data.html). Current and relevant city comparisons, as well as academic publications, and their data will be used to identify the most recent indicators of the knowledge society and the informational city. Because gathering first-hand data is expensive both in terms of time and money, other researchers tend to reuse data, for example, to compare the cities on different levels or to identify interconnections of indicators (see for example Bruno, Esposito, Genovese, and Gwebu (2011) or Stock (2011)). In this work, I will, accordingly, also reuse data collections of existing reports and investigations.

Even though the existing city indices already cover a lot of important indicators useful for investigating the knowledge society and prototypical informational world cities, an online content investigation of official city websites, institutional websites and business websites (if they are in any relation to the investigated topic) will be taken into account, as well. The web data was retrieved through simple online search engines such as Google or Bing. The languages that are used for this data collection are German, English, and Spanish. Of course, this will lead to better results for cities which have one of these languages as lingua franca and exclude information that is only available, for instance, in Chinese. In contrast to academic publications or official reports, online content is dynamic and multidimensional, containing text, links, audio data, and videos, and therefore needs to be investigated qualitatively (Volpers, 2013; Welker et al., 2010). Through this, it is possible to identify those topics which are of interest in the field of knowledge society or informational world city research which have so far not been received attention in academic research or official reports. The presented methods, content analysis as well as data and document investigations are, of course, all part of empirical sociological research methods (Atteslander, 2010).

4.6 Retrieval of patents and scientific publications

Informetrics as the main field of information science was used to identify the knowledge and innovation output of informational world cities. Accordingly, a corpus of data was created, with patents as innovation indicator. Following van Winden, van den Berg, and Pol (2007), patenting is one of the drivers of the knowledge economy. Hence, within information science "patent informetrics" can be used to investigate the productivity (Stock & Stock, 2015) by measuring the number of patents granted within a certain time period. To this end, patent information from the Derwent World Patent Index (DWPI) was retrieved using the host STN International. This host offers the possibility to adjust the retrieval by country (/PAA.CNY) and city (/PAA.CTY). Ambiguous city names like London, UK, and London, US, can be separated. However, while the DWPI covers 52 sources from around the world, no patent information is available for the United Arab Emirates (Thomson Reuters, n.d.-a).

In addition, the knowledge output of informational world cities was quantified in terms of science, technical, and medicine (STM) publications. Informational cities offer an enhanced knowledge infrastructure comprised of knowledge institutions, research centers, and universities. They all produce scientific publications and, according to Stock (2011), in cities of the knowledge society those publications dominate in quantity. The knowledge flow can also be investigated based on the citation of these publications; however, in this study, I concentrated on the investigation of informational world cities as centers of knowledge and therefore measured the productivity based on the knowledge output. For this investigation, Web of Science, a citation database covering more than 12,000 journals and 150,000 conference proceedings (Thomson Reuters, n.d.-b), was used. Similar to the patent information database, it is possible to search for city and country names within the address field and the year of publication.

4.7 Limitations of the results

The results presented in this investigation are subject to several limitations. Firstly, first-hand data and statistics were not gathered for every hypothesis, and, as such, the investigation is partly based on secondary data and their respective reliability – if possible, data investigations by independent researchers and organizations were used. Also, more than one data survey was compared to discuss diverse findings.

Secondly, not all data is available on the city level and not all data is available for the same time period. Thus, in some cases a comparison is only feasible on the national or regional level. However, the frame of reference should not be mixed within a data survey. Therefore, some investigations did not comprise all 31 cities. In addition, the presented figures, indices, and surveys are for the most part based on a data evaluation of a certain time period or of a snapshot at the time of their creation. To identify the development and trends, the presented research findings and methods would have to be applied to longer time periods and on larger sets of data than only 31 cities.

Thirdly, the presented case studies and examples of projects are limited to those which have been mentioned by the interviewed experts, identified through own experience or have been described in other publications already. Thus, for example, the online participation platform *Frankfurt gestalten* for the city of Frankfurt is mentioned, but even though other cities will have introduced similar projects, they may not be mentioned. In addition, the interviews were conducted between June 2010, starting in Singapore, and April 2014, ending with a telephone interview with an expert from Vancouver. Hence, the experience and statements of the experts interviewed need to be considered in retrospect. Also, the interviews were conducted exclusively in English or German. Language barriers during the interviews and research may have resulted in further, unrecognized limitations.

Finally, the corpus of investigated cities was limited to world cities that have already been mentioned in the literature as a creative, knowledge, digital or smart city. Therefore, cities that have not yet been recognized by another study or are only of small or medium size were not considered in the present work. The results will therefore not reflect the development in and of such cities.

References

Atteslander, P. (2010). *Methoden der empirischen Sozialforschung [42/5000 Methods of empirical social research]*. Berlin, DE: Erich Schmidt Verlag.

Bruno, G., Esposito, E., Genovese, A., & Gwebu, K. L. (2011). A critical analysis of current indexes for digital divide measurement. *The Information Society, 27*, 16–28. https://doi.org/10.1080/01972243.2010.534364

Diekmann, A. (2014). *Empirische Sozialforschung: Grundlagen, Methoden, Anwendungen [Empirical social research: Basics, methods, applications]*. Reinbek, DE: Rowohlt-Taschenbuch-Verlag.

Dornstädter, R., Finkelmeyer, S., & Shanmuganathan, N. (2011). Job-Polarisierung in informationellen Städten [Job polarization in informational cities]. *Information – Wissenschaft & Praxis, 62*(2–3), 95–102.

Eisenhardt, K. M., & Graebner, M. E. (2007). Theory building from cases: Opportunities and challenges. *Academy of Management Journal, 50*(1), 25–32. https://doi.org/10.5465/AMJ.2007.24160888

Fietkiewicz, K. J. (2013). *Emerging informational cities in Japan. An empirical investigation of Tokyo, Kyoto, Yokohama, and Osaka*. Unpublished master's thesis, Heinrich-Heine University Düsseldorf, DE.

Fietkiewicz, K. J., & Pyka, S. (2014). Development of informational cities in Japan: A regional comparison. *International Journal of Knowledge Society Research, 5*(1), 69–82. https://doi.org/10.4018/ijksr.2014010106

Fietkiewicz, K. J., & Stock, W. G. (2015). How "smart" are Japanese cities? An empirical investigation of infrastructures and governmental programs in Tokyo, Yokohama,

Osaka and Kyoto. In *2015 48th Hawaii International Conference on System Sciences* (pp. 2345–2354). IEEE. https://doi.org/10.1109/HICSS.2015.282

Fischer, H. (2003). *Ethnographie, ethnographische Feldforschung und die teilnehmende Beobachtung. Ein kurzer Überlick [Ethnography, ethnographical field research, and participating observation. A brief overview]*. Norderstedt, DE: GRIN Verlag.

Flick, U. (2009). *An Introduction to Qualitative Research* (4th ed.) London, UK: SAGE Publications.

Förster, T., Lamerz, L., Mainka, A., & Peters, I. (2014). The tweet and the city: Comparing twitter activities in informational world cities. In *Proceedings of the 3. DGI 2014 Conference: Informationsqualität und Wissensgenerierung*. (pp. 101–118). Franfurt a.M., DE: DGI.

Förster, T., & Mainka, A. (2015). Metropolises in the twittersphere: An informetric investigation. *ISPRS International Journal of Geo-Information*, 4, 1894–1912. https://doi.org/10.3390/ijgi4041894

Friedmann, J. (1986). The world city hypothesis. *Development and Change*, *17*(1), 69–83. https://doi.org/10.1111/j.1467-7660.1986.tb00231.x

Geertz, C. (1983). *Dichte Beschreibung: Beiträge zum Verstehen kultureller Systeme [Density description: contributions to the understanding of cultural systems]*. Franfurt a.M., DE: Suhrkamp.

Glaser, B. G., & Strauss, A. (1998). *Grounded Theory: Strategien qualitativer Forschung [Grounded theory: Strategies of qualitative research]*. Bern, CH: Huber.

Khveshchanka, S., Mainka, A., & Peters, I. (2011). Singapur: Prototyp einer informationellen Stadt [Singapore: Prototype of an informational city]. *Information – Wissenschaft & Praxis*, *62*(2–3), 111–121.

Kosior, A., Barth, J., Gremm, J., Mainka, A., & Stock, W. G. (2015). Imported expertise in world-class knowledge infrastructures: The problematic development of knowledge cities in the Gulf Region. *Journal of Information Science Theory and Practice*, *3*(3), 17–44.

Lichtman, M. (2013). *Qualitative research for the social sciences*. Los Angeles, CA: SAGE Publications.

Mainka, A., Fietkiewicz, K. J., Kosior, A., Pyka, S., & Stock, W. G. (2013). Maturity and usability of e-government in informational world cities. In W. Castelnovo & E. Ferrari (Eds.), *Proceedings of the 13th European Conference on e-Government University* (pp. 292–300). Como, IT: Academic Conferences and Publishing International Limited Reading.

Mainka, A., Hartmann, S., Meschede, C., & Stock, W. G. (2015). Mobile application services based upon open urban government data. In Proceedings of the iConference 2015: Create, Collaborate, Celebrate. 24.-27. March 2015, Newport Beach, CA.

Mainka, A., Hartmann, S., Orszullok, L., Peters, I., Stallmann, A., & Stock, W. G. (2013). Public libraries in the knowledge society: Core services of libraries in informational world cities. *Libri*, *63*(4), 295–319. https://doi.org/10.1515/libri-2013-0024

Mainka, A., Hartmann, S., Stock, W. G., & Peters, I. (2015). Looking for friends and followers: A global investigation of governmental social media use. *Transforming Government: People, Process and Policy*, *9*(2), 237–254. https://doi.org/10.1108/TG-09-2014-0041

Malinowski, B. (1922). *Argonauts of the Western Pacific*. New York, NY: E. P. Dutton & Co.

Mattissek, A., Pfaffenbach, C., & Reuber, P. (2013). *Methoden der empirischen Humangeographie [Methods of empirical humanography]* (2nd ed.). Braunschweig, DE: Westermann.

Meier, C. (2001). Ethnografie [ethnography]. In G. Schwabe, N. Streitz, & R. Unland (Eds.), *CSCW-Kompendium: Lehr- und Handbuch zum computergestützten kooperativen Arbeiten* (pp. 46–53). Berlin, Heidelberg, DE: Springer.

Melhuus, M., Mitchell, J. P., & Wulff, H. (Eds.). (2012). *Ethnographic practice in the present* (11th ed.). New York, NY: Berghahn Books.

Moed, H. F. (2009). New developments in the use of citation analysis in research evaluation. *Archivum Immunologiae et Therapiae Experimentalis, 57*(1), 13–18. https://doi.org/10.1007/s00005-009-0001-5

Murugadas, D., Vieten, S., Nikolic, J., Fietkiewicz, K. J., & Stock, W. G. (2015). Creativity and entrepreneurship in informational metropolitan regions. *Journal of Economic and Social Development, 2*(1), 14–24. Retrieved from http://hdl.handle.net/10125/41496

Murugadas, D., Vieten, S., Nikolic, J., & Mainka, A. (2015). The informational world city London. *Journal of Documentation, 71*(4), 834–864. https://doi.org/10.1108/JD-06-2014-0090

Nowag, B., Perez, M., & Stuckmann, M. (2011). Informationelle Weltstädte – Indikatoren zur Stellung von Städten im „Space of Flow" [Informational world cities – Indicators of the position of cities in the "space of flow"]. *Information – Wissenschaft & Praxis, 62*(2–3), 103–109.

Oktay, J. S. (2012). *Grounded Theory*. New York, NY: Oxford.

Parasumaran, A., Zeithaml, V. A., & Berry, L. L. (1988). SERVQUAL: A multiple-item scale for measuring consumer perceptions of service quality. *Journal of Retailing, 64*(1), 12–40.

Sassen, S. (2001). *The global city: New York, London, Tokyo* (2nd Ed.). Princeton, NJ: Princeton University Press.

Schmitz, L. (n.d.). Feldforschung und ethnografisches Interview mit Auswahl-Karten [Field research and ethnographic interview with selection cards]. Retrieved from https://soz-kult.fhduesseldorf.de/members/liloschmitz/feldforschung_und_ethnografisches_interview

Stallmann, A. (2014). *Sind Helsinki, Oslo und Stockholm Vorreiter der Wissensgesellschaft? Eine empirische informationswissenschaftliche Untersuchung [Are Helsinki, Oslo and Stockholm a pioneer of the knowledge society? An empirical study]*. Unpublished master's thesis, Heinrich-Heine University Düsseldorf, DE.

Stock, W. G. (2011). Informational cities: Analysis and construction of cities in the knowledge society. *Journal of the American Society for Information Science and Technology, 62*(5), 963–986. https://doi.org/10.1002/asi

Stock, W. G., & Stock, M. (2015). *Handbook of information science*. Berlin, DE; Boston, MA: De Gruyter Saur.

Strauss, A., & Corbin, J. (1996). *Grounded Theory: Grundlagen qualitativer Sozialforschung [Grounded theory: Fundamental qualitative social research]*. Weinheim, DE: Beltz.

Taylor, P. J. (2004). *World city network: A global urban analysis*. London, UK: Routledge. https://doi.org/10.4324/9780203634059

Thomson Reuters. (n.d.-a). Intellectual property & science. Country coverage and kind codes. Retrieved from http://ip-science.thomsonreuters.com/support/patents/coverage/

Thomson Reuters. (n.d.-b). Intellectual property & science is now known as clarivate analytics. Retrieved from http://thomsonreuters.com/en/products-services/scholarly-scientific-research/scholarly-search-and-discovery/web-of-science-core-collection.html

UNDP. (2014). *Human development report 2014*. New York, NY. Retrieved from http://hdr.undp.org/en/content/human-development-report-2014

van Winden, W., van den Berg, L., & Pol, P. (2007). European cities in the knowledge economy: Towards a typology. *Urban Studies, 44*(3), 525–549. https://doi.org/10.1080/00420980601131886

Volpers, H. (2013). Inhaltsanalyse [Content analysis]. In K. Ulauf, S. Fühles-Ubach, & M. Seadle (Eds.), *Handbuch Methoden der Bibliotheks- und Informationswissenschaft: Bibliotheks-, Benutzerforschung, Informationsanalyse* (pp. 413–424). Berlin, DE; Boston, MA: De Gruyter.

Welker, M., Wünsch, C., Böcking, S., Bock, A., Friedemann, A., Herbers, M., & Schweitzer, E. J. (2010). Die Online-Inhaltsanalyse: methodische Herausforderung, aber ohne Alternative [Online content analysis: Methodical challenge, but without alternative]. In M. Welker & C. Wünsch (Eds.), *Die OnlineInhaltsanalyse: Forschungsobjekt Internet* (pp. 9–30). Cologne, DE: Halem.

5 Identifying prototypical cities of the knowledge society

Urban researchers have investigated many cities with regard to different characteristics and published their findings in journals, conference proceedings, or reports. These publications build the basis of the identification process of prototypical informational world cities. Informational cities in turn may be considered the prototypical cities of the knowledge society. According to Stock (2011), informational world cities are a complex phenomenon that is determined by a diverse collection of characteristics. Many of their indicators are inherited from world or global city research, which is why the literature review starts with the definition of world cities. As a starting point, world cities should not be determined by the number of people living in a city but rather by their status in a global system, according to the world and global city researchers John Friedmann, Peter J. Taylor, and Saskia Sassen. Following Friedmann (1986) cities are not defined by political-administrative boundaries but through their economic interactions with other cities. A similar approach can be found in Taylor's (2004) definition of world cities: For him, world cities are never isolated but characterized through their linkages with other cities and regions. According to Taylor (2004), cities interlinking with other places is referred to as "cityness." Sassen (2001, p. 209) refers to the term "cityness" as the process happening within a city, with the components "complexity, incompleteness, and making."

The status of a city can change over time. Therefore, the formation of global or world cities is referred to as a process (Derudder, Hoyler, Taylor, & Witlox, 2012). The cities New York, London, and Tokyo belong to the group with the highest amount of interactions or, in other words, have the highest degree of cityness in the late twentieth century which makes them, according to Sassen (2001), true global cities. Friedmann (1986) identifies cities within a global network which he considers to be primary or secondary world cities based on the economic flows investigated. Of course, New York, London, and Tokyo belong to the primary world cities, as well. Like Friedmann, Taylor (2004) also investigates economic flows of cities to define the network of world cities. According to his research group GaWC, world cities can be categorized as alpha, beta, and gamma world cities (GaWC, 2014), and those with the potential to become a world city in the future. Finally, the basic idea of informational cities goes back to Castells' (1996) theory of the network society and his definition of prototypical cities, in which spaces of flows inherit a central role. As the flow of power can be represented through economic flows (Nowag, Perez, & Stuckmann, 2011), world cities that are defined through their economic flows and their interlinking connectivity are a good starting point for further investigation. All cities that were investigated in this thesis have been identified as a world or global city at least by one urban researcher (see Appendix II Literature Review Cities).

As mentioned above, informational world cities serve as prototypical cities of the knowledge society and require additional criteria to be met. According to Stock (2011), informational cities are characterized by their combination of further characteristics as they have been established in knowledge, creative, digital, and smart city research. Those topics are of interest in academic research as well as in economic analysis and may also impact political decision-making. Therefore, cities that are prominent for a special development towards an informational world city can be detected in the literature, e.g. in journals like *Urban Studies*, or at conference presentations, e.g. at the IEEE International Smart Cities Conference (ISC2-2015). They may also result in specific projects, e.g. in a project for Horizon 2020 named "Smart Cities and Communities." The main difficulty in this area is the overabundance of terms used to describe this process. In the following, I will present the identification process of prototypical informational world cities using literature review to highlight trends, examples, and studies. For this, I separate the findings into two categories: digital and smart cities on the one hand, as both are focused on ICT infrastructure and sustainability, and creative and knowledge cities on the other, as both are characterized by the cognitive infrastructure. The characteristics and definitions of the city types have much overlap and the definition as informational world city will therefore be an umbrella term to combine them all.

Following the method to identify prototypical informational world cities, two conditions have to be fulfilled (see chapter 4 Methods): First, the investigated city has to be identified as a world city according to the flow theory, and second the city has to have been investigated as a knowledge, creative, digital or smart city. All in all, 138 publications were considered. Figure 5.1 demonstrates the amount

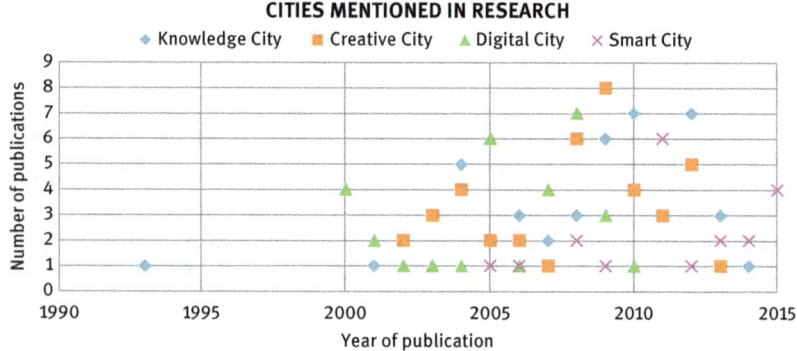

Figure 5.1: A Literature review of city research which uses at least one of the 31 informational cities as a case study or example (n = 138).

Table 5.1: A literature review of city research which uses at least one of the 31 informational cities as a case study or example (n = 138).

	Knowledge City	Creative City	Digital City	Smart City
1. Amsterdam (The Netherlands)	✓	✓	✓	✓
2. Barcelona (Spain)	✓	✓	✓	✓
3. Beijing (China)	✓	✓	✓	✓
4. Berlin (Germany)	✓	✓	✓	✓
5. Boston (USA)	✓	✓	✓	✓
6. Chicago (USA)	✓	✓	✓	✓
7. Dubai (UAE)	✓	✓	✓	✓
8. Frankfurt (Germany)	✓	✓	✓	✗
9. Helsinki (Finland)	✓	✓	✓	✓
10. Hong Kong (China, SAR)	✓	✓	✓	✓
11. Kuala Lumpur (Malaysia)	✓	✓	✓	✗
12. London (UK)	✓	✓	✓	✓
13. Los Angeles (USA)	✓	✓	✓	✗
14. Melbourne (Australia)	✓	✓	✓	✓
15. Milan (Italy)	✓	✓	✓	✗
16. Montréal (Canada)	✓	✓	✓	✓
17. Munich (Germany)	✓	✓	✓	✗
18. New York (USA)	✓	✓	✓	✓
19. Paris (France)	✓	✓	✓	✗
20. San Francisco (USA)	✓	✓	✓	✓
21. São Paulo (Brazil)	✓	✓	✗	✗
22. Seoul (South Korea)	✓	✓	✓	✓
23. Shanghai (China)	✓	✓	✓	✓
24. Shenzhen (China)	✓	✗	✓	✓
25. Singapore (Singapore)	✓	✓	✓	✓
26. Stockholm (Sweden)	✓	✓	✗	✓
27. Sydney (Australia)	✓	✓	✓	✓
28. Tokyo (Japan)	✓	✗	✓	✗
29. Toronto (Canada)	✓	✓	✓	✓
30. Vancouver (Canada)	✓	✓	✓	✓
31. Vienna (Austria)	✓	✓	✓	✓

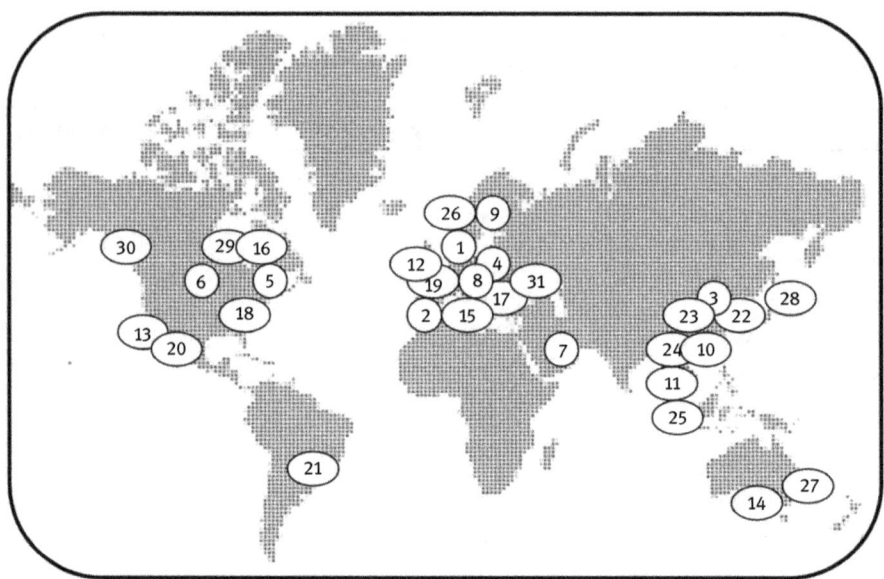

Figure 5.2: Informational world cities on the world map. Cf. numbers in Table 5.1.
Source: Mainka et al. (2013, p. 298).

of publications found in each year by city type. Using this method, a total of 31 cities were identified as prototypical informational world cities (see Table 5.1). The identified cities represent metropolises around the world (Figure 5.2). The literature review will be presented in this chapter in detail by giving an overview first of the cities identified and second of the topics investigated thus far in research regarding the city types under investigation.

5.1 Digital and smart cities

With regard to the relevance of ICT in a city, different terminologies arose in research literature and city projects. Anthopoulos and Fitsilis (2013) analyze the historical evolution of ICT-related city projects and set up a list of the terminology used. Thus, we have to extend the search of "digital cities" to the terms "broadband cities," "wireless cities," "u-cities," and "web-cities." Furthermore, they mention the terms "eco-city" and "smart city." Both terms have to be used with caution, as different approaches inform these terms. A smart city is, amongst others, related to sustainability and a green infrastructure (Abdoullaev, 2011). This includes, of course, parts of the concepts of the knowledge, creative, and digital city, but is not necessary related to world city characteristics. According to an interview with

Saskia Sassen (Meister, 2012), the city Songdo in South Korea, which was built as a smart city or u-city based on ICT networks and business solutions, is not a world city because it lacks cityness. In the following, world cities that are investigated in digital and smart city case studies and analyses will be presented.

5.1.1 ICT infrastructure

To become a digital city, the basic infrastructure has to be established. Events like the Olympic Games or the World Expo have a positive influence on the ICT infrastructure development. Beijing, for example, has established a "Metropolitan Area Network (MAN)" which offers broadband connectivity to public information services and free Wi-Fi for residents, planned to be available for three years (Anthopoulos & Fitsilis, 2010). Song, Zhang, and Zeng (2009) investigate governmental wireless city strategies and state that Beijing and Shanghai have already finished the implementation of wireless hotspots which will serve as a public network for tourists and citizens alike. Free Wi-Fi access has become an important aspect of public spaces, and efforts in this area have been made by different stakeholders, e.g. by the municipality in Toronto, by communities like New York Wireless, or Île Sans Fil in Montréal (Hampton & Gupta, 2008). However, my personal experience has shown that the accessibility and quality vary wildly both within each city and between cities.

So-called "ubiquitous cities" (u-City) go one step further. A u-City is a fully connected city in which all things, institutions, and services are embedded. D. Shin, Nah, Lee, Yi, & Won (2008) identified the topics home, work, education, traffic, health, environment, and government as actual projects in uCity plans. Cities that have adopt such plans are, for example, Hong Kong, Dubai, Seoul, Singapore, and Amsterdam. Their approaches differ from each other but are all grounded in ICT networks. In Seoul, for example, we can find an enhanced ICT infrastructure with traffic sensors and a high smartphone penetration rate (J. H. Lee, Hancock, & Hu, 2014). More than 97.5% of the population is connected to the national broadband network and about 57% use a fiber connection (J. H. Lee et al., 2014). According to Al-Hader and Rodzi (2009), the implementation of smart ICT infrastructure to manage waste, water, electricity, etc. is the basis of a smart city as we can find it in Dubai or Singapore.

5.1.2 ICT networks

From the city government perspective, the ICT infrastructure was used to enable communication between the municipal institutions, or additionally to connect themselves with citizens, organizations, and the economy. The term "digital city" was therefore often used to describe online platforms for communication between different stakeholders. Examples are the "digital city Amsterdam," "Virtual London," "Rete Civica di Milano" (Milan community network), "virtual Helsinki," or "Digital City Shanghai" (Aurigi, 2000; Benini, Cindio, & Sonnante, 2005; Ding, Lin, & Sheng, 2005; Ishida, 2000). Buzzwords such as "cyber cities," "virtual cities," or "civic networks," amongst others, have become popular (Ishida, Aurigi, & Yasuoka, 2005). Another aspect of digital city networks are 3D representations of the city's physical space. Early adopters of 3D representations are Chicago (developed by a private consulting agency) and Melbourne (developed in conjunction with the virtual design studio and the University of Melbourne) (Deren, Qing, & Xiafei, 2000). Further visualization models have followed, for example, in Amsterdam, Berlin, Boston, Dubai, Frankfurt, Helsinki, Hong Kong, London, Los Angeles, Munich, New York, Paris, San Francisco, Singapore, Sydney, Tokyo, Toronto, Vancouver, and Vienna (Horne, Thompson, & Podevyn, 2007). According to Horne, Thompson, and Podevyn (2007), the use of these visualizations can be categorized into "planning and design," "infrastructure and facility service," "commercial sector and marketing," and "promotion and learning of information on cities."

Initiatives that bring together citizens, business and government are often investigated under the heading e-government or e-participation (Palvia & Sharma, 2007). Digital space has been opened in some cities to stimulate political interest and debates. One example is the 2006 municipal election in Milan with the portal ComunaliMilano2006.it. This platform offered profile pages and open forums to discuss and inform about the candidates (De Cindio, 2009). Another example is Shanghai, where the IT industry has opened the way for an enhanced information infrastructure within the city and the government. This has resulted in better public service, for example in the areas of healthcare, fire departments, or public security, by developing one elaborate e-government system (Lagerkvist, 2010).

One of the pioneers of open innovators with regard to its digital city development is Amsterdam. In 1994, a free internet network was launched by the DDS (De Digitale Stad) initiative (Besselaar, 2001; Couclelis, 2004; Riemens & Lovink, 2002). This initiative opened up access to the internet for the public. According to Riemens and Lovink (2002, p. 327), the aim of this initiative "was to democratize the use of the Net and create a digital public domain." DDS was initiated by a grassroots project with funding by the city government. The idea was to bring communities together and to open space for political debates, as well. The digital city approach was used to help users to identify themselves with this virtual communication platform and to enhance the sense of orientation using city metaphors. In other cases, the digital city was a digital simulation of the physical city.

Nevertheless, due to the lack of ongoing funding, the foundation was not able to survive, eventually becoming a business and losing its independent character.

The intention of many digital city initiatives was to build communities. Those "virtual communities" were not intended to replace the physical communities, but rather to extend the space for communication. According to Schuler (2005), this is the case for many cities and towns in Europe. The EU has funded these communities with the goal of stimulating the communication between citizens, institutions, the government, and the economy. According to Couclelis (2004), best-practice examples are the digital cities Amsterdam and Helsinki. Schuler (2005) emphasizes community networks in Milan (wrcm.dsi.unimi.it), Amsterdam (www.dds.nl) and Barcelona (www.bcnet.upc.es/ravalnet) as successful examples. Further, Ishida (2000) mentions "Virtual Los Angeles" with its communicative approach and aim to involve the community in decision-making processes as a successful example from the US.

The European Union has established a funding for digital initiatives called the "TeleCities support program" (Ishida et al., 2005), whose goal was to enable partnerships between European cities and to enhance government-to-citizen communication. These projects have often been implemented in a top-down manner and lacked social inclusion. Thus, many projects were not able to establish themselves and withered. One of the few successful exceptions of the TeleCity project was Vienna. In their approach, the citizens and the government both took on responsibilities of the project and had to work cooperatively. Similar projects and activities were initiated between the years 1990 and 2000 in many European, American, and Asian cities. According to Ishida et al. (2005), these initiatives have served the needs of the information society, as most digital city platforms have been more information- than communication-oriented.

The idea behind these initiatives is to re-design decision-making processes. Citizens are to become a partner in policy and city debates. Montréal is a city that has started to use open innovation processes (Leydesdorff & Deakin, 2011). Open innovation on the city level means to learn from all parties and stakeholders. In a city, these are citizens, private businesses, universities, organizations, and all others who live or work in the city. In a creative city, like in Montréal, open innovation can build creative communities that are able to develop new strategies and policies which do not only depend on economic indicators. This is a further form of human capital, which is an important factor in a smart and sustainable city (Leydesdorff & Deakin, 2011).

5.1.3 Strategic master plan

Becoming a digital or a smart city is, for many cities, a key point of their visionary plans for future development. Take, for example, Malaysia's "vision 2020" which was set up in 1991 (Yigitcanlar & Sarimin, 2010). This vision includes the development of the "Multimedia Super Corridor" (MSC), which is located near the capital Kuala Lumpur and the international airport. The main objective was to establish a competitive knowledge economy based on an enhanced ICT infrastructure. Another early adopter in Asia is Singapore with its IT2000 Master Plan called "Singapore One: One Network for Everyone" (Ishida et al., 2005). Furthermore, Singapore has launched its iN2015 masterplan in 2006, whose aim it was to foster the digital infrastructure and to turn Singapore into a global and intelligent nation (Zhu, 2011).

Seoul has benefited from the nationwide ICT plans which were established early on in Korea, beginning with the Cyber Korea master plan in the late 1990s and followed by the E-Korea and U-Korea plans (S.-H. Lee, Yigitcanlar, Han, & Leem, 2008). The key goals have been high-speed internet nationwide, the establishment of e-government and e-commerce, and a virtual city as a communication platform between different stakeholders. The U-Korea plan called for the ICT network and physical space to be joined, which is why Wi-Fi hotspots were established in large and metropolitan areas. The actual plans focus on the development of U-cities with advanced U-infrastructures, which should enhance urban planning and urban life within an ubiquitously connected environment. In the past, Seoul was dominated by car manufacturing and known as a street-centered and cement-heavy city. Its shift towards a knowledge society with leisure time facilities and public involvement is manifested through the redevelopment of the Cheonggyecheon highway, which was demolished in favor of public space along the Cheonggyecheon creek in central Seoul (Choi & Greenfield, 2009). Today, Seoul has set itself the goal to become a ubiquitous city with the highest rate of broadband penetration, with multimedia boards/touch screens, and LED streetlights including hotspot functionality for Wi-Fi connection (Choi & Greenfield, 2009).

Asian cities as Beijing, Shanghai, Chengdu, Hangzhou, and Wuhan have also adopted master plans with the goal of becoming a smart city (Zhu, 2011). Beijing includes smart home, digital life, smart healthcare, smart traffic, dynamic navigation, and mobile payment in its vision of the future. For Shanghai, high-end industry, e-commerce, e-government, intelligent city management, information safety, and wireless broadband connectivity are major topics in its smart city master plan. In addition, due to its big port, Shanghai had gained in importance as China's center for import and export in the nineteenth century (Lagerkvist, 2010). This has positively affected further physical connectivity, foreign investment, and governmental intervention, which has in turn promoted the cities' ICT sector and the whole economy. Today, many of the "Fortune Global 500" companies have offices in Shanghai.

A further example is Shenzhen. The city has implemented a science and technology development plan to become a hub for high-tech industries and R&D with the aim to improve its competitiveness and to meet the needs of the future (de Jong, Yu, Chen, Wang, & Weijnen, 2013). The master plan of Shenzhen is called "Eco-2-Zone" and refers firstly to the Special Economic Zone with benefits for business and industries and secondly to the Ecological Zone which represents the city's initiatives towards a green city (de Jong et al., 2013; Kang et al., 2014). The ambition of Shenzhen is to attract knowledge workers with their knowledge-eco-city initiatives. One of their main strategies is to promote Shenzhen as a "living laboratory" for knowledge workers who are likely to develop a smart city. Furthermore, a government-to-government cooperation was established to implement green technology together with the expertise of the "Dutch Ministries of Infrastructure & the Environment and Economic Affairs, Agriculture & Innovation, the cities of Amsterdam and Eindhoven and a number of research institutes" (de Jong et al., 2013, p. 217). In China, it is a common practice to enable special economic zones or "sweeteners" like tax reduction in the first years of business to become an attractive space for the desired industry. Beijing, for example, has established the Capital Recreation District (CRD) which is already known as Cyber Recreation Industry due to the benefits it offers for the ICT industry and its workforce (Keane, 2009). Dubai is also following this approach in several economic districts such as Dubai Internet City (Vanolo, 2013).

To focus on establishing the knowledge economy with enhanced activities in ICT and related fields is also emphasized in other cities through the development of digital media city districts such as, "for example, Poblenou (Barcelona @22), Fashion City and World Jewelry Centre (Milan), Orestad (Copenhagen), Digital Corridors (Malaysia), [and] Digital Media City (Seoul)" (Evans, 2009, p. 1007). Hospers (2003a) defines the fashion districts of Milan and Paris as cultural-technological cities, and for Gdaniec (2000), the cultural/technology district Poblenou in Barcelona has resulted in a positive image in which the city is represented as a successful cultural and innovative place. The knowledge-intensive economy and the ability to attract talents is an argument that Barcelona is a prototype of a smart city for Bakıcı, Almirall, and Wareham (2013). Additional policies and reforms according to smart city initiatives in Barcelona are based on its political willingness (Cohendet, Grandadam, & Simon, 2011).

In many other cities, clusters of related business firms have evolved due to strategic master plans, but also due to the economic motivation of firms. This is true for knowledge-, digital- and creative-based industries, the financial sectors, as well as for educational institutions. Evidence for this development in the digital sector is provided by the investigation of Kotkin and DeVol (2001). Firms of the digital or information economy have to stay where the talents are. In the US, those cities are, for example, Boston with its Fort Point Channel called "cyber district," Lower Manhattan (Silicon Alley) in New York City, San Francisco with its "South of Market" district, and Chicago's "Bucktown" neighborhood. In these areas, companies of the information age as well as knowledge workers and creatives have settled down and reshaped the district or neighborhood. The reasons are the same as in the evolution of creative clusters: cheap rents, plenty of space, and a vibrant community. Further examples are the "digital media city" in Seoul, and London's development of a digital hub at Queen Elizabeth Olympic Park, which should cluster business, technology, media, education and data at one place to improve innovation (Angelidou, 2015; S.-H. Lee et al., 2008). Other cities do not establish an entire district, but rather build a "high-tech park" for ICT industries and R&D. One example of this is Shanghai's "Zhangjiang Hi-Tech Park" (Sigurdson, 2005).

New York City has also developed a smart city master plan which focuses on its human and social capital (Angelidou, 2015). In its "Digital Roadmap," the focus lies on the four topics industry, open government, engagement, and access, with the stated approach to humanize technology. Similarly, London focuses on creativity and technology in its smart city development plan (Angelidou, 2015), the main topics being "collaboration and engagement," "open data and transparency," "technology innovation," and "efficiency and resource management" with the overall goal to "improve the lives of Londoners."

According to Ben Letaifa (2015), smart city strategies depend on actual circumstances in a region or city. Thus, London is concerned most about smart transportation and mobility due to the increasing attractiveness of the financial sector located in the dense city center. Montréal is working on smart health (due to its aging population), smart transportation (because of its poor transport infrastructure), and smart grid technologies that are driven by the major ICT providers located in the city. Finally, Stockholm is constructing a smart traffic management intended to replace the 100-year-old infrastructure.

Some cities also call themselves "smart city" because they embed ICT in their developing process. Hollands (2008) has identified San Francisco, Amsterdam, and other cities as forerunners of this trend, with Vancouver and Montréal as further adopters. For Hollands (2008), the main aspects of a smart city are the ability of learning and innovation, a creative population, institutions of knowledge production, and digital networks for communication.

5.1.4 Economy and labor

Digital cities are also defined as hubs of the information and communication technology economy. One example is Helsinki, which has become an important player in the global network of ICT firms due to the success of Nokia (Roper & Grimes, 2005). In 2001 Nokia's global market share was 31%, the firm was active in R&D and on the global trade market. Because of this, Helsinki was dubbed "the telecommunication capital" of Europe (Yigitcanlar, 2009). But a more diversified strategy in ICT markets than just investing in mobile phones is critical for the future (Roper & Grimes, 2005; Van den Berg, Braun, & van Winden, 2001). A further example is Los Angeles' digital city development, which is highly correlated with the film industry. According to Kotkin and DeVol (2001), Los Angeles dominates film production services for the internet. Furthermore, it is a hub for animation and feature films as well as video games.

According to Blythe (2005), increasing e-commerce and innovative solutions like the electronic signature, which allows verifying online transactions in the private sector as well as in government interactions, have made Hong Kong into one of the "most wired" cities in the world. In addition, Hong Kong's neighbor Shenzhen has made a rapid development towards being a global player in the ICT industry as well (C. C. Wang, 2013). It was Chinas first established "Special Economic Zone" which has opened up Shenzhen as a marketplace for global cooperation. This and further governmental plans have pushed the growth of ICT manufacturing and software developing industries in Shenzhen. Nevertheless, Shenzhen cannot be compared with Silicon Valley or other high-performing and innovative cities of the IT industry, since it lacks educated and creative talents (C. C. Wang, 2013). Ding et al. (2005) have investigated the digital city initiatives in Shanghai and concluded that one reason why Shanghai lags behind Hong Kong or the US is the complicated procedure necessary to register a corporation, which is why startups, a very common feature in the IT industry, especially in IT service and software development, are less prevalent in Shanghai.

To become an important player in the digital or smart economy, Glaeser and Berry (2006) argue that a high percentage of educated adults are one important growth factor. They indicate that Boston or the San Francisco Bay Area are examples which possess a high percentage of highly skilled workers and hence higher income rates. This also positively influences the incomes of low-skilled workers in that region. They demonstrate that cities with a lower percentage of educated adults also had substantially less growth and a lower income rate up to the year 2000. Glaeser and Berry (2006) argue that in most industries that emerged from the 1980's until today, human capital has become an important resource. They claim that educated managers and entrepreneurs tend to hire a highly skilled workforce. Two examples support this tendency: First, the founder of Apple, Steve Jobs, started his high-tech company in Cupertino, California, which is located in Silicon Valley. Even today, Apple is still located in Cupertino. The second example is the founder of Gateway, Ted Waitt, who started his firm in Sioux City in Iowa.

He was forced to move with his company to a place with a higher amount of highly skilled workers.

According to Glaeser and Berry (2006), for a city that wants to become an attractive smart city and "stay smart" it is important to offer an excellent school system and to invest in education. This attracts educated parents and educates the next generation for the future. In addition, the city should invest in safety. One example is Boston, which was able to rapidly reduced crime rates until 1990 and in consequence attracted workers with college degrees. Affordable housing for young, educated people and students are necessary as well as limited tax rates to encourage economic growth.

5.1.5 Smart and sustainable city

Abdoullaev (2011) has defined the "trinity city" as a real smart city which combines the three features "Information Cyber City, Intelligent/Knowledge City and Ecological/Clean City." This merges the ICT network, the human population, and ecologically sustainable cities in one place. According to his investigation, cities that focus on a smart environment are Amsterdam, Stockholm, and Sydney. New cities that follow a smart city master plan and have been built from scratch are, for example, Dubai Waterfront, Dubai Central, Putrajaya (next to Kuala Lumpur), Songdo and Incheon next to Seoul, and Langfang located south of Beijing (Abdoullaev, 2011; Zhu, 2011).

Another definition is put forth by Komninos (2011). He joins the enhanced ICT infrastructure and the rise of the knowledge and innovation economy under the term smart city. In his opinion (Komninos, 2011, p. 184), examples are Cyberport Hong Kong and Smart City Amsterdam:

> Cyberport should not be seen as the usual technology district or technology park. It is an ecosystem that nurtures talent in the media industry, turning skills and talent into startups. It amplifies the skills and creativities of the Hong Kong population using experimental digital infrastructure and open platforms. The objectives are technology diffusion, up-skilling and the enhancement of human capabilities. Cyberport is a creative community supplied with advanced communication and media infrastructure and digital connectivity.

Amsterdam has also established an advanced ICT infrastructure. Smart devices (crowd data) and sensors in the city help to analyze and enhance citizens' everyday lives and organizational practices (Komninos, 2011). The government has opened its doors and minds to establish open innovation methods in decision-making processes. Finally, the main goal for Amsterdam is to become a sustainable city that is able to reduce CO_2 emissions permanently.

Amsterdam is a green city which is not exclusively built on ICT, as their transportation infrastructure is focused on bicycles. There are more bicycles in the city than inhabitants, which could be counted as sustainable since the city is independent of petroleum (Gilderbloom, Hanka, & Lasley, 2009). Further plans to make the city more sustainable and raise the quality of life are being developed by the Municipality of Amsterdam. According to Gilderbloom et al. (2009), they are concerned amongst others with noise pollution, air quality, climate change, green space, and recreation spaces. These measures are accompanied by the Amsterdam Smart Meter Service, which is intended to help in reducing energy consumption (J. H. Lee et al., 2014). The reduction of emissions is also the aim of the Urban Eco-Map in San Francisco (J. H. Lee et al., 2014). On this map, citizens can look at visualizations of the emissions produced in their neighborhood and compare them with other neighborhoods. In addition, they can learn about actions to reduce their own ecological footprint. J. H. Lee et al. (2014) have found that more than 44% of San Francisco services (smartphone applications or websites) have a direct or indirect impact on the environment. Many services combine technology like sensors and civic participation. In Vienna, for example, the project "Smart City Wien" has developed the "Smart Energy Vision 2050" which offers advice on how the climate goals could be met to, step by step, become a smart and sustainable city (Hofstetter & Vogl, 2011).

In the case of Seoul, a participatory approach has helped to reduce 10% of CO_2 emissions annually (C40 Cities, 2011), introducing a one-day-no-driving campaign (J. H. Lee et al., 2014). Citizens were able to participate by registering online at no-driving.seoul.go.kr and choosing a day between Monday and Friday on which they would only use public transportation. After registering, they received an e-Tag based on RFID technology to monitor if the car was not driving in the city. This campaign was introduced in 2003 and includes monetary tax incentives for participating citizens (Seoul Solutions, 2015). Further real-time monitoring is used in other public sectors like "the public drainage system, Han River bridges, the fire service, public parking, garbage trucks, some public buildings and SMC air pollution" (J. H. Lee et al., 2014, p. 94). GPS and real-time information are also used in San Francisco, e.g. for real-time public transportation navigation (Routesy San Francisco) or parking space sensors to adjust parking fees response to actual demand (J. H. Lee et al., 2014). Furthermore, metering systems are integrated into water, energy, and gas networks. Sensors have also been integrated in the Royal Seaport of Stockholm. Real-time consumption of energy and material production are monitored by a "Smart Urban Metabolism" framework (Shahrokni, Lazarevic, & Brandt, 2015), analyzing and visualizing energy flows within the seaport.

Different rankings determine certain nations and cities to be digital or smart. Smart communities have for example been nominated by the Intelligent Communities Forum (ICF) since 2006. Cities are then considered for the "ICF's annual Intelligent Community Award." Each year, 21 finalists are chosen by a committee. The finalists are scored according to an indicator framework which is based on a questionnaire. Finally, the top seven cities are announced as best practice examples "in broadband deployment and use, workforce development, innovation, digital inclusion and advocacy that offer lessons to regions, cities, towns and villages around the world. They are charting new paths to lasting prosperity for their citizens, businesses and institutions" (Intelligent Community Forum (ICF), 2014, para. 2). They have indeed identified cities that show the political willingness to become smart, e.g. Hong Kong, Shanghai, Seoul, Singapore, Stockholm, San Francisco, Toronto, Vancouver and many more. The case of Singapore is rather unique: Singapore is a city-state island with the strong political willingness to become an intelligent island. That Singapore is making significant progress towards this objective is evident through international surveys, in which Singapore is highly ranked because of the e-government status and due to its ambitions to make IT accessible and available across the whole island (Baum, Yigitcanlar, Mahizhnan, & Andiappan, 2008).

5.1.6 Smart city applications

Many cities use ICT and sensors around the city to implement innovative city services. In recent years, this has increasingly involved user-generated data. In combination with open data offered by the government, smart city applications have been developed by citizens as well as the private sector (Mainka, Hartmann, Meschede, & Stock, 2015b). The different types and approaches to develop smart city services in Seoul and San Francisco have been investigated by J. H. Lee et al. (2014). They determine that Seoul has implemented services for diverse topics such as health, welfare, and education. San Francisco is more focused on public administration and transportation. Most applications in Seoul are built by the city's IT department, but in 2011 they launched an open data portal and have tried to follow a more participatory development strategy since (J. H. Lee et al., 2014). Due to the early adoption of open data by San Francisco MTA (public transportation) many transportation applications have already been developed (J. H. Lee et al., 2014).

Mobile applications are moving more and more into focus in smart city development. Open data portals and hackathons are therefore the ideal backbones for innovative solutions (Mainka, Hartmann, Meschede, & Stock, 2015a; Mainka et al., 2015b). The City of Chicago, for example, offers an open data portal but does not develop its own mobile apps. According to John Tovla, chief technology officer of Chicago, "volunteers and private sector will probably innovate quicker and certainly at a scale that government cannot match" (Walravens, 2015, p. 219). The aim of such applications is called public value. Walravens (2015) has created a business model grid to identify and compare the public value of apps. Apps that are of high public value are, for example, "I Amsterdam QR spots," "London Bike App," and "Berlin Neighborhood." All three apps pursue different goals but can be identified as apps that enhance the quality of life in a city.

5.2 Creative and knowledge cities

According to Lor and Britz (2007), "usable content" and "human capacity" are pillars of the knowledge society. For both we have to determine the cognitive infrastructure in a city. For Stock (2011, p. 970) the creative city – with its "[c]reative-artistic activities and the result thereof" – and the knowledge city – with its "[s]cientific-technical-medical activities and the result thereof" – represent characteristics of this infrastructure. Knowledge and creative activities are part of the development of soft infrastructures, but are not as easy to measure through hard facts as roads or ICT (Setunge & Kumar, 2010; Stock, 2011). Etzkowitz and Klofsten (2005) see the academic facilities, municipal institutions, and enterprises as responsible for the development of the needed infrastructure. In the same vein, content and shared knowledge are of high importance as well. Lifelong learning has to be included as part of the cognitive infrastructure and is equally important for institutions and municipalities (Stock, 2011). For Evans (2009) and Foord (2008), the creative economy can be subsumed under the knowledge economy. In the following, I will highlight the topics which are investigated in the literature with regard to both city types in case studies and investigations.

5.2.1 Historic development

Different historical backgrounds may positively influence a city's development towards a creative or knowledge city. For instance, cities that are historically known as cultural hubs now have a high potential to become creative cities. Musterd and Gritsai (2009) have investigated European cities and identify this potential for, among others, Amsterdam, Munich, Milan, and Barcelona. The capital function of a city is seen as an advantage to become a creative or knowledge city: "Capitals are often culturally rich, with major museums, art galleries, universities and other important seats of research and learning located there" (Brown, Redmond, & Miquel, 2013, p. 62). According to Hospers (2003a), the creativity of a city increases during unstable times, for instance during the two World Wars in Berlin, or in 1600 in Amsterdam, or during the nineteenth century in Vienna, London, and Paris.

At the same time, historically established creative cities will not lose their role. Hospers (2003a) mentions, for example, Los Angeles, New York, and London as cities which have been the first to agglomerate cultural and creative activities and gain success through this. For Hospers, this is the Matthew effect, summarized by the principle "success breeds success." Further cities with a cultural/creative background are Vienna and Berlin (Hall, 2004). Vienna, for example, has established itself as a knowledge hub with excellent universities, a high number of scientists and students, and an increasing number of patents (Trippl, 2012).

5.2.2 Economic transformation

Another factor is the past economic specialization of a city: Cities that are characterized through industrialization have to overcome bigger challenges to restructure their economic orientations. Musterd and Gritsai (2009) indicate that a good starting point for the establishment of the creative and knowledge industry are high-skilled activities, engineering, and high-tech activities, but a service-oriented economy is also advantageous. Within Europe the cities of Amsterdam, Barcelona, Helsinki, and Munich are examples.

For instance, Amsterdam has become a creative and knowledge city because of the city's open and tolerant image and due to the creative industry which now is located there (Gilderbloom et al., 2009; Musterd, 2004; Pethe, Hafner, & Lawton, 2010): "A tradition of innovative economic talent, combined with a high degree of openness, has resulted in the distinctive Amsterdam atmosphere, a major pull factor for knowledge workers and creative individuals" (Musterd & Deurloo, 2006, p. 81). This has resulted in a high metropolitan quality of life and an international atmosphere which attracts talents (van den Berg, Pol, van Winden, & Moets, 2005). Other cities are prominent because they are known to be diverse and tolerant as well, such as Barcelona and Munich (Musterd & Gritsai, 2009; Pethe et al., 2010). These cities are a home for the creative class. In North America, for example, 14% of the nation's artists live in the cities New York and Los Angeles (Kotkin & DeVol, 2001). To attract creatives and talents, a city's tolerance can be an indicator. Florida and Gates (2001) have identified an increasing correlation between the Gay Index and the growths of regional ICT clusters. This does not mean that all talents work in the ICT industry but that talent from abroad who are members of a minority may feel more comfortable in these places (Florida, 2014). Such tolerant and creative regions are, among others, San Francisco's Bay Area, New York's Greenwich Village, semi-autonomous Hong Kong, Greater Amsterdam, Toronto and Vancouver (Heywood, 2008).

One city that has undergone a transformation from a car manufacturing center (dominated by SEAT) to a vibrant, creative, and knowledge-driven city is Barcelona (Edvinsson, 2006, 2011). Today, the city benefits from

> newly constructed and remodeled city areas, positive image, attractive world-class events, and high performing schools and research organizations. What might be the most interesting is the explicit focus and vision of shaping the context for the new workers of tomorrow rather than the present focus groups of politics. The artists, designers, food and restaurant entrepreneurs, biomedical researchers, educational entrepreneurs, all working with intangible and intellectual capital have been attracted to this city. (Edvinsson, 2006, p. 8)

An extreme transformation of the city's economic orientation, as in the case of Barcelona, can help the creative and knowledge industry to flourish, other prominent examples being Singapore, Hong Kong and Dubai (Edvinsson, 2006). Further cities that Edvinsson (2006) recognizes as hubs of talents of the creative and knowledge economy are for example Boston, New York, and Stockholm. In an investigation of European cities van Winden, van den Berg, and Pol (2007) identify Amsterdam, Munich, and Helsinki as cities that are able to attract talents and the knowledge-based industry. According to van den Berg et al. (2005), Munich, which is mostly known for its car manufacturer BMW, has a very diversified economic base and a large number of knowledge-intensive companies and institutions which they call the "Munich Mix." This kind of business landscape shapes the foundation of the knowledge economy, which is the employer for a high amount of workers with high qualifications. In the case of Helsinki, Inkinen and Vaattovaara (2010) argue that the city's advantages are the generous welfare, excellent educational system, and the development of the ICT infrastructure which has helped Helsinki to rise above the Nordic periphery. Hu (2012) has investigated Sydney's transformation into a knowledge city. Today, Sydney is a hub for advanced producer service firms, which is the main aspect of global cities, but creative industries, cultural and media services are located in Central Sydney as well, characterizing Sydney as a knowledge city. Rocco (2012, p. 391) argues that "providers of services are providers of knowledge" and has investigated the knowledge economy of São Paulo, which has also attracted national and international producer service firms. A prominent example of the creative economy is Berlin. According to an investigation by Kunzmann and Ebert (2007), branches of the cultural or creative economy are represented above average in 43 of Berlin's 192 districts. They also identify a higher concentration of creative economy activities near the city center. Another example is the development of the knowledge economy in Dubai. Dubai benefitted from the oil boom and became able to rebuild its landscape into a prosperous and vital city. According to Alraouf (2008), Dubai has built up a successful mix of the knowledge economy and service industry.

Some cities are acknowledged as a digital city although their economic orientation is not prominent in IT and engineering fields. This is true for example in the case of New York. Currid (2006) has investigated the distribution of occupations according to different economic sectors for the years 2000 and 2004, demonstrating that New York has just a marginal proportion of occupations in IT and engineering. Currid (2006) argues that New York's ongoing status as a global center is perhaps the effect of its vital and diverse industries, most importantly the creative economy. Occupations in arts and culture dominate in New York and have a higher potential to grow than the financial sector (Currid, 2006).

5.2.3 Strategic master plan

In the literature, many cities that are on their way to becoming a knowledge city or have already gained the status of a knowledge city are investigated according to their "Knowledge-Based Urban Development," resting on the three pillars environment, economy and society (Yigitcanlar, O'Connor, & Westerman, 2008). This includes many and diverse strategic decisions and the inclusion of industry and society in governmental urban planning. The result of this kind of planning is usually the manifestation of a master plan or vision for the future of the city. In addition, a strategic master plan for a city with the goal to become a creative or knowledge city may have a positive effect to face future challenges. As positive examples, Barcelona with its vision for 2015 and Melbourne's City Plan 2010 are highlighted in a report from Price Waterhouse Coopers, *Cities of the Future* (Bolz et al., 2005). Both action plans are concerned with a wider range of critical aspects, such as environmental quality, social equity, education, economy, culture, and more. Today, it can be observed how successful these plans have been implemented and manifested. Melbourne, for example, has received the "Most Admired Knowledge City Award" (MAKCI) by the World Capital Institute with regard to their outstanding city development (World Capital Institute, 2012). Currently, Melbourne is working on their plans for 2030 (Victorian Government Department of Sustainability and Environment Melbourne, 2003).

As a further example, Edvinsson (2006) mentions the impressive transformation of Malaysia and its capital Kuala Lumpur. Today, the city has taken many steps towards their 2020 vision, which includes becoming an attractive place for living and working, and services on a world class level. Edvinsson (2006) calls Kuala Lumpur a prototypical intelligent city which is able to attract the creative class. A strategic plan to develop a knowledge city is acknowledged as fundamental to be successful (Yigitcanlar, 2009). Following Yigitcanlar (2009), other cities that endeavor to become a knowledge city are Dubai and Shanghai. More cities, like San Francisco, Seoul, Shanghai, and Singapore, have also set the goal of becoming a knowledge city, with the focus on "people's skills and abilities" (Reffat, 2010). Berlin is mentioned as a city which has also established a master plan to encourage the city's development towards a knowledge city. According to Franz (2009), Berlin is a city which is rich in knowledge and poor in capital. World cities that mentioned in the literature as having already successfully completed the transition towards a knowledge city or creative region due to their strategic plans and visions are Barcelona, Berlin, Beijing, Boston, Dubai, Helsinki, Hong Kong, Kuala Lumpur, London, Melbourne, Montréal, Munich, Seoul, Shanghai, Shenzhen, Singapore, Stockholm and Vancouver (Carrillo, Yigitcanlar, García, & Lönnqvist, 2014; Durmaz, Yigitcanlar, & Velibeyoğlu, 2008; Dvir & Pasher, 2004; Foord, 2008; Heng & Low, 1993; Hospers, 2003a, 2003b; Kong & O'Connor, 2009; Lange, Kalandides, Stöber, & Mieg, 2008; Y.-S. Lee & Hwang, 2012; Metaxiotis & Ergazakis, 2012; D. Wang, Wu, Li, & Wang, 2012; C. Wong, 2008; C. Y. L. Wong, Millar, & Choi, 2006; Yigitcanlar, 2009, 2012; Yigitcanlar et al., 2008; Yusuf & Nabeshima, 2005; Zhao, 2010).

Master plans that endeavor to establish the knowledge economy and additionally set a focus on cultural activities are acknowledged as important for the development of a creative city. According to Evans (2009), many cities have started to try and replicate successful creative city or creative economy approaches. For example, the "Creative London" commission was adopted by the following councils/commissions: Creative New York, Creative Amsterdam, Create Berlin, Design Singapore, and Design Toronto. Landry (2008) emphasizes the cities Barcelona and Frankfurt as creative cities because their visionary plans are highly focused on cultural development. In some master plans the creative city development includes the formation of "creative clusters" which "helps to shape places or urban sites as 'creative milieu'" (Chen, 2012, p. 439). The development of new urban space through special economic zones (SEZ), reuse of industrial estates and warehouses for artistic activity, and strengthening the development of links to universities and traditional creative institutions are core implementations of the creative cluster in Shanghai (Chen, 2012). According to Wu (2005), the political interventions to build clusters, such as creative or R&D clusters, actively contribute to developing local creative hubs. A further example is Boston, as evidenced by its creative growth in the software industry and biotech industry, which is university-based and located there as well. Those clusters have emerged in many Asian cities as well, examples being Singapore and Beijing, which have both been developed as top-down initiatives by the government (Keane, 2009; Kong, 2009). In Singapore's master plans the explicit development of the creative economy belongs to the key innovations (Hornidge, 2007). According to Porter (1990), cluster theory is not a recent development, as industries have always tended to cluster. The aim is to create vertical (supplier to the buyer) or horizontal (host, vendors, and suppliers – business to business) links. Formal as well as informal interactions are seen as a generator of innovation because of the direct or indirect idea and knowledge exchange.

Landry (2008) mentions further examples of governmental plans that were established years ago, such as the "pedestrianization" of Munich in the 1970s or the regeneration of the waterfront in Barcelona in the 1990s. Both reinventions were made in preparation of the Olympic Games and are helpful factors for the development of the creative city. Landry and Bianchini (2006) emphasize Montréal, Singapore, and Milan due to their pedestrianization and car congestions, which exclude cars from the city center as forerunners of a livable city. These programs have made the redesigned places into a magnet for citizens and tourists alike, enhancing the quality of life.

The city's shape and design is also a part of urban planning. A positive example which is mentioned by Landry and Bianchini (2006) because of its creativity is Melbourne, for its skyline which looks metropolitan but has no high-rising buildingsas, for example, Manhattan does. Another example includes the Metro Stations in Stockholm. The stations have been decorated and designed by artists. Landry and Bianchini (2006) argue that this conveys a positive feeling with regard to public transportation and, additionally, offers another tourist attraction.

The strategic master plan "22@Barcelona" focuses on the IT sector but is also important for the establishment of the creative city because it attracts creative talents to work in Barcelona (Cohendet et al., 2011; Marti-Costa & Miquel, 2012). The redevelopment of the city district Poblenou has succeeded in turning it into a vital creative neighborhood, which is a positive result of the knowledge city master plan by the government. For Barcelona's government, cultural and creative activities are embedded in their knowledge city ideology.

Landry (2008) has investigated the development of creative cities. For him, many creative innovations were established because of necessity or scarcity without any master plan whatsoever. Examples are the high-rise buildings in New York, which were developed due to increasing immigration, or the floating car parks in Amsterdam, which were established because of the city's limited space. Another innovation he mentions is the car sharing initiative in Berlin called "STATTAUTO" (lit. 'instead of car', but also homophone and near-homograph to

Stadtauto ('city car'). Compared to the average car use, the shared cars have driven twice as many kilometers and have mostly carried two persons instead of the statistical average of 1.4. For Hospers (2003a) the underground train system was a creative solution for the traffic in Paris, Stockholm, or London. This has had a positive effect on the further growth of these cities.

5.2.4 Face-to-face facilities

Communication is also a major aspect of the creative city. People have to be able to meet each other in the physical world, requiring a diversity of locations spread throughout the city. One example which has a lot of those facilities is Vienna. There are countless *Kaffeehäuser* ('coffeehouses') which are open from early morning until the late evenings and provide spaces for meeting and collaboration. Hospers (2003a) has called this the "café factor." According to Landry (2008), the café culture has traditionally been a common feature in Central Europe, and can also be observed in cities such as Berlin or Munich.

For Landry (2008, p.126) communication and partnerships between cities are also important in creative cities: "Networking and creativity are intrinsically symbiotic, as the greater the number of nodes in a system the greater its capacity to reflexive learning and innovation." Communities can come together either physically or virtually. He is focused on community networking and provides examples such as Virtual Helsinki, Copenhagen, Amsterdam, or Manchester.

Network communities which focus on urban development within a city are a further example for creative cities. Landry (2008) describes the case of Chicago, in which a partnership of organizations and volunteers have developed ideas for the future of the city. The project is called "Imagine Chicago." Furthermore, Chicago has the advantage of being geographically centralized (Reffat, 2010). Therefore, it is a good place for conventions and conferences where people can meet. This is also a magnet for professionals of the knowledge economy. Accordingly, Chicago is an excellent location for knowledge exchange.

In his discussion on human capital on the national, regional, and business level, Pawlowsky (2011) emphasizes cities that have established knowledge centers which enhance the communication between politics, business, and citizens. One example is Dubai, which has passed the competing Singapore in as a knowledge city due to projects that have fostered education and human capital. For Pawlowsky (2011) the establishment of a high-tech campus with impressive architecture has been a major asset in this regard. Perhaps the city most noted for its scientific development and education is Boston, with its globally known universities MIT and Harvard (Evans, 2009; Reffat, 2010). For Hospers (2003a) those hubs may also be referred to as cultural-intelligent cities, mentioning Boston, Toulouse, and Heidelberg as examples. In addition, Barcelona has been mentioned as a developed knowledge city due to its orientation towards a knowledge economy (Pawlowsky, 2011; Walliser, 2004). In Helsinki a university-government corporation called the "Culmitaum Innovation Oy Ltd." aims to increase the collaboration between universities and firms and to foster the production of technological innovations (Stachowiak, Pinheiro, Sedini, & Vaattovaara, 2013). A similar approach is followed by Frankfurt (am Main): The city has established the "House of Finance" located at the Goethe University, which concentrates on public-private partnerships focusing on science, politics, and business (Szogs, 2011). In Milan, the "Politecnico di Milan Acceleratore d'Impresa" supports the establishment of startups based on academic research and services for enterprises (Stachowiak et al., 2013). Further knowledge centers, for example, have been established in Hong Kong (Future Nest), Tokyo (KDI Future Center), New York (Metrotech), Singapore (One-North), Shanghai (The Zhangjiang Hi-Tech Park), Amsterdam (Amsterdam Science Park ASP), Sydney (The Macquarie Technology Business Precinct), Melbourne (The LaTrobe Research and Development Precinct), Toronto (MaRS) and Helsinki (Helsinki Science Park) (Bugliarello, 2004; Edvinsson, 2011; Evans, 2009; Pawlowsky, 2011; Sigurdson, 2005; van den Berg et al., 2005; Yigitcanlar & Martinez-Fernandez, 2010). Those centers are often called "Knowledge Innovation Zones," "Science Parks," or "Knowledge Parks."

5.2.5 Knowledge output

Focusing on science and knowledge output in terms of academic papers, the cities Tokyo, London, San Francisco, Boston, New York, Paris, Los Angeles, Amsterdam (including surrounding region), Beijing, Moscow, and Osaka have led the city ranks in the last two decades according to the bibliometric investigation of Matthiessen, Winkel Schwarz, and Find (2006) (listed in descending order). A further research hub is Berlin. According to Franz (2009), Berlin has a high concentration of publicly and privately funded research institutions with a high quality related to the number of graduate students and the ability of professors to apply for funding.

The role of universities is not limited to academic output: They are important as teaching institutions as well as mediators between different stakeholders in developing a knowledge city (Powell, 2012). Examples are the "UN Global Cities" program invented by the RMIT in Melbourne or the "Camp for Social Innovation" by the Alto University in Helsinki. Those initiatives try to meet the actual needs of urban and social development.

Research output may also be used as an indicator of "knowledge flow," which represents the connectivity between knowledge cities (Haustein, 2012). According to the investigations of Matthiessen et al. (2006) the cities London, New York, Los Angeles, the San Francisco Bay Area, Boston, Baltimore, and Philadelphia are the seven largest research nodes.

5.2.6 Knowledge economy and labor market

For Kotkin and DeVol (2001), scientists as well as professionals working in IT jobs belong to the group of knowledge or creative workers. In their investigation of the North American labor force, they identified an increase of IT professionals in the cities San Francisco, Seattle and New York. To attract highly skilled professionals, the city has to emphasize the quality of life. In Amsterdam, for example, the strategic plan includes the improvement of amenities for culture, arts, leisure, tourism, education, and more (Romein & Trip, 2012), because an attractive living climate improves the business climate in a creative city. Strategic plans or campaigns may help to attract talents to the city, e.g. Barcelona's initiative "Talencia" (Wesselmann, Meyer, & Lisowski, 2012).

Pareja-Eastaway, Chapain, and Mugnano (2013) have investigated the "city branding" initiatives by Barcelona and Amsterdam. Both cities aim to attract foreign talent by setting the focus on their creative- and knowledge-intensive economy. An important aspect is that these efforts are made in cooperation with citizens, institutions, and economic actors. This can also be promoted by established networks in a city. Streit and Lange (2013) mention, for example, the network "CREATE Berlin," which helps to connect creative people in the city. The City of Amsterdam has founded the "Bureau Broedplaatsen," a service that helps artists and creatives to find affordable space for their activities (Streit & Lange, 2013). Local cultural activities or tourist attractions also play a role for creative cities. Helsinki was emphasized as one example of cultural innovation by Landry (2008): The city is known for its lightening festival "Vailon Voimat," which takes place in winter to overcome the dark time of the year and has resulted in increased cultural cooperation with other cities such as Barcelona and London.

According to Landry (2008), city branding is also related to unique creativity. For him, examples like the creative interventions in the public space and world-class architecture create a unique image, e.g. in Barcelona by prominent artists such as Miro, Serra, Pollock, and Ellsworth. Barcelona is a living art city with a lot of design and beautiful architecture (Landry, 2006).

Amenities of higher education may attract many young people which will become the next creatives and knowledge workers. Pareja-Eastaway, Bontje, and d'Ovidio (2010) have investigated the potential of Amsterdam, Milan and Barcelona to attract young and high-skilled workers. They see these three cities as creative knowledge centers which are recognized internationally, with a broad spectrum of higher education institutions but also with a high cost of living. Similar findings are described by Baum, Yigitcanlar, and O'Connor (2008) for the cases of Sydney, Melbourne, and Canberra. All cities have a creative industry which is concentrated in the city center and all cities have the same problem of costs of living which are too high for the creative class.

To investigate the knowledge workforce in Melbourne, Johnson (2012) has analyzed the 2006 ABS Census data for industry and deployment, counting 244,000 people as part of the knowledge economy workforce, split into (Johnson, 2012, p. 280):
1. Telecommunications, IT and Media (around 31 000 jobs or 13 per cent)
2. Finance and Banking (111 000 or 45 per cent)
3. R&D/Higher education (71 000 or 29 per cent)
4. Design-related industries (e.g. architecture) (27 000 or 11 per cent)
5. Cultural industries (e.g. performing arts) (6 000 or 2 per cent).

The data shows that just a small part (13%) of the workforce has a job in design-related or cultural industries, but a large number of people is working in knowledge-intensive jobs like R&D and higher education as well as in the finance sector. The case of Los Angeles shows another aspect of the creative economy: Los Angeles is known for its film and entertainment production activities. For Landry (2008) this is the best example of a successful match of cultural and technological creativity.

5.2.7 Creative milieu

It is not always possible to promote or plan the knowledge or creative city. Looking back to around 1900, so-called "creative" or "bohemian milieus" evolved in city areas which were formerly dedicated, for example, to industrialization or commerce (Bontje & Musterd, 2009; Wojan, Lambert, & Mcgranahan, 2007). Old manufacturing spaces were rented or occupied by artists, as for example in the city district Soho in New York (Vivant, 2010; Zukin & Braslow, 2011), where during the deindustrialization of Manhattan, artists started to repurpose the space for artistic activities. This turned Soho into a flourishing neighborhood, but caused commercial galleries, luxury shops, cafés and restaurants to enter the district and in turn effectively banish the artists responsible for the rejuvenation, as they could not afford the rising rent costs and more expensive lifestyle. Today, we still see examples of those squatters or cheap buildings for rent, e.g. *Squat Chez Robert* and *Electron Libre* in Paris, a former film academy, parts of harbour and shipyards areas, the *Westergasfabriek* (Western Gas Factory) in Amsterdam, the former village of Gracia in Barcelona, the *Cable Factory* in Helsinki, the *Domagk* (a former army barrack) in Munich, the art district *TianZiFang* in Shanghai, or the recently closed art house *Tacheles* in Berlin (Bontje & Musterd, 2009; Brake, 2012; Hitters & Richards, 2002; Uitermark, 2004; Vivant, 2010; Wei & Jian, 2009). For Evans (2009) creative quarters evolved from prior bohemian quarters as in the cases of London, Paris or New York or also from the garages of Silicon Valley. In the case of Canadian cities (Montréal, Toronto, and Vancouver) "bohemians" have also occupied inner city districts which used to be industrial warehouses or inner city neighborhoods (Gertler, 2004; Ley, 2003). Today, we can still find creative neighborhoods in Barcelona (Bairro de Grácia Bairro do Raval) or São Paulo (Vila Madalena) (Costa & Oliveira, 2009), for example. Evans (2009) adds new districts of commerce like the Fashion City in Milan to the category of a creative milieu. Further creative clusters which arose due to policy interventions are the Dashanzi arts district (yishu qu) in Beijing, the m50 art district in Shanghai, the West Kowloon Cultural Centre in Hong Kong, the Creative Gateway at King's Cross London, Liberty Village in Toronto, or Wicked Park in Chicago (Catungal, Leslie, & Hii, 2009; Foord, 2008; Henry, 2010; Kong & O'Connor, 2009; Lloyd, 2002; J. Wang, 2009). Of course, others arose without policy intervention, such as the film/TV production industry in Los Angeles, the fashion and furniture industries in Milan, and the fashion industry in New York. Evans (2009) also locates hubs of design in London and New York.

It is not unusual that such creative milieus grow up to creative clusters or districts (Cinti, 2007). If there is, for example, one successful creative site, e.g. the cable factory in Helsinki, other creative, innovative, cultural or artistic amenities may choose to settle down in the neighborhood as well. To create or promote a creative district may have the following goals: to reuse degraded space, enhance the city image, attract tourists, or conserve heritage and culture. Different approaches and best practices of how to establish a creative cluster have been discussed in the literature (see for a review Cinti, 2007).

Cohendet, Grandadam, and Simon (2010) have investigated the layers of a creative milieu in a city in a case study of Montréal. In this city, for example, the circus Cirque du Soleil and the game developer Ubisoft are important for the city's creative milieu. The Cirque du Soleil has established a National Circus School and is the initiator of the International Circus Festival. Ubisoft has also initiated a festival, the Ubisoft Street Festival, and additionally organizes the International Game Summit. Schools and educational amenities for creatives also attract the creative class. In Montréal the National Circus School recruits artists from around the world (Cohendet et al., 2011).

In another investigation Cohendet et al. (2011, p. 154) analyzed Barcelona's creative cluster and identified five main institutions:

> (i) the Pompeu Fabra University, focusing on communications related training, research and production; (ii) the Engineering schools of UPC-Barcelona and the new School of Industrial Engineering of Barcelona; (iii) the Barcelona Media Innovation Centre (CIBM), which conducts research, innovation and experimental production projects in the field of communications and audiovisual production; (iv) the new Barcelona Digital ICT Technology Center, which aims to contribute to the development of the Information Society and the growth of the ICT sector; and (v) the 22@Living Lab, led by 22@ and the Barcelona Digital Foundation, which forms part of a network of different urban laboratories operated by the public and private sectors, aimed at developing new ICT based mobile technology products and service.

Creativity, culture, and heritage are also important for the tourism and visitor economy (Evans, 2009). However, cities today are confronted with a new type of tourists (Maitland, 2010): This type of tourist wants to learn more about the daily life of citizens. One example of this is London: The London Eye and Big Ben are the main tourist attractions, but some tourists want to make their own way through the city, to get in touch with locals. According to Maitland (2010), especially gentrified areas, where the "yuppies" live, are the new tourist spots for the "cosmopolitan consuming class" and "transnational elites." Thus, London's new hotspot is Islington, which is a neighborhood of the wealthy middle class with a lot of shopping opportunities.

Another way to express the creativity of a community in a city are events which are organized by independent groups. One example mentioned by Landry and Bianchini (2006) is from Shanghai: Families living in houses in the Putuo district sing from their balconies in a competition and are judged by a jury staying in the courtyard. This event has become a festival which is now celebrated with special lighting and food in the district near the estate.

5.2.8 Knowledge city benchmarks

E. Ergazakis, Ergazakis, Metaxiotis, and Charalabidis (2009) present in their research paper a framework of characteristics which should be met by successful knowledge cities, such political support, an advanced library network, tolerant and open society, e-government services and other factors. Prosperous examples for them are Barcelona, Montréal, Munich, Stockholm, Singapore, Dublin, and Delft. In a prior publication, K. Ergazakis, Metaxiotis and Psarras (2004) have also highlighted the cities Melbourne – due to its City Plan 2010 which has focused on knowledge as economic driver – and São Paulo – due to its developed research network and its role in connecting the knowledge community virtually.

Another benchmark for knowledge cities is the Most Admired Knowledge City Award (MAKCi), which was established as a cooperation between the World Capital Institute and Teleos in 2007, creating a framework based on eight categories: (1) identity, (2) intelligence, (3) financial, (4) relational, (5) human individual, (6) human collective, (7) instrumental-material, and (8) instrumental-intangible (World Capital Institute, 2008). Experts with expertise in knowledge management or knowledge-based development nominate cities and eventually vote to determine the most admired knowledge city. Cities that have been elected by 10% or more belong to the finalists. Because cities come in vastly different shapes and sizes, cities can be nominated either as knowledge metropolis or knowledge city region since 2009.

Following finalists were selected in the last awards (Table 5.2):

Table 5.2: Finalists of MAKCi 2007–2012.

2012	Austin	Bilbao	Brisbane	**Melbourne**	Montréal	Ottawa	Seoul	Singapore
2011	Austin	Bangalore	Holon	Manizales	**Melbourne**	Nuremberg	**Singapore**	
2010	**Barcelona**	Manchester	**Melbourne**	Ottawa	**Singapore**			
2009	**Barcelona**	**Boston**	Istanbul	Manchester	**Melbourne**	**Montréal**	Shenzhen	Valencia
2008	Bangalore	Manchester	**Montréal**	Ottawa	**Singapore**	Valencia		
2007	**Barcelona**	Bilbao	Boston	Ottawa	**Singapore**			

Source: World Capital Institut (2007, 2008, 2009, 2010, 2011, 2012).

In 2012 the last report was published online. More current information can only be found at a Wiki run by an independent knowledge management group ("Most admired knowledge cities," 2016). Accordingly, further awards were given to Boston and Melbourne in 2013, Ottawa in 2014, Montréal and Vienna in 2015, and to Melbourne and Dublin in 2016. Further finalists in 2016 were Tokyo, Seoul, and Vancouver. Since the information about the MAKCI Award and the investigation process of the winning city and region are not available since 2013, the reliability and trustworthiness of the award has been somewhat affected. Nevertheless, the initiators of this award are highly cited researchers in the field of knowledge cities (e.g. Francisco Javier Carrillo, WCI President, Blanca Garcia, WCI Executive Director Awards Program, Tan Yigitcanlar, WCI Executive Director Events Program, and Kostas Metaxiotis, WCI Executive Director Editorial Program).

5.3 Conclusion of the identification process

In this chapter, the review of the investigated literature has been described in detail. The aim was to identify a corpus of cities that are either already an informational world city, or well on their way to becoming one. Hence, the initial precondition of this investigation was that the city had to be mentioned as a hub in world city research. This condition is fulfilled by all cities listed in Table 5.1. As a consequence, all cities that may also constitute a good example of an informational city, but are not explicitly regarded as world cities, will be excluded from this investigation.

Additionally, the cities had to be mentioned at least in one of the following research topics: knowledge, creative, digital, or smart city. The literature investigated here was retrieved from Web of Science, Scopus, and Google Scholar. All retrieved publications were evaluated manually and were therefore limited to the researchers' own language proficiency (German, English, and Spanish), with the majority of the publications written in English. All of the cities have been identified as a knowledge city or developing knowledge city. All cities except for Shenzhen and Tokyo have also been determined to be a creative city. These exceptions may well have come about due to the language limitations. Furthermore, all cities except for São Paulo and Stockholm have been mentioned in the literature as digital cities. This absence of any mention should not be taken to mean that both cites are disconnected from the internet, they merely have not been mentioned as examples of a digital city or investigated in a case study on a related topic. Finally, 23 out of the 31 cities have been mentioned or investigated in the literature as a smart city. This may be caused by the inconsistent use of the term smart city (Hollands, 2008), whereas topics like sustainability and mobile applications based on urban data have been considered.

The literature review has reflected that many of the topics that are part of the city's development towards an informational city are hard to be confined to one category, as there are always interrelations. Thus, for example, the gaming industry belongs to the category entertainment and is included in the creative economy. But it is also increasingly related to ICT and therefore might also be counted as value added in the technology sector. Overall, each city type has garnered a lot of research attention in recent years, as the literature review in this chapter has demonstrated. Therefore, all of them have to be analyzed in an investigation that is aiming to identify interrelations of city development in the twenty-first century.

References

Abdoullaev, A. (2011). A smart world: A development model for inteligent cities. In *IEEE 11th International Conference on Computer and Information Technology* (pp. xxxv–xxxvii).

Al-Hader, M., & Rodzi, A. (2009). The smart city infrastructure development & monitoring. *Theoretical and Empirical Researches in Urban Management*, *2*(11), 87–94.

Alraouf, A. A. (2008). Emerging Middle Eastern knowledge cities: The unfolding story. In T. Yigitcanlar, K. Velibeyoglu, & S. Baum (Eds.), *Creative urban regions: Harnessing urban technologies to support knowledge city initiatives* (pp. 240–259). Hershey, PA; New York, NY: IGI Global.

Angelidou, M. (2015). Smart cities: A conjuncture of four forces. *Cities*, *47*, 95–106. https://doi.org/10.1016/j.cities.2015.05.004

Anthopoulos, L. G., & Fitsilis, P. (2010). From digital to ubiquitous cities: Defining a common architecture for urban development. In *2010 Sixth International Conference on Intelligent Environments* (pp. 301–306). IEEE. https://doi.org/10.1109/IE.2010.61

Anthopoulos, L. G., & Fitsilis, P. (2013). Using classification and roadmapping techniques for smart city viability's realization. *Electronic Journal of E-Government*, *11*(2), 326–336.

Aurigi, A. (2000). Digital city or urban simulator? In T. Ishida & K. Isbister (Eds.), *Digital Cities* (pp. 33–44). Berlin, Heidelberg, DE: Springer.

Bakıcı, T., Almirall, E., & Wareham, J. (2013). A smart city initiative: The case of Barcelona. *Journal of the Knowledge Economy*, *4*(2), 135–148. https://doi.org/10.1007/s13132-012-0084-9

Baum, S., Yigitcanlar, T., Mahizhnan, A., & Andiappan, N. (2008). E-government in the knowledge society: The case of Singapore. In T. Yigitcanlar, K. Velibeyoglu, & S. Baum (Eds.), *Creative Urban Regions* (pp. 132–147). Hershey, PA; New York, NY: IGI Global. https://doi.org/10.4018/978-1-59904-838-3.ch008

Baum, S., Yigitcanlar, T., & O'Connor, K. (2008). Creative industries and the urban hierarchy: The position of lower tier cities in the knowledge economy. In T. Yigitcanlar, K. Velibeyoglu, & S. Baum (Eds.), *Knowledge-Based urban development: Planning and applications in the information era. IGI Global, Information Science Reference, United States of America, Pennsylvania, Hershey* (pp. 42–57). Hershey, PA; New York, NY: IGI Global.

Ben Letaifa, S. (2015). How to strategize smart cities: Revealing the SMART model. *Journal of Business Research*, *68*(7), 1414–1419. https://doi.org/10.1016/j.jbusres.2015.01.024

Benini, M., Cindio, F. De, & Sonnante, L. (2005). Virtuose, a virtual community open source engine for integrating civic networks and digital cities. In P. van den Besselaar & S. Koizumi (Eds.), *Digital cities III. Information technologies for social capital: cross-cultural perspectives* (pp. 217–232). Berlin, Heidelberg, DE: Springer.

Besselaar, P. Van Den. (2001). E-community versus e-commerce: The rise and decline of the Amsterdam digital city. *AI & Society*, *15*, 280–288. https://doi.org/10.1007/BF01208709

Blythe, S. E. (2005). Hong Kong electronic signature law and certification authority regulations: Promoting e-commerce in the world's most wired city. *North Carolina Journal of Law & Technology*, *7*(1).

Bolz, U., Castilla Porquet, M., Sivertsen, T., Ford, A., Rakel, J., & Sturesson, J. (2005). *Cities of the future: Global competition, local leadership*. New York, NY.

Bontje, M., & Musterd, S. (2009). Creative industries, creative class and competitiveness: Expert opinions critically appraised. *Geoforum*, *40*(5), 843–852. https://doi.org/10.1016/j.geoforum.2009.07.001

Brake, K. (2012). Reurbanisierung. [Reurbanisation.] In K. Brake & G. Herfert (Eds.), *Reurbanisierung. Materialität und Diskurs in Deutschland* (pp. 258–286). Wiesbaden, DE: VS Verlag für Sozialwissenschaften. https://doi.org/10.1007/978-3-531-94211-7

Brown, J., Redmond, D., & Miquel, M. P. i. (2013). Capitalising on position. In S. Musterd & Z. Kovács (Eds.), *Place-making and Policies for Competitive Cities* (pp. 59–76). Oxford, UK: John Wiley & Sons. https://doi.org/10.1002/9781118554579.ch5

Bugliarello, G. (2004). Urban knowledge parks, knowledge cities and urban sustainability. *International Journal of Technology Management, 28*(3–6), 388–394.

C40 Cities. (2011, November 7). Seoul car-free days have reduced CO2 emissions by 10% annually. Retrieved from http://www.c40.org/case_studies/seoul-car-free-days-have-reduced-co2-emissions-by-10-annually

Carrillo, F. J., Yigitcanlar, T., García, B., & Lönnqvist, A. (2014). *Knowledge and the city: Concepts, applications and trends of knowledge-based urban development*. New York, NY: Routledge.

Castells, M. (1996). *The rise of the network society*. Malden, MA: Blackwell.

Catungal, J. P., Leslie, D., & Hii, Y. (2009). Geographies of Displacement in the creative city: The case of Liberty Village, Toronto. *Urban Studies, 46*(5–6), 1095–1114. https://doi.org/10.1177/0042098009103856

Chen, Y. (2012). Making Shanghai a creative city: exploring the creative cluster. In A. Romein & P. Nijkamp (Eds.), *Creative knowledge cities. Myths, visions and realities* (pp. 437–464). Cheltenham, UK; Northampton, MA: Edward Elgar Publishing.

Choi, J. H., & Greenfield, A. (2009). To connect and flow in Seoul: Ubiquitous technologies, urban infrastructure and everyday life in the contemporary Korean city. In M. Foth (Ed.), *Handbook of research on urban informatics: The practice and promise of the real-time city* (pp. 21–36). Hershey, PA; New York, NY: IGI Global.

Cinti, T. (2007). Cultural clusters and districts: The state of the art. In P. Cooke & L. Lazzeretti (Eds.), *Creative cities, cultural clusters and local economic development* (pp. 70–92). Cheltenham, UK; Northampton, MA: Edward Elgar Publishing.

Cohendet, P., Grandadam, D., & Simon, L. (2010). The anatomy of the creative city. *Industry & Innovation, 17*(1), 91–111. https://doi.org/10.1080/13662710903573869

Cohendet, P., Grandadam, D., & Simon, L. (2011). Rethinking urban creativity: Lessons from Barcelona and Montréal. *City, Culture and Society, 2*(3), 151–158. https://doi.org/10.1016/j.ccs.2011.06.001

Costa, P., & Oliveira, A. R. (2009). *From "creative cities" to "urban creativity"? Space, creativity and governance in the contemporary city* (No. 2009/80). Madrid, ES.

Couclelis, H. (2004). The construction of the digital city. *Environment and Planning B: Planning and Design, 31*(1), 5–19. https://doi.org/10.1068/b1299

Currid, E. (2006). New York as a global creative hub: A competitive analysis of four theories on world cities. *Economic Development Quarterly, 20*(4), 330–350. https://doi.org/10.1177/0891242406292708

De Cindio, F. (2009). Moments and modes for triggering civic participation at the urban level. In M. Foth (Ed.), *Handbook of research on urban informatics* (pp. 97–113). Hershey, PA; New York, NY: IGI Global. https://doi.org/10.4018/978-1-60566-152-0.ch007

de Jong, M., Yu, C., Chen, X., Wang, D., & Weijnen, M. (2013). Developing robust organizational frameworks for sino-foreign eco-cities: Comparing sino-Dutch Shenzhen low carbon city with other initiatives. *Journal of Cleaner Production, 57*, 209–220. https://doi.org/10.1016/j.jclepro.2013.06.036

Deren, L., Qing, Z., & Xiafei, L. (2000). Cybercity: Conception, technical supports and typical applications. *Geo-Spatial Information Science, 3*(4), 1–8. https://doi.org/10.1007/BF02829388

Derudder, B., Hoyler, M., Taylor, P. J., & Witlox, F. (2012). Introduction: A relational urban studies. In B. Derudder, M. Hoyler, P. J. Taylor, & F. Witlox (Eds.), *International handbook of globalization and world cities* (pp. 1–4). Cheltenham, UK; Northampton, MA: Edward Elgar Publishing.

Ding, P., Lin, D. D., & Sheng, H. H. (2005). Digital city Shanghai: Concepts, foundations, and current state. *Digital Cities III. Information Technologies for Social Capital: Cross-Cultural Perspectives*, 141–165.

Durmaz, B., Yigitcanlar, T., & Velibeyoğlu, K. (2008). Creative cities and the film industry: Antalya's transition to a Eurasian film centre. *The Open Urban Studies Journal, 1*(1), 1–10. https://doi.org/10.2174/1874942900801010001

Dvir, R., & Pasher, E. (2004). Innovation engines for knowledge cities: an innovation ecology perspective. *Journal of Knowledge Management, 8*(5), 16–27. https://doi.org/10.1108/13673270410558756

Edvinsson, L. (2006). Aspects on the city as a knowledge tool. *Journal of Knowledge Management, 10*(5), 6–13. https://doi.org/10.1108/13673270610691134

Edvinsson, L. (2011). Einige wichtige Fragestellungen zum Intellektuellen Kapital der Zukunft [Some important questions about the intellectual capital of the future]. In S. Jeschke, I. Isenhardt, F. Hees, & S. Trantow (Eds.), *Enabling Innovation* (pp. 357–362). Berlin, Heidelberg, DE: Springer. https://doi.org/10.1007/978-3-642-24299-1_33

Ergazakis, E., Ergazakis, K., Metaxiotis, K., & Charalabidis, Y. (2009). Rethinking the development of successful knowledge cities: An advanced framework. *Journal of Knowledge Management, 13*(5), 214–227. https://doi.org/10.1108/13673270910988060

Ergazakis, K., Metaxiotis, K., & Psarras, J. (2004). Towards knowledge cities: Conceptual analysis and success stories. *Journal of Knowledge Management, 8*(5), 5–15. https://doi.org/10.1108/13673270410558747

Etzkowitz, H., & Klofsten, M. (2005). The innovating region. Toward a theory of knowledge-based regional development. *R&D Management, 35*(3), 243–255.

Evans, G. (2009). Creative cities, creative spaces and urban policy. *Urban Studies, 46*(5–6), 1003–1040. https://doi.org/10.1177/0042098009103853

Florida, R. L. (2014). The creative class and economic development. *Economic Development Quarterly, 28*(3), 196–205. https://doi.org/10.1177/0891242414541693

Florida, R. L., & Gates, G. (2001). Technology and tolerance: To high-technology growth. The Brookings Institution. Retrieved from http://www.urban.org/sites/default/files/alfresco/publication-pdfs/1000492-Technology-and-Tolerance.PDF

Foord, J. (2008). Strategies for creative industries: An international review. *Creative Industries Journal, 1*(2), 91–113.

Franz, P. (2009). Knowledge City Berlin? Potenziale und Risiken einer Stadtentwicklungsstrategie mit dem Fokus Wissenschaft [Knowledge City Berlin? Potentials and risks of a city development strategy with a focus on science]. In U. Matthiesen & G. Mahnken (Eds.), *Das Wissen der Städte* (pp. 95–110). Wiesbaden, DE: VS Verlag für Sozialwissenschaften. https://doi.org/10.1007/978-3-531-91648-4_7

Friedmann, J. (1986). The world city hypothesis. *Development and Change, 17*(1), 69–83. https://doi.org/10.1111/j.1467-7660.1986.tb00231.x

GaWC. (2014). GaWC – The world according to GaWC 2012. Retrieved from http://www.lboro.ac.uk/gawc/world2012t.html

Gdaniec, C. (2000). Cultural industries, information tecnology and the regeneration of port-industrial urban landscapes. Poblenou in Barcelona – a virtual city? *GeoJournal*, *50*(4), 379–387. https://doi.org/10.1023/A:1010804102645

Gertler, M. S. (2004). *Creative cities: What are they for, how do they work, and how do we build them?* Ontario, CAN.

Gilderbloom, J. J. I., Hanka, M. M. J., & Lasley, C. C. B. (2009). Amsterdam: Planning and policy for the ideal city? *Local Environment*, *14*(6), 473–493. https://doi.org/10.1080/13549830902903799

Glaeser, E. L., & Berry, C. R. (2006). Why are smart places getting smarter? (Policy Briefs) Cambridge, MA. Retrieved from https://www.hks.harvard.edu/sites/default/files/centers/rappaport/files/brief_divergence.pdf

Hall, P. (2004). Creativity, culture, knowledge and the city. *Built Environment*, *30*(3), 256–258.

Hampton, K. N., & Gupta, N. (2008). Community and social interaction in the wireless city: Wi-fi use in public and semi-public spaces. *New Media & Society*, *10*(6), 831–850. https://doi.org/10.1177/1461444808096247

Haustein, S. (2012). *Multidimensional journal evaluation: Analyzing scientific periodicals beyond the impact factor*. Berlin, DE: Walter de Gruyter. https://doi.org/10.1515/9783110255553

Heng, T. M., & Low, L. (1993). The intelligent city: Singapore achieving the next lap. *Technology Analysis & Strategic Management*, *5*(2), 187–202. https://doi.org/10.1080/09537329308524129

Henry, R. (2010). The emergence of a creative enterprise distirct – Toronto´s Liberty Village. *PLan Canada*, 22–26. Retrieved from http://www.lclmg.org/lclmg/Portals/0/Museums Documents/Plan Canada Magazine.pdf

Heywood, P. (2008). The place of knowledge-based development in the metropolitan region. In T. Yigitcanlar, K. Velibeyoglu, & S. Baum (Eds.), *Creative urban regions: Harnessing urban technologies to support knowledge city initiatives* (pp. 1–23). Hershey, PA; New York, NY: Information Science Reference.

Hitters, E., & Richards, G. (2002). The creation and management of cultural clusters. *Creativity and Innovation Management*, *11*(4), 234–247. https://doi.org/10.1111/1467-8691.00255

Hofstetter, K., & Vogl, A. (2011). "Smart city Wien": Vienna's stepping stone into the European future of technology and climate. In M. Schrenk, V. V. Popovich, & P. Zeile (Eds.), *REAL CORP 2011: Change for stability: Lifecycles of cities and region* (pp. 1373–1382). Essen, DE: CORP Association.

Hollands, R. G. (2008). Will the real smart city please stand up? *City*, *12*(3), 303–320. https://doi.org/10.1080/13604810802479126

Horne, M., Thompson, E. M., & Podevyn, M. (2007). An overview of virtual city modelling: Emerging organisational issues. In *CUPUM '07 10th international conference on computers in urban planning and urban management*. Igassu Falls, BRA: http://nrl.northumbria.ac.uk/2142/.

Hornidge, A.-K. (2007). *Knowledge society: Vision and social construction of reality in Germany and Singapore* (Doctoral dissertation). Retrieved from https://www.ssoar.info/ssoar/handle/document/32331.

Hospers, G.-J. (2003a). Creative cities: Breeding places in the knowledge economy. *Knowledge, Technology & Policy*, *16*(3), 143–162. https://doi.org/10.1007/s12130-003-1037-1

Hospers, G.-J. (2003b). Creative cities in Europe. *Intereconomics*, *38*(5), 260–269. https://doi.org/10.1007/BF03031728

Hu, R. (2012). Clustering: Concentration of the knowledge-based economy in Sydney. In T. Yigitcanlar, K. Metaxiotis, & F. J. Carrillo (Eds.), *Building prosperous knowledge cities: Policies, plans and metrics* (pp. 195–212). Cheltenham, UK; Northampton, MA: Edward Elgar Publishing.

Inkinen, T., & Vaattovaara, M. (2010). Creativity and knowledge-based urban development in a Nordic welfare state. In K. Metaxiotis, F. J. Carrillo, & T. Yigitcanlar (Eds.), *Knowledge-Based Development for Cities and Societies* (pp. 196–210). Hershey, PA: IGI Global. https://doi.org/10.4018/978-1-61520-721-3.ch012

Intelligent Community Forum (ICF). (2014). Top 7 intelligent communities of the year. Retrieved from http://www.intelligentcommunity.org/top7

Ishida, T. (2000). Understanding digital cities. In T. Ishida & K. Isbister (Eds.), *Digital cities: Experiences, technologies and future perspectives* (Vol. 1765, pp. 7–17). Berlin, Heidelberg, DE: Springer.

Ishida, T., Aurigi, A., & Yasuoka, M. (2005). World digital cities: Beyond heterogeneity. In P. van den Besselaar & S. Koizumi (Eds.), *Digital cities III. Information technologies for social capital: cross-cultural perspectives* (pp. 188–203). Berlin, Heidelberg, DE: Springer.

Johnson, K. (2012). Commuting: The geography of Melbourne`s knowledge economy. In T. Yigitcanlar, K. Metaxiotis, & F. J. Carrillo (Eds.), *Building prosperous knowledge cities: Policies, plans and metrics* (pp. 279–308). Cheltenham, UK; Northampton, MA: Edward Elgar Publishing.

Kang, Y., Lei, Z., Ca, C., Yuming, G., Hao, L., Ying, C., ... Hart, T. (2014). *Comparative study of smart cities in Europe and China*. Retrieved from http://projects.sigma-orionis.com/choice/wp-content/uploads/2015/01/Smart_City_report-Final-Draft-March-2014.pdf

Keane, M. (2009). The capital complex: Beijing's new creative clusters. In L. Kong & J. O´Connor (Eds.), *Creative economies, creative cities. Asian-European perspectives* (pp. 77–95). London, UK; New York, NY; a.o.: Springer. https://doi.org/10.1007/978-1-4020-9949-6_6

Komninos, N. (2011). Intelligent cities: Variable geometries of spatial intelligence. *Intelligent Buildings International*, *3*(3), 172–188. https://doi.org/10.1080/17508975.2011.579339

Kong, L. (2009). Beyond networks and relations: Towards rethinking creative cluster theory. In L. Kong & J. O'Connor (Eds.), *Creative economies, creative cities. Asian-European perspectives* (Vol. 98, pp. 61–75). London, UK; New York, NY; a.o.: Springer. https://doi.org/10.1007/978-1-4020-9949-6_5

Kong, L., & O'Connor, J. (2009). *Creative economies, creative cities. Asian-European perspectives*. New York, NY: Springer.

Kotkin, J., & DeVol, R. C. (2001). *Knowledge-value cities in the digital age*. Santa Monica, CA: Milken Institute.

Kunzmann, K. R., & Ebert, R. (2007). Kulturwirtschaft, kreative Räume und Stadtentwicklung in Berlin [Culture, creative spaces and urban development in Berlin]. *disP: The Planing Review / Netzwerk Stadt Und Landschaft, ETH Zürich*, *43*(171), 64–79.

Lagerkvist, A. (2010). The future is here: Media, memory, and futurity in Shanghai. *Space and Culture*, *13*(3), 220–238. https://doi.org/10.1177/1206331210365247

Landry, C. (2006). *The art of city making*. London, UK: Earthscan.

Landry, C. (2008). *The creative city: A toolkit for urban innovators* (2nd ed.). New York, NY: Earthscan Publications.

Landry, C., & Bianchini, F. (2006). *The creative city*. London, UK: Demos.

Lange, B., Kalandides, A., Stöber, B., & Mieg, H. A. (2008). Berlin's creative industries: Governing creativity? *Industry & Innovation*, *15*(5), 531–548. https://doi.org/10.1080/13662710802373981

Lee, J. H., Hancock, M. G., & Hu, M.-C. C. (2014). Towards an effective framework for building smart cities: Lessons from Seoul and San Francisco. *Technological Forecasting and Social Change*, *89*, 80–99. https://doi.org/10.1016/j.techfore.2013.08.033

Lee, S.-H., Yigitcanlar, T., Han, J.-H., & Leem, Y.-T. (2008). Ubiquitous urban infrastructure: Infrastructure planning and development in Korea. *Innovation: Management, Policy & Practice*, *10*(2–3), 282–292. https://doi.org/10.5172/impp.453.10.2-3.282

Lee, Y.-S., & Hwang, E.-J. (2012). Global urban frontiers through policy transfer? Unpacking Seoul's creative city programmes. *Urban Studies*, *49*(13), 2817–2837. https://doi.org/10.1177/0042098012452456

Ley, D. (2003). Artists, aestheticisation and the field of gentrification. *Urban Studies*, *40*(12), 2527–2544. https://doi.org/10.1080/0042098032000136192

Leydesdorff, L., & Deakin, M. (2011). The triple-helix model of smart cities: A neo-evolutionary perspective. *Journal of Urban Technology*, *18*(2), 53–63. https://doi.org/10.1080/10630732.2011.601111

Lloyd, R. (2002). Neo-bohemia: Art and neighborhood redevelopment in Chicago. *Journal of Urban Affairs*, *24*(5), 517–532. https://doi.org/10.1002/9780470752814.ch16

Lor, P. J., & Britz, J. J. (2007). Is a knowledge society possible without freedom of access to information? *Journal of Information Science*, *33*(4), 387–397. https://doi.org/10.1177/0165551506075327

Mainka, A., Hartmann, S., Meschede, C., & Stock, W. G. (2015a). Mobile application services based upon open urban government data. In Proceedings of the iConference 2015: Create, Collaborate, Celebrate. Newport Beach, CA.

Mainka, A., Hartmann, S., Meschede, C., & Stock, W. G. (2015b). Open government: Transforming data into value-added city services. In M. Foth, M. Brynskov, & T. Ojala (Eds.), *Citizen's right to the digital city: Urban interfaces, activism, and placemaking* (pp. 199–214). Singapore, SG: Springer.

Mainka, A., Hartmann, S., Orszullok, L., Peters, I., Stallmann, A., & Stock, W. G. (2013). Public libraries in the knowledge society: Core services of libraries in informational world cities. *Libri*, *63*(4), 295–319. https://doi.org/10.1515/libri-2013-0024

Maitland, R. (2010). Everyday life as a creative experience in cities. *International Journal of Culture, Tourism and Hospitality Research*, *4*(3), 176–185. https://doi.org/10.1108/17506181011067574

Marti-Costa, M., & Miquel, M. P. I. (2012). The knowledge city against urban creativity? Artists' workshops and urban regeneration in Barcelona. *European Urban and Regional Studies*, *19*(1), 92–108. https://doi.org/10.1177/0969776411422481

Matthiessen, C. W., Winkel Schwarz, A., & Find, S. (2006). World cities of knowledge: Research strength, networks and nodality. *Journal of Knowledge Management*, *10*(5), 14–25. https://doi.org/10.1108/13673270610691143

Meister, F. (2012). Die Global City ist ein brutaler Ort [The global city is a brutal place]. *Woz*, (25), 15–17.

Metaxiotis, K., & Ergazakis, K. (2012). Formulating: An integrated strategy for the development of knowledge cities. In T. Yigitcanlar, K. Metaxiotis, & F. J. Carrillo (Eds.), *Building prosperous knowledge cities: Policies, plans and metrics* (pp. 149–174). Cheltenham, UK; Northampton, MA: Edward Elgar Publishing.

Most admired knowledge cities. (2016). In *DACH KM Wiki*. Retrieved from http://dachkm.org/wiki/index.php?title=Kategorie:Most_Admired_Knowledge_Cities

Musterd, S. (2004). Amsterdam as a creative cultural knowledge city: Some conditions. *Built Environment, 30*(3), 225–234. https://doi.org/10.2148/benv.30.3.225.54307

Musterd, S., & Deurloo, R. (2006). Amsterdam and the preconditions for a creative knowledge city. *Tijdschrift Voor Economische En Sociale Geografie, 97*(1), 80–94. https://doi.org/10.1111/j.1467-9663.2006.00498.x

Musterd, S., & Gritsai, O. (2009). Creative and knowledge cities: Development paths and policies from a European perspective. *Built Environment, 35*(2), 173–188. https://doi.org/10.2148/benv.35.2.173

Nowag, B., Perez, M., & Stuckmann, M. (2011). Informationelle Weltstädte – Indikatoren zur Stellung von Städten im „Space of Flow" [Informational world cities – Indicators of the position of cities in the "space of flow"]. *Information – Wissenschaft & Praxis, 62*(2–3), 103–109.

Palvia, S. C. J., & Sharma, S. S. (2007). E-government and e-governance: Definitions/domain framework and status around the world. In *International Conference on E-governance* (pp. 1–12). https://doi.org/10.3991/ijac.v5i1.1887

Pareja-Eastaway, M., Bontje, M., & D'Ovidio, M. (2010). Attracting young and high-skilled workers: Amsterdam, Milan and Barcelona. In S. Musterd & A. Murie (Eds.), *Making competitive cities* (pp. 192–207). Chichester, UK: Wiley-Blackwell.

Pareja-Eastaway, M., Chapain, C., & Mugnano, S. (2013). Successes and failures in city branding policies. In S. Musterd & Z. Kovács (Eds.), *Place-making and policies for competitive cities* (pp. 149–171). Oxford: John Wiley & Sons. https://doi.org/10.1002/9781118554579.ch10

Pawlowsky, P. (2011). Wissen 2010 – Intellektuelles Kapital als Motor des Wohlstands [Knowledge 2010 – Intellectual capital as a driver of prosperity]. In S. Jeschke, I. Isenhardt, F. Hees, & S. Trantow (Eds.), *Enabling innovation* (pp. 329–355). Berlin, Heidelberg, DE: Springer.

Pethe, H., Hafner, S., & Lawton, P. (2010). Transnational migrants in the creative knowledge industries: Amsterdam, Barcelona, Dublin and Munich. In S. Musterd & A. Murie (Eds.), *Making competitive cities* (pp. 163–191). Chichester, UK: Wiley-Blackwell.

Porter, M. . (1990). *The competitive advantage of nations.* London, UK: Macmillan.

Powell, J. A. (2012). Enterprising: Academics, knowledge capital and towards PASCAL universities. In T. Yigitcanlar, K. Metaxiotis, & F. J. Carrillo (Eds.), *Building prosperous knowledge cities: Policies, plans and metrics* (pp. 241–259). Cheltenham, UK; Northampton, MA: Edward Elgar Publishing.

Reffat, R. M. (2010). Essentials for developing a prosperous knowledge city. In K. Metaxiotis, F. J. Carrillo, & T. Yigitcanlar (Eds.), *Knowledge-based development for cities and societies: Integrated multi-level approaches* (pp. 118–130). Hershey, PA: IGI Global.

Riemens, P., & Lovink, G. (2002). Local networks: Digital city Amsterdam. In S. Sassen (Ed.), *Global networks, linked cities* (pp. 327–345). London, UK: Routledge.

Rocco, R. (2012). Location patterns of advanced producers service firms: The case of São Paulo. In A. Romein & P. Nijkamp (Eds.), *Creative knowledge cities. Myths, visions and realities* (pp. 385–413). Cheltenham, UK; Northampton, MA: Edward Elgar Publishing.

Romein, A., & Trip, J. J. (2012). Theory and practice of the creative city thesis: Experiences from Amsterdam and Rotterdam. In M. Van Geenhuizen & P. Nijkamp (Eds.), *Creative knowledge cities. Myths, visions and realities* (pp. 23–52). Cheltenham, UK; Northampton, MA: Edward Elgar Publishing.

Roper, S., & Grimes, S. (2005). Wireless valley, silicon wadi and digital island – Helsinki, Tel Aviv and Dublin and the ICT global production network. *Geoforum, 36*(3), 297–313. https://doi.org/10.1016/j.geoforum.2004.07.003

Sassen, S. (2001). *The global city: New York, London, Tokyo* (2nd Ed.). Princeton, NJ: Princeton University Press.

Schuler, D. (2005). The Seattle community network: Anomaly or replicable model? In P. van den Besselaar & S. Koizumi (Eds.), *Digital cities III. Information technologies for social capital: cross-cultural perspectives* (pp. 17–42). Berlin, Heidelberg, DE: Springer.

Seoul Solutions. (2015). Weekly no-driving day: A voluntary program to reduce traffic volume in Seoul. Retrieved from https://www.seoulsolution.kr/content/weekly-no-driving-day-voluntary-program-reduce-traffic-volume-seoul?language=en

Setunge, S., & Kumar, A. (2010). Knowledge infrastructure: Managing the assets of creative urban regions. In T. Yigitcanlar (Ed.), *Sustainable urban and regional infrastructure development: Technologies, applications and management* (pp. 102–117). Hershey, PA: Information Science Reference.

Shahrokni, H., Lazarevic, D., & Brandt, N. (2015). Smart urban metabolism: Towards a Real-time understanding of the energy and material flows of a city and its citizens. *Journal of Urban Technology, 22*(1), 65–86. https://doi.org/10.1080/10630732.2014.954899

Shin, D., Nah, Y., Lee, I.-S., Yi, W. S., & Won, Y.-J. (2008). Security protective measures for the ubiquitous city integrated operation center. In *Third international conference on broadband communications, information technology & biomedical applications* (pp. 239–244). IEEE. https://doi.org/10.1109/BROADCOM.2008.65

Sigurdson, J. (2005). *Shanghai – from development to knowledge city* (EIJS Working Paper Series No. 217). Stockholm,SWE. Retrieved from http://econpapers.repec.org/paper/hhseijswp/0217.htm

Song, J., Zhang, J., & Zeng, J. (2009). Study of WAPI infrastructure and wireless city operation in China. In *International conference on management and service science* (pp. 1–3). IEEE. https://doi.org/10.1109/ICMSS.2009.5302831

Stachowiak, K., Pinheiro, R., Sedini, C., & Vaattovaara, M. (2013). Policies aimed at strengthening ties between universities and cities. In S. Musterd & Z. Kovács (Eds.), *Place-making and policies for competitive cities* (pp. 263–291). Oxford: John Wiley & Sons. https://doi.org/10.1002/9781118554579.ch16

Stock, W. G. (2011). Informational cities: Analysis and construction of cities in the knowledge society. *Journal of the American Society for Information Science and Technology, 62*(5), 963–986. https://doi.org/10.1002/asi

Streit, A., & Lange, B. (2013). Governance of creative industries. In S. Musterd & Z. Kovács (Eds.), *Place-making and policies for competitive cities* (pp. 293–311). Oxford: John Wiley & Sons. https://doi.org/10.1002/9781118554579.ch17

Szogs, G. M. (2011). Future Center – Internationale Impulse für die Innovations- und Wissenskommunikation [Future Center – International impetus for innovation and knowledge communication]. In *Enabling innovation* (pp. 421–435). Berlin, Heidelberg, DE: Springer. https://doi.org/10.1007/978-3-642-24299-1_40

Taylor, P. J. (2004). *World city network: A global urban analysis*. London, UK: Routledge. https://doi.org/10.4324/9780203634059

Trippl, M. (2012). Innovation networks in a cross-border context: The case of Vienna. In M. Van Geenhuizen & P. Nijkamp (Eds.), *Creative knowledge cities* (pp. 273–302). Cheltenham, UK; Northampton, MA: Edward Elgar Publishing. https://doi.org/10.4337/9780857932853.00018

Uitermark, J. (2004). The co-optation of squatters in Amsterdam and the emergence of a movement meritocracy: A critical reply to Pruijt. *International Journal of Urban and Regional Research, 28*(3), 687–698. https://doi.org/10.1111/j.0309-1317.2004.00543.x

Van den Berg, L., Braun, E., & van Winden, W. (2001). *Growth Clusters in European Metropolitan Cities.* Aldershot, UK; Ashgate.

van den Berg, L., Pol, P. M. J., van Winden, W., & Moets, P. (2005). *European cities in the knowledge economy: The cases of Amsterdam, Dortmund, Eindhoven, Helsinki, Manchester, Munich, Münster, Rotterdam and Zaragoza.* Aldershot, UK; Burlington, VT: Ashgate Publishing, Ltd.

van Winden, W., van den Berg, L., & Pol, P. (2007). European cities in the knowledge economy: Towards a typology. *Urban Studies, 44*(3), 525–549. https://doi.org/10.1080/00420980601131886

Vanolo, A. (2013). Smartmentality: The smart city as disciplinary strategy. *Urban Studies, 51*(5), 883–898. https://doi.org/10.1177/0042098013494427

Victorian Government Department of Sustainability and Environment Melbourne. (2003). *Melbourne 2030. Planing for sustainable growth.* https://doi.org/10.1080/08111140309955

Vivant, E. (2010). The (re)making of Paris as a bohemian place? *Progress in Planning, 74*(3), 107–152. https://doi.org/10.1016/j.progress.2010.05.002

Walliser, A. (2004). A place in the world: Barcelona's quest to become a global knowledge city. *Built Environment, 30*(3), 213–224. https://doi.org/10.2148/benv.30.3.213.54300

Walravens, N. (2015). Qualitative indicators for smart city business models: The case of mobile services and applications. *Telecommunications Policy, 39*(3–4), 218–240. https://doi.org/10.1016/j.telpol.2014.12.011

Wang, C. C. (2013). *Upgrading China's Information and Communication Technology Industry.* Co-Publishing with Zhejiang University Press, China. Distribute Worldwide Except Mainland China. https://doi.org/10.1142/8502

Wang, D., Wu, Z., Li, Y., & Wang, Y. (2012). The eco-knowledge city theory and its practice in Shenzhen of China. *International Journal of Asian Business and Information Management, 3*(3), 42–55. https://doi.org/10.4018/jabim.2012070106

Wang, J. (2009). "Art in capital": Shaping distinctiveness in a culture-led urban regeneration project in Red Town, Shanghai. *Cities, 26*(6), 318–330. https://doi.org/10.1016/j.cities.2009.08.002

Wei, L. W., & Jian, H. (2009). Shanghai's emergence into the global creative economy. In L. Kong & J. O´Connor (Eds.), *Creative economies, creative cities. Asian-European perspectives* (pp. 167–171). London, UK; New York, NY; a.o.: Springer. https://doi.org/10.1007/978-1-4020-9949-6_11

Wesselmann, S., Meyer, C., & Lisowski, R. (2012). Researching: Key factors for the success of knowledge cities in Germany. In T. Yigitcanlar, K. Metaxiotis, & F. J. Carrillo (Eds.), *Building prosperous knowledge cities: Policies, plans and metrics* (pp. 96–110). Cheltenham, UK; Northampton, MA: Edward Elgar Publishing.

Wojan, T. R., Lambert, D. M., & Mcgranahan, D. a. (2007). Emoting with their feet: Bohemian attraction to creative milieu. *Journal of Economic Geography, 7*(August), 711–736. https://doi.org/10.1093/jeg/lbm029

Wong, C. (2008). Knowledge economy in transition: The case of Singapore. In T. Yigitcanlar, K. Velibeyoglu, & S. Baum (Eds.), *Knowledge-based urban development* (pp. 58–81). Hershey, PA: Information Science Reference. https://doi.org/10.4018/978-1-59904-720-1.ch004

Wong, C. Y. L., Millar, C. C. J. M., & Choi, C. J. (2006). Singapore in transition: From technology to culture hub. *Journal of Knowledge Management, 10*(5), 79–91. https://doi.org/10.1108/13673270610691198

World Capital Institute. (2007). *2007 most admired knowledge cities (MAKCi) report*. Retrieved from http://www.worldcapitalinstitute.org/makci/history

World Capital Institute. (2008). *2008 most admired knowledge cities (MAKCi) framework* (2nd.). World Capital Institut. Retrieved from http://www.worldcapitalinstitute.org/makci/history

World Capital Institute. (2009). *2009 most admired knowledge cities (MAKCi) report*. Retrieved from http://www.worldcapitalinstitute.org/makci/history

World Capital Institute. (2010). *2010 most admired knowledge cities (MAKCi) report*. Retrieved from http://www.worldcapitalinstitute.org/makci/history

World Capital Institute. (2011). *2011 most admired knowledge cities (MAKCi) report*. Retrieved from http://www.worldcapitalinstitute.org/makci/history

World Capital Institute. (2012). *2012 most admired knowledge cities (MAKCi) report*. Retrieved from http://www.worldcapitalinstitute.org/makci/history

Wu, W. (2005). Dynamic cities and creative clusters. *Group*. https://doi.org/http://dx.doi.org/10.1596/1813-9450-3509

Yigitcanlar, T. (2009). Planning for knowledge-based urban development: Global perspectives. *Journal of Knowledge Management, 13*(5), 228–242. https://doi.org/10.1108/13673270910988079

Yigitcanlar, T. (2012). Comparing: Knowledge-based urban development of Vancouver, Melbourne, Manchester and Boston. In T. Yigitcanlar, K. Metaxiotis, & F. J. Carrillo (Eds.), *Building prosperous knowledge cities: Policies, plans and metrics* (pp. 327–351). Cheltenham, UK; Northampton, MA: Edward Elgar Publishing.

Yigitcanlar, T., & Martinez-Fernandez, C. (2010). Making space and place for knowledge production. In K. Metaxiotis, F. J. Carrillo, & T. Yigitcanlar (Eds.), *Knowledge-based development for cities and societies* (pp. 99–117). Hershey, PA: IGI Global. https://doi.org/10.4018/978-1-61520-721-3.ch006

Yigitcanlar, T., O'Connor, K., & Westerman, C. (2008). The making of knowledge cities: Melbourne's knowledge-based urban development experience. *Cities, 25*(2), 63–72. https://doi.org/10.1016/j.cities.2008.01.001

Yigitcanlar, T., & Sarimin, M. (2010). Orchestrating knowledge-based urban development. In K. Metaxiotis, F. J. Carrillo, & T. Yigitcanlar (Eds.), *Knowledge-based development for cities and societies* (pp. 281–295). Hershey, PA: IGI Global. https://doi.org/10.4018/978-1-61520-721-3.ch017

Yusuf, S., & Nabeshima, K. (2005). Creative industries in East Asia. *Cities, 22*(2), 109–122.

Zhao, P. (2010). Building knowledge city in transformation era: Knowledge-based urban development in Beijing in the context of globalisation and decentralisation. *Asia Pacific Viewpoint, 51*(1), 73–90. https://doi.org/10.1111/j.1467-8373.2010.01415.x

Zhu, Z. (2011). Smart cities and industries. *Industry*, 14–16.

Zukin, S., & Braslow, L. (2011). The life cycle of New York's creative districts: Reflections on the unanticipated consequences of unplanned cultural zones. *City, Culture and Society, 2*(3), 131–140. https://doi.org/10.1016/j.ccs.2011.06.003

6 Case study investigation of 31 informational world cities

Many theories and case studies exist regarding the development of cities of the knowledge society. As established in the previous chapters, I consider informational world cities to be defined by diverse characteristics. How these characteristics could be measured is described in chapter 3 ("Measuring cities of the knowledge society"). However, cities are not all the same: In Asia, for example, politics have a significantly larger influence on economic development than in western countries (Chow, 2004). Religion and leisure time also have different significance in different cultures (Haller, Hadler, & Kaup, 2013). Finally, while there are many differences, globalization has led to a loss of identity, especially in world cities– a small number of architects is responsible for shaping cities around the world, and shopping malls offer the same brands almost everywhere (Madanipour, 2013). In addition, cities of the twenty-first century are confronted with a continuously increasing density on the one hand and with a shift in economic values on the other. Production and manufacturing is increasingly replaced by creative and knowledge industries, especially in informational world cities.

In this chapter, I will fill the theory of informational world cities with life. The main data collection is based on expert interviews and first-hand experiences.[1] The interviews are split into quantitative data (SERVQUAL) and qualitative statements. As described in chapter 4 ("Methods"), 31 cities are investigated. Unfortunately, no interview partner could be found for Beijing. Furthermore, the quantitative data has to be read with caution since the data represents only the personal opinion of the experts and is not representative for the whole city or even society. In addition, data from articles, websites, reports and rankings will round out the investigation of the cities. The aim of this chapter is not to identify "the best" informational city but rather to identify best practice examples with regard to different developments. The findings are structured according to the twelve hypotheses developed in chapter 3 ("Measuring cities of the knowledge society"), grouped as "Infrastructure," "Political will," and "World city."

---

[1] *Additional note:* To offer a better reading flow in this chapter, the references for the expert interviews are shortened by an abbreviation for each city following by a number for each interview partner, e.g. for a personal communication in Amsterdam with the first expert the abbreviation AM 1 would be used. The associated reference in full APA style (AM 1, personal communication, Month Day, Year) is listed in Appendix I.

6.1 Infrastructures

The infrastructures discussed in this chapter are divided into two main parts: the digital infrastructure and the cognitive infrastructure. The former is described mainly by hard location factors, for instance by the internet connectivity and the number of users, while the latter is acknowledged as a soft location factor because the focus lies on the human capital – the creative class.

6.1.1 ICT infrastructure

The cities investigated (31 in total) are highly diverse and in different stages of their development. Following the development of the fifth Kondratieff cycle, the ICT infrastructure should be more important than past infrastructure (automotive traffic). Therefore, many companies of the information market have likely settled down in informational world cities. During the field study and interviews hints and arguments of the growing importance of companies of the information market and ICT infrastructure have been investigated in light of the first two hypotheses:

H1 Informational world cities are hubs for companies with information market activities, e.g. telecommunication companies.

H2 The ICT infrastructure in an informational world city is more important than automotive traffic infrastructure.

H1 Hub of companies with information market activities
The first hypothesis is closely related to the investigations of world city networks in which the presence of headquarters (HQ) and branches of an economic sector are counted. This has already been adapted by researchers, for example through the investigations of internet backbone capacity, the number of internet domain names, internet exchange points (IXPs), number of content providers such as Google and of cloud computing providers such as Dropbox (Malecki, 2012, p. 117). Thus, a city with many company HQs with information market activities (called ICT in the following) is, in theory, a hub of this economic sector on a global scale. But whether this really is an important factor for a city on its way to becoming an informational world city will be investigated according to the following hypothesis:

H1 Informational world cities are hubs for companies with information market activities, e.g. telecommunication companies.

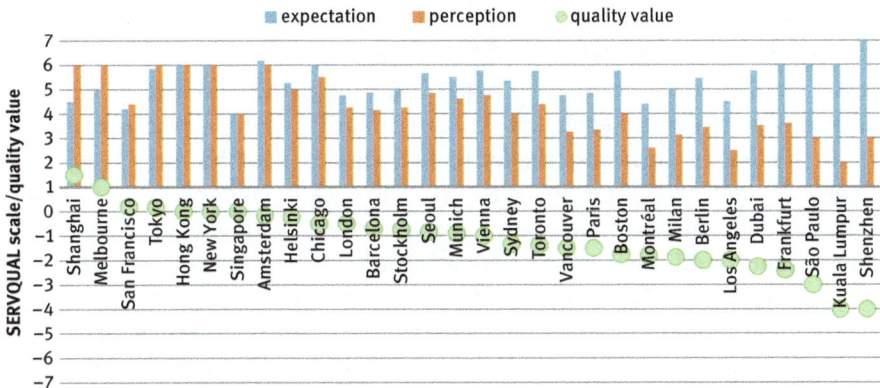

Figure 6.1: Quantitative interview results according to SERVQUAL (quality value = perception − expectation) for H1 (Informational world cities are hubs for companies with information market activities, e.g. telecommunication companies).

Based on the interview findings, several factors influence the importance of being an ICT hub in relation to becoming or being an informational world city. Most of the interview partners agreed that being an ICT hub is of advantage. 37 interviewees stated that it is "important" for becoming an informational world city, 15 that it is "helpful" and 20 experts do not see any necessity to be a physical hub of the ICT market. Further, relational aspects have been stated as important, e.g. research and development in ICT, creativity, and synergy effects with other sectors. Startups within the ICT sector empower innovation and are seen as more important than big telecommunication HQs (stated by nine interview partners). Experts which stated that it is not important to be an ICT hub argued that the presence of an enhanced ICT infrastructure is the driving factor of economic growth. This will be discussed with regard to the second hypothesis in this chapter. The SERVQUAL evaluation reveals that most experts, accumulated per city, have a higher expectation that an informational world city has to be an ICT hub compared with the actual perception for the particular city (Figure 6.1), with an overall quality value of $Q = -1.12$. This indicates that the ICT sector in most cities is acknowledged to not be as developed as it could be. Cities that are stated as advanced ICT hubs are Shanghai, Melbourne, San Francisco, Tokyo, Hong Kong, New York, and Singapore.

ICT hubs

Prominent ICT hubs are for example the San Francisco Bay Area (including Silicon Valley) or Singapore due to its vast development driven by ICT. Many facts and studies provide evidence for this popularity. For example, following the investigation by Forman, Goldfarb, and Greenstein (2016), the proportion of patents related to computers and communication has increased from 10% to over 30% annually between 1980 and 2005 in the San Francisco Bay Area. Furthermore, the Bay Area is more active in patenting than other US cities such as New York or Boston (Forman et al., 2016). Many firms of the information and communication economy are headquartered in this region, e.g. Adobe, Apple, Facebook, Google, Twitter, and others. While San Francisco's information economy evolved bottom-up, in Singapore, policies and governmental willingness to establish an "Intelligent Island" based on ICT were decisive for its success (Choo, 1995). As early as in the late 1970s, Singapore's government acknowledged the importance of ICT for economic growth (Hornidge & Kurfürst, 2010). One vanguard of Singapore's success is its port which aims to be the world's busiest port (Pike & Tomaney, 2010). According to Qiang, Rosotto, and Kimura (2009), this success story is based on the enhanced ICT infrastructure and its constant improvement, e.g. the implementation of a wireless broadband infrastructure.

Based on the interviews, many experts of the cities investigated mentioned that their city is an ICT hub at least for the continent or region. Thus, Vienna was mentioned as the third biggest IT location in Europe (VI 5). The experts stated that the ICT industry is more important than tourism, even though most people would know Vienna because of its sightseeing spots. A special example is Paris. One interviewee stated that the Paris ICT hub is globalized within the French-speaking world due to the language preference of most French people (PA 2). Helsinki is another example for profiting from the ICT market. The city has in the past benefited from the mobile phone producer Nokia, but today, while Nokia is in trouble, the city is not. However, the ICT infrastructure is still highly advanced due to former activities of Nokia. An important economic factor now is the gaming industry which is also technology-driven (HE 4). Consequently, information market activities are not only related to telecommunication but are based on information and communication technology.

A study of information market activities within the OECD countries has revealed that the focus has shifted from the computer, electronic, and optical products industries to the IT and other information services industries (OECD, 2014) (Figure 6.2). Since the demand for ICT goods has not decreased, it is obvious that production of physical goods is being outsourced to other countries. The increase in employment and value added in IT and other information service industries indicates the further economic growth of the IT sector.

When investigating a hub within the global network, the list of the Global Fortune 500 can be a good starting point (Alderson & Beckfield, 2012; Wall & van der Knaap, 2012). Looking at the listed corporations in the years 2000, 2005, 2010 and 2015, the number of telecommunication companies has declined from 21 in

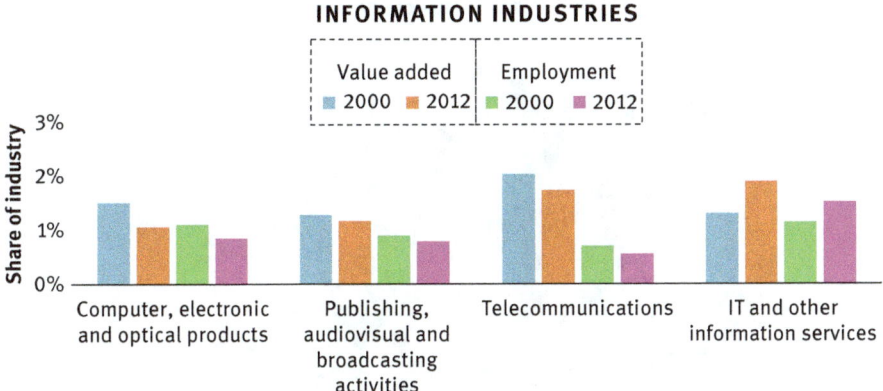

Figure 6.2: The relative size of information industries in the OECD. Percentage points of total value added and employment calculated. Source: OECD (2014).

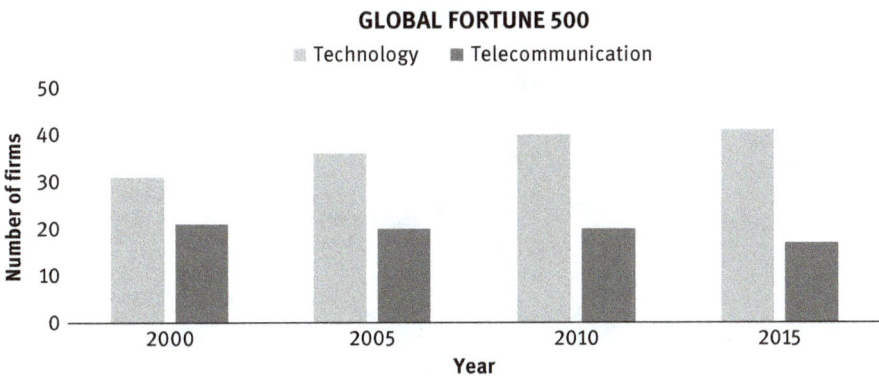

Figure 6.3: Amount of firms listed as Global Fortune 500 by industry "Technology"* and "Telecommunication." Data source: Global Fortune 500. *Technology industry was not available for 2000 and 2005, therefore the sectors "Computers, Office Equipment," "Computer services and software," "Information Technology Service," "Internet Service and Retailing," and "Electronics" have been accumulated.

2000 to 17 in 2015. On the other hand, there is an increase of firms of the technology industry from 31 to 41 firms as displayed in Figure 6.3.

Taking a closer look at the companies with the highest revenues in the world, both sectors still play a major role. Thus, the ocation of these companies' headquarters may acknowledge the city or the entire region as a hub within the technology or telecommunication sector. Which cities are home to these companies is displayed in Figure 6.4 ordered by revenue.

110 — 6 Case study investigation of 31 informational world cities

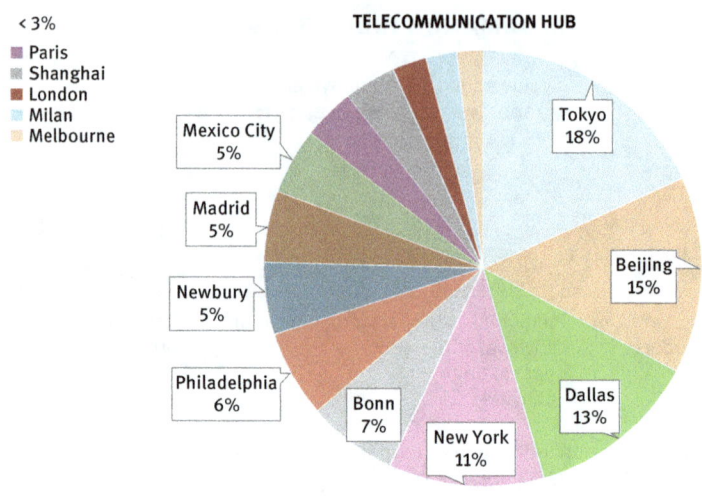

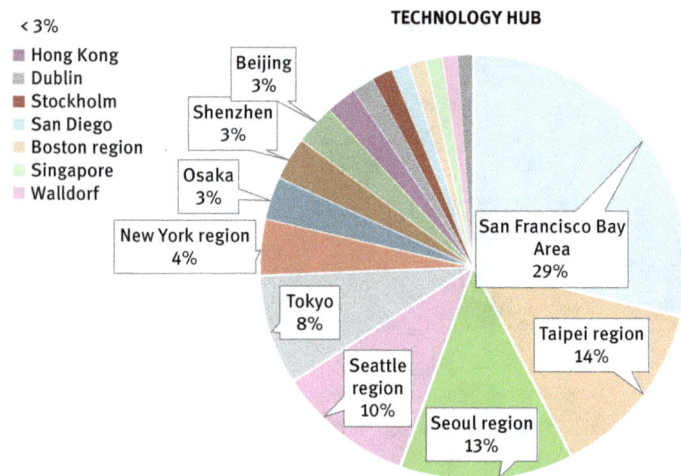

Figure 6.4: Telecommunication and technology hub according to headquarters locations of companies listed in the Global Fortune 500 in 2016 within the sector telecommunication or technology.

Synergy effects

It is possible to derive synergy effects through the combination of different sectors. One example for this is Frankfurt: The city is well-known for being a hub of the finance sector, but not necessarily for being an ICT hub. Finance service and the stock exchange are, however, highly dependent on ICT and fast internet connections. Thus, it is a major advantage for Frankfurt that the internet exchange hub DE-CIX is located there. Many banks have HQs or branches in Frankfurt. Finally, this economic situation has an impact on tech companies and developments related to the finance sector. For example, Deutsche Bank has outsourced its IT services to IBM (FR 4). Possibly as a result, IBM opened a branch in the Frankfurt region (Sossenheim) and, at the same time, established their main data center for the whole federal state of Hessen. Thus, the capital to establish the IT sector in Frankfurt came from the financial sector due to its high dependency on IT services (FR 4). Today, most IT services are clustered around Eschborn and Sossenheim, neighboring cities of Frankfurt.

A further advantage can be gained by offering adequate education at the same place where highly skilled personnel is needed. In Paris, for example, France Telecom (now renamed to Orange), the most important telecommunication company in France, has its headquarters: "situated in Paris and it's also linked to the polytechnic school, so the company can recruit new brains, new young people, freshly educated in the polytechnic school" (PA 3).

Barcelona is another example that is not particularly known for big ICT firms. Nevertheless, it is the World Mobile Congress Capital and hosts the Smart City Congress (BA 3). The World Mobile Capital Barcelona is an initiative to enhance the ICT development of the city, fostering "the digital empowerment of new generations, professionals and citizens; the digital transformation of industries; and the acceleration of digital innovation through entrepreneurship" ("Mobile World Congress," n.d.). Telefónica is one major company which performs R&D in the ICT sector in Barcelona, while the company HQ is located in Madrid (other branches of Telefónica exist, for example, in London or São Paulo). Another big player is Cisco, which is also involved in the smart city campus in Barcelona (BA 6). Starting projects under the banner of being or becoming a smart city always includes the increase of ICT, whether to increase society skills, R&D, government or industry (Albino, Berardi, & Dangelico, 2015). The driving force is the innovation and the positive economic impact that is ascribed to the ICT economy.

Further, ICT may influence or revolutionize creative work and culture, such as the gaming and animation picture industries, which are both parts of the entertainment sector and will be further investigated as part of the creative city. Nevertheless, cultural and creative activities can profit from synergy effects between themselves and firms of the ICT market. For example, "[i]n Paris, Google opened a cultural institute where artist can meet with computer scientist" (PA 5). 3D painting is one of the innovations that have been developed at this Google Cultural Institute (n.d.).

It is a difficult task to measure synergy effects between different sectors, even if it is between the finance and the ICT sector, or the impact of ICT in educational and cultural institutions and vice versa. To identify such synergy effects an investigation on a deep case study level for each city is required. Some examples have been described in the present work but they cannot be counted as hard numbers in a global comparison since every city has different cultural and economic requirements.

Entrepreneurship
Entrepreneurship and startups have been mentioned as important for innovation within the ICT sector. This is also the driving force of economic growth in Silicon Valley. Accordingly, the innovation industry in the San Francisco Bay Area grew by 174% whereas the rest of the economy grew only by 57% between the years 1994 and 2014 (B. L. Herrmann, Gauthier, Holtschke, Bermann, & Marmer, 2015). Startups are important job creators in North America. Kane (2010, p. 2) investigated the impact of startups related to the labor market and states that "without startups, there would be no net job growth in the U.S. economy." Following Murugadas, Vieten, Nikolic, Fietkiewicz, and Stock (2015), the impact of entrepreneurship has been determined to not be as significant globally as it is in the US.

Nevertheless, there is a global trend that cities strive to be entrepreneur-friendly, especially towards tech startups, and many cities have introduced their own "Silicon Anything" program to join this global development. For example, Silicon Alley in New York is not only a place but a name that joins the tech industry and entrepreneurship (NY 1). Some reports have started to rank the best cities to start a business according to the existing policies and environment in diverse cities, e.g. the Global Startup Ecosystem Ranking 2012 and 2015 (B. L. Herrmann et al., 2015; Bjoern Lasse Herrmann, Marmer, Dogrultan, & Holtschke, 2012), or City Initiatives for Technology, Innovation and Entrepreneurship (Gibson, Robinson, & Cain, 2015). Table 6.1 records all cities ranked by these reports for the year 2015. Interestingly, San Francisco is not the winner according to the report by Gibson, Robinson, and Cain (2015). They rank the top five cities which are, aside from New York, all located in Europe. This might be because San Francisco does not need to implement innovative initiatives for technology, as it is the best practice example. Following the report of Herrmann et al. (2015), the top four cities for startups are located in the US, followed by Tel Aviv in Israel. Other rankings exist, as well, but they are not globally oriented, for example the European Digital City Ranking (Nesta, 2015) or the country-level Global Entrepreneurship Index (Ács, Szerb, & Autio, 2016).

Conclusion: Hub of companies with information market activities
In summary, the question whether informational world cities are hubs for companies with information market activities did not meet with universal agreement

Table 6.1: Startup rankings according to Global Startup Ecosystem Ranking 2015 (B. L. Herrmann et al., 2015) on the left and City Initiatives for Technology, Innovation and Entrepreneurship (Gibson et al., 2015) on the right.

The Global Startup Ecosystem Ranking 2015		City Initiatives for Technology, Innovation and Entrepreneurship 2015	
City	Rank	City	Rank
Silicon Valley	1	New York	1
New York City	2	London	2
Los Angeles	3	Helsinki	3
Boston	4	Barcelona	4
Tel Aviv	5	Amsterdam	5
London	6		
Chicago	7		
Seattle	8		
Berlin	9		
Singapore	10		
Paris	11		
São Paulo	12		
Moscow	13		
Austin	14		
Bangalore	15		
Sydney	16		
Toronto	17		
Vancouver	18		
Amsterdam	19		
Montréal	20		

among the experts interviewed. Nevertheless, the SERVQUAL investigation has revealed that there is still space for improvement within the ICT sector in the most cities. During the interviews it was emphasized that there are different components of the information and communication technology market that play a role with regard to economic growth and therefore also for becoming a successful informational world city. ICT corporations do not only comprise telecommunication companies, but technology firms and information service corporations, as well. As investigated in the OECD countries, telecommunication activities declined but IT and information services increased (OECD, 2014). A similar development can be discerned by looking at the Global Fortune 500 firms filtered by industry. In 2015, 41 companies of the technology sector and 17 companies of the telecommunication sector were listed in the Global Fortune 500. In addition, synergy effects of different sectors as well as the impact of education and culture have been mentioned during the interviews. A positive impact of ICT on education has been identified, for instance, by a study from the telecommunication company Ericsson (2016). However, further independent research in this area is needed. Special attention has been paid toward the overall flourishing startup scene related to technology. Two rankings on a global scale have been published (Global Startup Ecosystem Ranking, and City Initiatives for Technology, Innovation and Entrepreneurship), which are diverse in their methodology. Nevertheless, they reflect those cities that are entrepreneurship-friendly and represent a new kind of hub within the global ICT sector.

H2 ICT Infrastructure vs. automotive traffic
With the second hypothesis, I will investigate whether the infrastructure of the fifth Kondratieff cycle (ICT infrastructure) is displacing the automotive infrastructure. In fact, no infrastructure has been completely eliminated in favor of another. For example, railways are an important part of mobility nowadays. But the focus has shifted. Therefore, the following hypothesis will be investigated in accordance with the expert interviews and secondary data:

H2 The ICT infrastructure in an informational world city is more important than automotive traffic infrastructure.

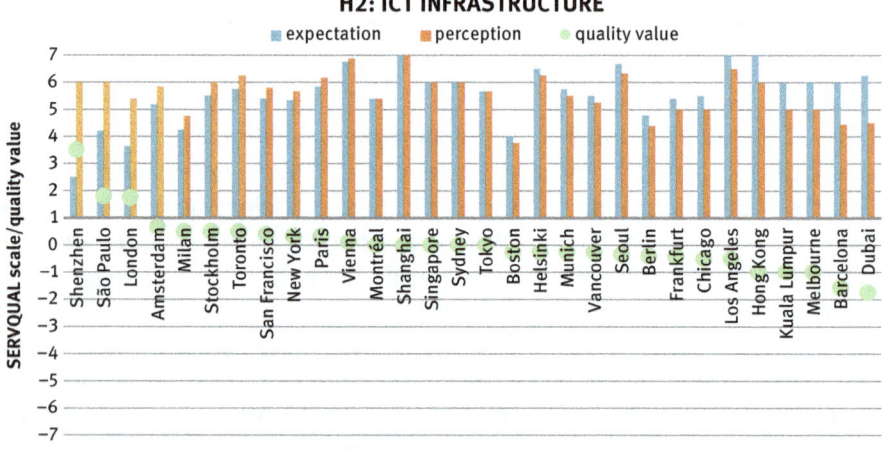

Figure 6.5: Quantitative interview results according to SERVQUAL (quality value = perception – expectation) for H2 (The ICT infrastructure in an informational world city is more important than automotive traffic infrastructure).

According to the experts, the expectation and perception are very diverse across the cities (Figure 6.5). The mean quality value according to SERVQUAL is near zero (0.03). Which infrastructure is in the spotlight largely depends on the current demand. For Shenzhen, São Paulo, and London, the interviewed experts have marked their perception very high in contrast to their expectation. São Paulo and London are both financial hubs that require advanced ICT infrastructure for every-day business. And Shenzhen is a city that has been built nearly from scratch and therefore has included an enhanced ICT infrastructure from the early beginnings. One expert stated that if you live in a city like Montréal where many streets crack due to the extreme climate (–30 °C in the winter and +30 °C in the summer), then everybody cares about road and street problems (MO 2). All in all, 38 interviewed experts stated that the ICT infrastructure is more important, with 13 emphasizing that cars are becoming less important and 15 that cars are very important for mobility in the city. Contrasting the digital infrastructure of ICT and the physical infrastructure of cars encouraged many of the experts to discuss the importance of Wi-Fi hot spots in the city and further the influence of ICT on mobility. Thus, I will in the following discuss the maturity of the ICT infrastructure, Wi-Fi connectivity, mobility in the city, and the influence of ICT on mobility.

ICT infrastructure maturity
With regard to the maturity of the ICT infrastructure, many rankings and indicators have been developed to compare different citied, e.g. the "ICT Development Index" on a national level (see chapter 3.1.1 ICT infrastructure). On city level, Ericsson has published the "Networked Society City Index" (NRI) for 2014 and 2016 (Ericsson, 2014, 2016). Indices by independent institutions are increasingly being published, e.g. by the International Technology Union (ISO/IEC JTC 1. InformationTechnology, 2015). With regard to actual indices and studies, ICT maturity calculations are based on broadband speed, accessibility, and quality, the amount of internet users as determined by the number of broadband subscribers, the amount and quality of Wi-Fi hotspots, political support, and information literacy. As it is evident that a positive correlation between the human development and the ICT maturity (Bruno, Esposito, Genovese, & Gwebu, 2011) as well as between the GDP and ICT indicators (Jin & Cho, 2015) exists on national level, it can be concluded that a city's economic success is also positively related to its maturity of the ICT infrastructure. Furthermore, Ericsson (2016) has identified positive correlations between ICT maturity and health and education. And according to the economic success, ICT maturity has a higher correlation with productivity (GDP per capita) than with economic competitiveness (R&D expenditure, patents, business start-ups, knowledge-intensive employment, and higher education attainment).

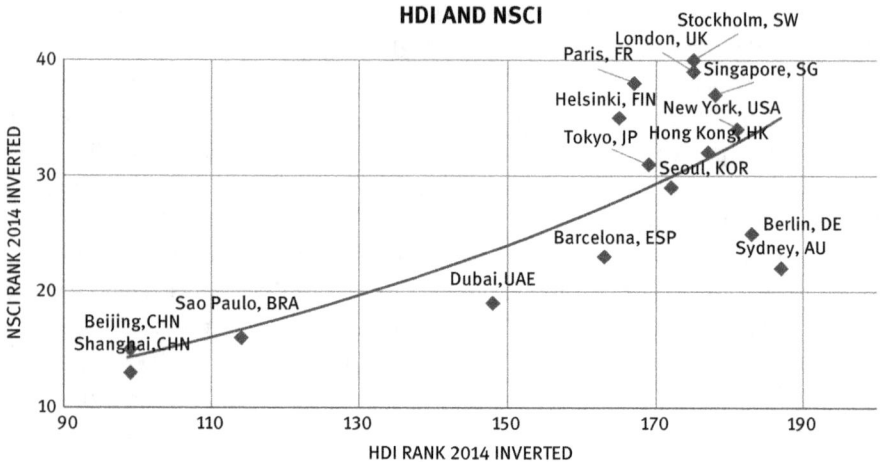

Figure 6.6: Correlation of the Human Development Index (HDI) and Networked Society City Index (NSCI) for the year 2014 with inverted ranking scores. Data source: Ericsson, 2014; UNDP, 2014.

In order to ascertain the relation of the national development and a city's ICT maturity I have correlated the Human Development Index (HDI) with the NSCI by applying the national score to cities (Figure 6.6). For example, São Paulo is assigned the HDI score of Brazil. In total, only 16 scores are available in both indices. The cities are presented with inverted ranking scores, e.g. Stockholm is ranked first amongst 40 cities and therefore received the value 40 in the NSCI. As assumed, the cities located in nations with a high HDI are ranked high in the NSCI (Pearson correlation R = 0.74). The cities at the top of both indices are New York, Singapore, and Stockholm, with São Paulo, Beijing, and Shanghai at the bottom. It should be kept in mind that the calculation was limited to 16 cities and that the NSCI was developed by the telecommunication corporation Ericsson headquartered in Stockholm – the city that is ranked first in the NSCI. However, the ICT infrastructure of the 31 cities that are investigated in the present work would probably result in high scores of ICT maturity across the board, since all of the cities are located in highly developed countries.

One issue that came to light during the interviews was accessibility. A world city does not have a population that consists solely of wealthy people. Social inequality in turn results in inequalities of access to ICT, frequently called the digital divide (Ravi S. Sharma, Ng, Dharmawirya, & Keong Lee, 2008). To bridge this divide is a challenging task for every city. One remarkable initiative comes from Boston: My interview partner there, Deb Socia, the head of Tech Goes Home (at the time of the interview, 2013) and now head of Next Century Cities, told me: "We are wiring the poorest neighborhoods of the city by our Tech Goes Home initiative" (BO 1). She is the winner of the NATOA Community Broadband Hero in 2013 and has received multiple other awards for her work in connecting all neighborhoods and closing the technology gap. The Tech Goes Home initiative connects people with low-income, unemployed and underemployed, people who do not speak English, and individuals with disabilities (TGH, 2016). People like Deb Socia and initiatives like Tech Goes Home are needed in this changing society to ensure that disadvantaged people are not left behind.

Public institutions like libraries are also important in bridging the gap between the information poor and the information rich. Libraries in the twenty-first century have to identify their new role as an information hub for society and at the same time face diverse problems like budget cuts (see chapter H7 Library as physical space). Thus, libraries are not all the same: "If you have the money the technology is phenomenal. [In Los Angeles] Beverly Hills and West-Hollywood have state of the art libraries with good broadband connection" (LA 1). In addition, the central library in Los Angeles offers a free Wi-Fi hotspot that is available around the clock, even the library is closed, which means that even on Sundays, people are spending time near the library with their private devices (LA 3). An investigation of public libraries in informational world cities demonstrated that in 30 out of the 31 cities public libraries do offer Wi-Fi, but there are differences in matters such as charging and log in barriers (Mainka, Hartmann, et al., 2013). Thus, for example, in Singapore the Wi-Fi access is offered in all public institutions and not merely in the library, whereas Dubai does not possess many public libraries, and the one we visited was more akin to a reading room for international newspapers and did not have Wi-Fi access.

The digital divide is not only caused by high costs for technology devices and telecommunication contracts, as the adoption rate of ICT also correlates with the personal ability to use the technologies (SG 5). In Singapore, for instance, classes for the elderly and for the youngest are introduced to help society to call attention to information technology and to promote information literacy (Mainka, 2011). So far, no standardized measure of information literacy on a global scale has been developed (Beutelspacher, 2014). Still, city-level courses in schools, libraries or other public institutions give us a hint as to whether the city is willing to bridge the digital divide. The Vancouver Public Library has, for example, introduced a special program for illiterate adults (VA 3). The library has acquired e-readers which are intended to animate the adults to learn to read and at the same time to learn how to use the technological equipment. In addition, "[o]ne advantage of an e-reader is that nobody around you will see which book you are reading" (VA 3), possibly eliminating another barrier.

As previously mentioned, all cities investigated are located in well-developed countries and offer an enhanced ICT infrastructure. Nevertheless, with the problem that cities will grow and increase in density in mind, the quality of all infrastructures has to be ensured with regard to the present and future needs of the population. Thus, an additional problem of ICT access can be slow connectivity rates, which at the very least hinder some functionalities if not outright exclude citizens from the network in worst-case scenarios. Therefore, the political willingness to support the digital infrastructure, be it through financial incentives to private providers or through direct municipal investments, is essential. The ICT infrastructure is, in most cases, not regulated by municipalities but by national or regional governments. Therefore, a direct comparison of a city's expenditure for ICT would not reflect the real investments that are made, with the exception of Singapore, which is a city-state. The government authority for ICT, the Infocomm Development Authority (IDA) states on their website: "IDA actively engages various stakeholders in collaborative efforts and initiatives to ensure that our work is closely aligned with the needs of the infocomm industry, business sectors and end-user communities" (IDA, n.d.-b). The IDA also continually develops future master plans that emphasize the increasing importance of ICT – currently, they are working on realizing their smart nation vision (IDA, n.d.-a). In a speech at the annual Industry Briefing in Singapore, Ms. Jacqueline Poh, Managing Director of the IDA, has stated that the government will make a $2.82 billion expenditure for ICT. The core investments will be in infrastructure-related programs like data centers and storage, cloud networking, routers, new PCs and laptops for government agencies, and enhancing Wi-Fi coverage at schools. As Singapore is a city state, the governmental decisions and expenditures are made by one authority. In comparison to other countries, Singapore is ranked first in the Network Readiness Index (NRI) in 2015, followed by Finland and Sweden. The NRI covers, amongst others, the indicators political and regulatory environment, business and innovation environment, ICT infrastructure and usage. Finally, the NRI has a high correlation with the HDI which also indicates the positive correlation of ICT indicators and economic success (Peña-López, 2006).

Ubiquitous connectivity through Wi-Fi hot spots
One indicator of the ICT infrastructure that is not considered in comparisons on a national level is the number of Wi-Fi hotspots. Hotspots for public internet connections are especially useful when many people are able to access them. As a result, many cities have started to offer publicly or privately supported Wi-Fi hotspots in city centers, metros, or in other busy areas. The approaches how public Wi-Fi is implemented varies between the cities. In some cases, the municipality is involved, in others private initiatives evolved to answer the public demand.

According to my own experience with an Android smartphone, many of the offered hotspots are sorely lacking. Internet connectivity that is too slow, bugs, and connection errors have often kept me from using the public Wi-Fi hotspots. In Paris, for example, to access the Wi-Fi via a smartphone, terms of use have to be accepted on a mobile website via the push of a button. Most open Wi-Fi hotspots offer entry pages of this kind as the providers are often liable for illegal internet activities like copyright infringement under local law. In the case of Paris' free public Wi-Fi, the confirm button did not work and therefore prevented me from accessing the Wi-Fi (own experience, December 12, 2013).

Another aspect of public Wi-Fi is the comfort of use. In Barcelona, the famous market street *La Rambla* offers a public Wi-Fi hotspot approximately every 250 meters. It is easy to connect to and the landing page works adequately. However, accessing the internet connection constantly while walking down the more than one kilometer-long street is not possible, as each hotspot requires a new login. The best connectivity was, of course, reached directly in front of a hotspot (own experience, December 4, 2013). Generally, Barcelona has invested a large amount to improve their ICT infrastructure and offer, for instance, about 590 Wi-Fi hotspots all around the city (Ajuntament de Barcelona, n.d.). Wi-Fi hotspots in a pedestrian zone naturally invite people to use the internet connection for such purposes as navigation tools or augmented reality functionalities with their smartphones. However, due to login restrictions, continuous connectivity is not possible today. As wearables, navigation systems, and other functions of mobile devices will increasingly enter our everyday life, the legal limitations on public internet access need to be overcome.

In some cities, the public transport provider also offers free Wi-Fi, as it is the case in New York City. According to a recent report, the Metropolitan Transport Authority will offer Wi-Fi in all 278 subway stations by the end of 2016 (Hawkins, 2016). Of course, when making use of public hotspots one cannot expect uninterrupted high-quality broadband access. However, my own experience in August 2014 was quite underwhelming. It was difficult to connect to the internet through the New York City metro station hotspots, and if the connection did succeed, it was very slow. Still, it is by no means a given that free Wi-Fi is available at public transportations today. As Wi-Fi hotspots in public areas do not only serve the needs of tourists and visitors, some telecommunication providers offer hotspots for their customers and, subject to additional costs, also for non-costumers. This is the case for example in Paris (and in other parts of France), where the hotspots are provided by the biggest telecommunication provider Orange and in Berlin, Frankfurt, and Munich (as in other German cities) by the provider Telekom. However, the fact that Telekom hotspots are available in most rail stations in Germany is likely to serve as an obstacle to ubiquitous free public Wi-Fi.

Wi-Fi connectivity in the city can also be achieved through private citizen initiatives, as for example in Berlin, where the initiative *Freifunk* ('free wireless') is encouraging all citizens to open their personal WiFi hotspot for the public (Freifunk – Berlin, n.d.). As of mid-2016, private citizens offering an open internet connection are as liable as other providers (BER 2). German law has been transformed with the goal to offer more opportunities for open Wi-Fi hotspots (BMWi, 2016), and, in the future, private internet sharing could enhance public Wi-Fi infrastructure in urbanized areas (BER 2).

There are still many municipalities that do not offer public Wi-Fi, or not to the same extent as Barcelona. In Milan, for instance, there is free Wi-Fi for the first 30 minutes (MI 3) and in Munich, there is only one place in the city center where free public Wi-Fi is available (MU 1, MU 4). Nevertheless, in cities where the municipal government does not offer public Wi-Fi, the private economy often meets that need to some extent. Thus, there are open Wi-Fi hotspots in countless Starbucks cafés, McDonald's branches and also in smaller restaurants and cafés that make municipal Wi-Fi ostensibly obsolete. Hence, to use an internet connection in London you do not need to enter a café. Often the hotspot can be reached from the street (own experience, June 24–30, 2013). One example which offers both to a high extent is Singapore. Next to a lot of public hotspots, private providers such as shopping malls, cafés, and restaurants offer nearly ubiquitous connectivity. During my stay in diverse cities, in most cases, the private providers like shopping malls or restaurants offered a more satisfactory internet connection than public hotspots at outdoor places. Among public institutions, internet connections in public libraries have most been acceptable.

Finally, another idea to offer free public internet connectivity on the go are media poles (Schumann & Stock, 2015). Among the investigated cities, Seoul and Milan have installed media poles experimentally. The idea is that pedestrians can use the touch screens at the poles like a huge smartphone with diverse applications, e.g. a restaurant guide or travel assistance. In Seoul, the screens can be switched between the languages Korean, Japanese, and English. 22 media poles that are twelve meters high each have been installed every 35 meters along the 760-meter long street *Gangnam daero* in 2009. They are also used as advertising platforms and Wi-Fi hotspots (see Figure 6.7). According to my own experience in July 2013, the media poles are largely ignored by pedestrians. In addition, the screen was working very slow and disconnected after a few minutes of use. In Milan, the media poles are called "multimedia totems" and offer first of all a smart navigation system for public transportation, and in addition Wi-Fi connection, power sockets to recharge devices, and NFC technology to purchase additional services. They were installed along with other smart city initiatives in preparation of the then-upcoming Expo

Figure 6.7: Media Poles in Seoul. A u-city street project. Left: One media screen in use by Prof. Wolfgang G. Stock. Middle: Street in Gangnam Districts with 22 media poles. Right: One multimedia information totem in Milan. Photos: Agnes Mainka.

2015 (Morandi, Rolando, & Di Vita, 2016). As in Seoul, I have not seen any person who was interested in or used these poles (own experience, November 5–7, 2013). These experiences are similar to the findings of an investigation of media screens in Oulu, Finland (Schumann & Stock, 2015). Due to the nearly ubiquitous penetration of smartphones, media screens are used less and less and will likely become obsolete, similar to public phones. Therefore, media screens will need to identify niches with additional features which are not served by a smartphone to survive.

Mobility

The fact that ICT is growing in importance in our everyday life has been described by one interview partner as follows: "In the future, the blood system of the city will not be the streets but the information networks. Because when you have information, you could act and maybe you don't need to move by transport" (BA 6). Even if the ICT infrastructure will be highly developed, it may not supplement mobility at all. In addition, the advantages of the ICT infrastructure are not recognized as an important service for the whole of society. For example, in São Paulo, "[t]he ICT infrastructure is better developed than the automotive infrastructure, but the most Paulistas care more about their cars" (SP 5).

With the growing density within cities, the issue of mobility gained in importance. Dargay and Gately (1999) have shown a disproportionate relation between economic growth and the amount of car owners per capita for the years from 1970 to 1990. In their study, published in 1999, they forecast an annual growth rate of 18% for the car ownership rate in China. Finally, in nearly all cities around the world, the growing number of cars has resulted in traffic jams, especially during rush hour. Today the problems are crowded cities, pollution, and high costs. Thus, many cities have to face these problems and develop ideas on how to reduce the negative effects of cars in the city, such as congestion fees for entering the city, high parking fees in the city center, and other governmental incentives or restrictions. The reallocation of the road network to sustainable modes of traffic has been discussed as early as the 2000s (Cairns, Atkins, & Goodwin, 2002). Today, while diverse smart city initiatives are facing the problems of climate change, approaches of sustainable transportation are becoming reality.

According to Kenworthy (2006), three types of mobility modes are acknowledged in cities: auto city, transit city, and walking city. He has investigated the correlation between population density and the preferred mode of mobility. As a result, he identifies thinly populated cities as auto cities, cities with 30 to 100 persons per ha as transit cities, and those with 100 or more per ha as walking cities. He concludes that being a transit or walking city makes the city sustainable and livable. A similar result was reached by the *Mobility in Cities Database. Synthesis Report* (UITP, 2015). In this report, the share of sustainable transport modes (walking, cycling, and public transport) was correlated with the population density. Accordingly, a positive correlation is the result. Nevertheless, in most world cities traffic is still a problem. To investigate the effect of the fifth Kondratieff, which is dedicated to the growing importance of ICT rather than to automotive traffic, different cases of car traffic reduction will be introduced in the following.

In Europe, many cities have introduced a green zone to reduce dust pollution in city centers. Thus, many cars that do not fulfill modern standards are not allowed to enter the city center ("Die Umweltplakette/Feinstaubplakette," n.d.). Other cities like London or New York have introduced congestions fees. In many more cities, the parking fees are exorbitantly high, preventing many people automatically from entering the city by car (LA 1). As high costs discourage people to drive a car, Singapore has introduced a tremendously high registration fee that is nearly the price of a new car (SG 12). Seoul has introduced the "No Driving Day" with monetary incentives for participating citizens. In the case of São Paulo, the municipality had to establish a "No Driving Day" without any incentives for the citizens, due to the overcrowded streets in and around the city center (SP 1). However, such an approach might lead people to buy a second car instead of using public transportation. Another approach was chosen in Helsinki, where "slow driving zones kill the car driving fun factor" (HE 1). Finally, to reduce car use in the city, alternative transportation modes need to be offered, which in conclusion could be interpreted as the loss of importance of cars in the city.

Figure 6.8: Public transportation modes. Examples of Dubai Metro (upper pictures) and trams in Hong Kong (bottom-left) and Barcelona (bottom-right). Photos: Adriana Kosior (top-left), Agnes Mainka (both right), Carsten Brinker (bottom-left).

The informational world cities investigated are all off different sizes and density. Nevertheless, all are global cities that have to deal with many people that enter the city center, be it for business, leisure, or as a visitor. Today, all of the 31 cities have a rather dense population and offer some kind of public transportation. Therefore, they are all at least Transit cities (Kenworthy, 2006). Due to the need for mass transport opportunities, diverse mobility modes from omnibuses on streets over subway railroad systems up to driverless metro trains on high-rise levels have evolved (Post, 2007). Figure 6.8 shows examples of the trams in Barcelona and Hong Kong, and of the self-driving high-rise metro in Dubai. According to many of the experts interviewed, public transportation is well-developed in the city center, but only to a lesser extent in the whole metropolitan area or region.

In Los Angeles, the main mode of public transportation is the bus. Buses have to share the road with the cars and are thus very slow, especially during rush hour (LA 2). Further, LA's autonomous urban development has resulted in a car-oriented city in which most places are best reached by freeway (Bratzel, 1995). Hence, due to the overcrowded streets in the core of the city, a lot of smaller centers have evolved everywhere in the LA metropolitan area. Today, the city is very widespread and only offers a subway system for the interior city center. However, "[t]he majority of the City of Los Angeles is considered a High Quality Transit Area. These areas have frequent access to some form of transit, whether it is light rail, bus, BRT [bus rapid transit], or subway" (Chamberlain & Riggs, 2016, p. 54). According to Chamberlain and Riggs (2016), the inhabitants of LA are not likely to live in walking distance of metro stations, but 97% live at least in a "bikeable" distance to a public transit station. Referring to a report of the United States Census Bureau (Wilson et al., 2012), LA is second after New York City according to the average density of population per square mile, which definitively characterizes the city as transit city.

To make public transportation attractive, a seamless infrastructure is important. An example how an infrastructure should not be established is Kuala Lumpur. In this city, two light rail provider and one monorail provider coexist. Each is independent and this has resulted in uncomfortable commutes (own experience). When one needs to change the train, this can often include a longer walk, since the providers do not share metro stations. Furthermore, pedestrians are not integrated into the road infrastructure, which makes it difficult to cross the street in some parts. Thus, the lack of traffic lights for pedestrians makes the use of public transportation dangerous and inconvenient.

As there exist a lot of different modes of transportation today, it has become possible to banish automotive traffic from city centers in favor of repurposing this space as public space, although this happens only rarely. One prominent example is the highway Cheonggyecheon in Seoul. The highway that crossed the city center was demolished in favor of revitalizing the former river underneath the street (Figure 6.9). Today, this space is an attraction for pedestrians, offering light entertainment in the evening. This was not the only step Seoul took towards a more sustainable city: Amongst others, since 2002 a total of 15 highways have been closed, which has resulted in a decrease of 3.6°C in the summer (Mesmer, 2014).

There are two examples that would also have the potential to redevelop a riverside. First, Kuala Lumpur has two rivers crossing the city (the Sungai Klang and Sungai Gombak), which are in need of improvement with regard to water quality and could be redesigned (Shaziman, Usman, & Tahir, 2010). And second, Singapore has built a four-lane expressway along a canal in the city center. Still, even in Seoul the space next to the river is a frequently used motorway and Singapore already has a large number of recreated public spaces along the Singapore River. Another example is New York City. At the famous entertainment area Times Square, a pedestrian plaza was realized which has reduced the space available for cars (NY 1). In both cases (Seoul and New York), the government was the main planner and decision maker. In other cities, the government is the biggest challenge with regard to such development plans: "The habit in the administration is the problem" (MO 6). The creation of car-free zones also needs to be accepted by

Figure 6.9: From highway to greenway. Photos: Carsten Brinker (bottom-left), Agnes Mainka (others).

the citizens (MU 1). In Munich, for example, which is home to the leading automobile manufacturer BMW, a debate on establishing a car-free zone in the old town area was initiated, but to the present day, no decision has been made. In other cities like Frankfurt "[a]utomotive traffic is important for local and global logistic [, as] Frankfurt is a logistic hub" (FR 2). Looking at Vienna, the city development plan includes the goal of reducing car traffic by 20% by offering new pedestrian zones and enhancing the number of journeys with public transportation by up to 40% (VI 5). One incentive is the cheap price of an annual public transport ticket, which is available for 356 €. This is half the price compared to Frankfurt (VI 5). But even in Vienna "there is a clash between the car driver faction and the car free zone faction" (VI 5, own translation). Thus, there are also experts that state that "the city wouldn't function without cars" (LO 1). Milan's government has gone one step further and offers free public transit vouchers for commuters who leave their vehicles at home (Peters, 2015). Further cities have already reached some goals with regard to car reduction rates, e.g. in Paris the number of car owners has been reduced from 60% in 2001 to 40% of inhabitants in 2015 (Peters, 2015). In Hong Kong "80% of people use public transport, only 20% own cars" (HK 1). One expert opined that "[t]he city should be a meeting place for people and not for traffic" (ST 1).

Beyond pedestrianization we can also see an additional change in behavior, as one expert from Toronto stated: "The younger generation is less car-oriented. In addition, an own car is too expensive for most inhabitants" (TOR 2). In New York it is also the case that most people do not own a car, since the fastest way of commuting is via the Metro (NY 1). Further, not all cities are developed for cars, for example "Vancouver has no freeway going through the city center. Compared to other North American cities this is unique. Some streets are designated only for bicycles" (VA 1). Similar examples are Amsterdam or Boston. However, in some situations a car is advantageous, for example to buy larger quantities at the supermarket (although the number of delivery services is growing as well). To satisfy such short-term needs, car sharing has found its niche in many cities. In many cases, the cars are very compact and powered electrically (Figure 6.10).

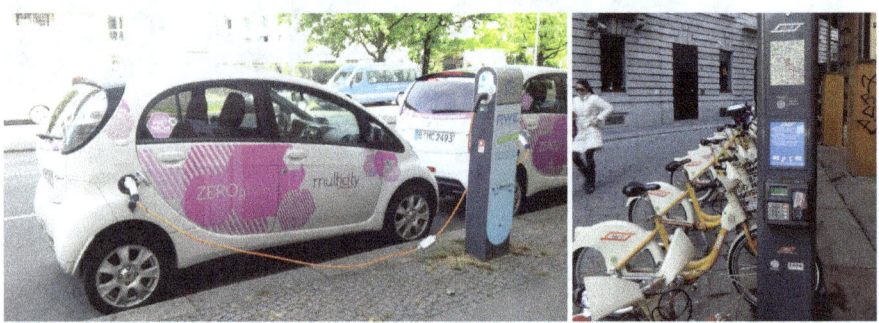

Figure 6.10: Sharing services. Left: Car-sharing in Berlin. Right: Bike-sharing in Milan. Photos: Agnes Mainka.

An equivalent system has been introduced for bikes in nearly all of the cities investigated, with the exception of Amsterdam, Hong Kong, Kuala Lumpur, and Sydney (Appendix III Bike sharing). However, "Amsterdam is not famous for fast connections via cars" (AM 1), and the city is a bike-friendly city. Due to this fact the city is sustainable and has a naturally lower level of emissions. Amsterdam has been offering bike rentals since quite some time before bike-sharing services were first launched. Figure 6.10 shows a typical bike-sharing station in Milan. According to the numbers of bikes in use and stations available, China has the largest bike sharing systems (EWContributor, 2015). Among the top cities, Shanghai offers 19,000 bikes and 600 stations. Paris has one of the most successful bike sharing systems in Europe (EWContributor, 2015), offering almost 20,000 bikes in roughly 1,800 stations ("Paris – Vélib'," n.d.). New York City was the first city to introduce a bike-sharing system in the USA. Most bike-share programs are a joint effort by local businesses and municipalities. But in New York, the bike sharing program was developed exclusively by private parties. Thus, this service is the most expensive in comparison with other global cities (EWContributor, 2015). Paris and New York have suffered from vandalism, resulting in negative revenues for the investors (EWContributor, 2015). However, only a sufficient amount of bikes and stations all over the city makes the sharing system successful. A bike-sharing density of 10 bike stations per square kilometer has been determined as ideal (NACTO, 2015). According to an interview partner from Berlin, the increasing importance of bikes and bike lanes in a city results in a loss of importance of cars (BER 7). An interview partner from Boston observes: "People always have ridden bicycles, but now there is much more support for bicycles, like bike racks for bike sharing. It's the second or third year that they are in the city. It's very popular in Boston. ... [B]eyond the automobile Boston is a great city for walking" (BO 4).

With regard to car reduction rates in cities, there are also some constraints that need to be considered. Thus, one interviewee stated: "It depends on what class you are living. If you are a worker and you have to get to work you need to be able to use transportation. If you are a knowledge worker and you are flexible and you make lots and lots of money then ICT infrastructure is more important" (LA 3). Public transportation is generally seen as a transportation mode for the masses and not for the upper classes. In Dubai, for example, there is a big gap between the rich and the poor. So in order to make public transportation attractive for all societal groups, a "first class" cabin at the metro was established. In addition, forward-looking urban planning is required. For example, in Vienna, planning of the new neighborhood "Aspern" directly integrated a metro line from the very beginning. Thus, people who would like to live in Aspern in the future do not have to wait until the neighborhood gets crowded before the infrastructure will follow, the infrastructure will already be there (VI 5). Finally, the increasing success of bike-sharing systems is not yet wholly integrated into most cities as a secure alternative to public transportation. In many cities, additional bike lanes are planned or under construction. Some cities like Barcelona are not entirely bicycle-friendly due to their hilly landscape or weather conditions unfavorable for riding a bike, e.g. the freezing winters of Toronto and Montréal. Of course, the weather cannot be changed, but geographical conditions like a hilly landscape in Barcelona could be overcome by, for example, offering e-bikes instead (BA 3). Other issues are, for instance, bike accidents. Bike lanes should be established in order to reduce this worry. In the case of Paris, users of the bike-sharing systems are actually involved in fewer traffic accidents than other cyclists: "In 2009, ... Vélib' riders were responsible for one-third of all bike trips in Paris but were involved in only one-fourth of all traffic crashes involving a bicycle" (Kazis, 2011).

Summing up all interviews, there is a split regarding the question whether cars are important or not. Most have stated that a car is required if one does not live in the city center, since outside of the core, public transportation is not developed enough. In London, for example, two experts mentioned independently of each other that the ICT and the automotive traffic needs improvement (LO 2, LO 3). Finally, whether a city is a "walking city" does not depend exclusively on the ratio of commuters that are using a sustainable mode of transport. According to a study by Duncan, Aldstadt, Whalen, Melly, and Gortmaker (2011), the "walk score" is a valid measure to investigate certain aspects of walkability in a city. This score was calculated for major cities in the US and is freely available online. Two indices already use the usage of sustainable modes of transport as part of their indicator catalog with regard to the development towards a sustainable city: "The Green City Index" by Siemens (2012) and the "Sustainable Cities Index" by Arcadis (2015). According to The Green City Index, the greenest city in Europe is Copenhagen, in Asia Singapore, and in North America San Francisco. Following the Sustainable Cities Index, the top three cities are Frankfurt, London, and Copenhagen. In their ranking (which is not separated by continent), San Francisco holds the 27th and Singapore the 10th rank.

Internet of Things – Smart Mobility
The examples above show that mobility is still important within a city and will not be replaced by ICT. However, ICT can enhance the efficiency of the use of physical space and mobility. For instance, the usage of sensors in the city that are connected to the internet is referred to as "Internet of Things" (IoT) (Kamel Boulos & Al-Shorbaji, 2014). This technology is, for example, used in Amsterdam, where smart mobility initiatives are being developed, for example in the recent cooperation between HERE (a software developer for navigation systems) and the Dutch Ministry of Infrastructure (Nokia, 2015) to test a new road messaging system for improve road safety and traffic flow. Mobility data or data gathered about the usage of physical space originates from installed sensors and as well from citizens that are using their smartphone and different apps (Castelnovo, 2015).

Among the top mobile apps based on urban data, navigation systems are downloaded the most (Mainka, Hartmann, Meschede, & Stock, 2015a). Through mobile apps, customers are able to request departure times of trains and buses, even though, some cities such as Montréal offer no real-time transit information via an app (MO 3). In addition, smartphones can replace ticket vending machines. For instance, Munich is generally considered to be a very conservative city and very cautious in introducing new services. Therefore, the city is not a forerunner, for instance, in e-tickets. The city waited for the technology to mature and until it was successfully tested in other cities before finally implementing it (MU 1).

However, there are many providers of public transportation in each city – bike sharing, car sharing, metro, and buses. ICT also increase the ease of use, for instance by offering a one-stop paying system (through an app) for all transport modes (Regio IT, n.d.). Furthermore, the combination of all public mobility modes in one app could, in the future, bridge the different providers to offer a nearly seamless public infrastructure. Such a service is offered in Helsinki in the development stage with the ambition to make car ownership obsolete (Peters, 2015).

Citizens as sensors and open urban data will have an impact on urban planning decisions in the future. It might become possible to identify the main bike routes and meet the demand if additional bike rakes or lanes are needed. One example is the running and biking app Strava (Davies, 2014). Originally this app was built to monitor the personal tracking of the customer's workout distances and times. Therefore, a lot of data was collected about when and where the app users travel. Today, Strava is cooperating with cities, London among others, to sell the data to urban planners who can now, for instance, more easily identify routes on which new bike lanes are needed.

According to Vekatraman (2014), a shift in managing traffic conditions is needed. Since in 2013 about US$ 101 billion have been spent to make cars more efficient and only US$ 3.56 billion to make traffic more efficient. He argues that it will not help drivers to own a car equipped with all kinds of technology, as they will not become more efficient because the traffic management is the true limiting factor. Thus, using data from apps may help to track routes from cyclist as well as from commuters that are using any other mode of transport. One example that evolved in Boston is the "Street Bump" app (www.streetbump.org). While driving a car, the app automatically tracks when the car is passing a pothole and records that data. The stored data can be used to help local governments to identify where the street conditions are worst. Finally, accumulating all real-time data (traffic data, transit service data, routes and city maps, weather data) could in future be used to supply citizens with information about which mode of transport would be most efficient, for example sharing a taxi with other people one morning and commuting by bike on another (Busher, Doody, Webb, & Aoun, 2014). And from the urban planning perspective, this data may be used to enhance the efficiency in urban transit by identifying the mobility demands in more detail. However, neighborhoods with poor data coverage should not be ignored in future city development plans.

Conclusion of ICT infrastructure
According to the theory of economic cycles by Kondratieff, the twenty-first century is the century of information and communication technology (fifth Kondratieff cycle). In line with this new technology, a new network has evolved – the internet.

As economic development and innovation are concentrated in cities of the knowledge society, the impact of ICT will be enormous (Castells, 1989). Therefore, the question arises whether the importance of ICT has increased in comparison with the previous economic cycle driven by automotive traffic dependent on the network of streets. According to the experts interviewed this question is very controversial. Some see opportunities to enhance sustainability in the city, but the inclusion of the whole region is lacking. Most public mobility options are developed for the inner city center and not for the surrounding area. Therefore, the change in behavior which was acknowledged as an important step towards a sustainable city cannot grow without a sufficient offer of alternatives to automotive traffic.

Today, many indices exist that use core indicators of the digital infrastructure such as internet broadband speed, accessibility, and quality. Thus, the digital infrastructure is seen as a foundation for economic innovation and success. Finally, cities located in a nation that is highly ranked by the Human Development Report and the Network Readiness Index will generally offer a good ICT infrastructure. In addition, a higher GDP rate correlates with an enhanced ICT infrastructure. Public institutions are bridging the digital divide in society by supporting information literacy programs for citizens, including the poor, disabled, youngsters, or seniors. Therefore, political willingness to support the ICT market and the social inclusion of the information poor are advantageous.

In addition, continuous access to the internet is envisioned either for people through Wi-Fi hotspots or to add things by way of IoT technology. Diverse opportunities how free Wi-Fi can be accessed have emerged in cities. From the view of the municipality, the cheapest way is if private businesses such as restaurants and cafés offer access to internet hotspots. Telecommunication companies such as Telekom in Germany or Orange in France offer additional hotspots for their customers. Providing a lot of free hotspots as in Barcelona or investing into media poles as Seoul and Milan have done can additionally add a smart city image (Morandi et al., 2016), even if the technology is not completely satisfactory. In line with the increasing sharing economy, Wi-Fi hotspots can also be shared by private citizens as in the case of the *Freifunk* initiative in Berlin. Overall, the open access to internet hotspots has not yet reached its limits and could, if efficiently included, enhance the ICT infrastructure within cities.

ICT is also used to open new opportunities in managing urban traffic. It is hardly news that too many cars in the city cause high pollution, costs, and traffic jams, but with the dawn of smart city initiatives, sustainable modes of transportation are in the focus of urban planners and the special role of the increasing amount of sharing services and real-time information services (based on IoT) is being acknowledged. ICT in combination with public transportation services can make commuting via sustainable modes of traffic more comfortable. The possibilities of using technology in improving transit conditions or enhancing the ease of use are still in their infancy, e.g. tracking common cyclist routes to adjust the construction of bike lanes according to the demands of the user, or introducing a one-stop payment system for all kinds of public transportation including bike- and car-sharing services. Finally, governmental incentives may enhance the public willingness to switch from the car to more sustainable modes of transportation.

6.1.2 Cognitive infrastructure

As informational world cities are prototypical cities of the knowledge society, the cognitive infrastructure is related to institutions of the creative and knowledge city. Hence, in this subchapter, I will investigate whether the aspects of both city types are prevalent in the 31 cities analyzed. Knowledge flows are represented through the scientific-technical output, e.g. scientific articles published and patents granted. Both are the output of knowledge institutions and knowledge-intensive companies. Synergy effects and innovation will be encouraged foremost by their cooperation. Therefore, the following hypothesis will be investigated:

H3   Science parks or university clusters that cooperate with knowledge-intensive companies are important in an informational world city.

In order to attract the "creative class" that is the workforce of the knowledge and creative economy, the city's ability to become a creative hub is acknowledged as an advantage for economic success. Thus, the 31 cities will also be investigated with respect to whether they can be characterized as a creative city:

H4   An informational world city needs to be a creative city.

Cities that are recognized as creative cities are home to so-called "creative milieus." They offer space for face-to-face communication and collaboration:

H5   Physical space for face-to-face interaction is important for an informational world city.

In addition, the main resource of the knowledge society is the human capital. To encourage life-long learning and to offer access to all kinds of information is a special role of digital libraries:

H6   A fully developed content infrastructure, e.g. supported by digital libraries, is a characteristic feature of an informational world city.

Due to the digital access to information in the twenty-first century, libraries as spaces inherit a new role. They shift from being mere depositories of publications to being places for cooperation, and as such offer additional space for face-to-face meetings:

H7   Libraries are important in an informational world city as a physical place for face-to-face communication and interaction.

In the following, I will investigate each hypothesis separately as part of the cognitive infrastructure.

H3 Science parks and university clusters

Following Stock (2011), the knowledge city can be investigated according to its output of scientific and knowledge work, which he determines through the amount of STM publications and patents registered. However, the origin of these publications are knowledge institutions, which therefore can be defined as the infrastructure that is required to be successful in knowledge-related activities. In addition, today's economic success is based on human capital. Thus, private R&D in cooperation with scientific institutions can contribute to this. To identify the importance of science parks and knowledge clusters in informational world cities, the experts have been interviewed about the following hypothesis:

H3   Science parks or university clusters that cooperate with knowledge-intensive companies are important in an informational world city.

Taking a look at the SERVQAL results, the total quality value is slightly negative $Q = -0.44$ (Figure 6.11). Nevertheless, the overall expectations on this indicator are rather high, but only in four cities the experts are convinced that the perception of science parks and universities exceeds the expectations: "[T]his [question] does go to the heart of an informational city ... But really the question is, 'Do you have that added proportion to your size of city?' That, to me, is the issue. And Vancouver probably leans towards maybe having them slightly more" (VA 1). But overall, the Vancouver experts revealed that there is still room for more university clusters and science parks. It is not surprising that a city like Boston has achieved a positive quality value with the maximum perception score (7), since some of the best universities are located within the Greater Boston Area (e.g. Harvard and MIT). In contrast, a city like Melbourne received a negative quality value, even though the city is acknowledged as a knowledge city according to

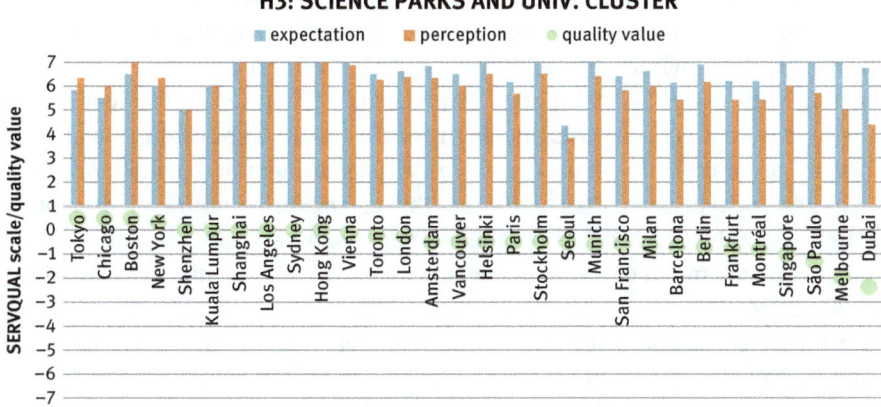

Figure 6.11: Quantitative interview results according to SERVQUAL (quality value = perception − expectation) for H3 (Science parks or university clusters that cooperate with knowledge intensive companies are important in an informational world city).

the Most Admired Knowledge City award (City of Melbourne, 2013). Accordingly, in this city more science parks and university clusters should be developed. Furthermore, a physical agglomeration of private businesses and universities is seen as important for a good partnership between them (77 experts). For instance, in Seoul, all three interview partners agreed that the existence of science parks or clusters are not as important. The Universities in Seoul, as in the whole of South Korea, produce a high number of qualified people, but the economic system is not comparable to the innovation production in western countries (SE 2). Most academics are looking for a job at Samsung or LG after they graduate. This low contribution to economic development by university clusters in South Korea has also been confirmed by Sohn and Kenney (2007). Overall, the expectation that science parks and university clusters are an important characteristic is rather high across all cities investigated.

The interview partners have also argued that the base for strong R&D is education (16 experts). According to the experts interviewed, it is important to be a strong university city (MO 2), leading many experts to state during the interview the number of universities and colleges that are located in their city or region (18 experts). Additionally, the city should attract students to graduate and eventually settle down in the region (mentioned by 6 experts).

Boston, for instance, is a successful example of institutions of higher education. The institutions are not located in the city center but are easy and fast to reach from there. The metro from Boston city center to Harvard University takes approximately 15 minutes. MIT and Harvard are both located in Cambridge and both are among the world's leading universities in several rankings (O'Neill, 2016). Following the investigation of Von Zedtwitz and Heimann (2006), Cambridge and its environs are acknowledged as a central knowledge cluster of Boston, with about 50 institutions of higher education and a medical center. The high number of scientists and engineers in the Greater Boston area attracts further corporations to set up their own R&D center here or to collaborate with local research institutes. In addition, the established culture of promoting entrepreneurship, for example in the form of student startup competitions, has enhanced the growth of the economy and consequently the number of startups in that region. Von Zedtwitz and Heimann (2006) conclude that network events and the personal character of the networks (everybody knows everybody) are an especially important driver of the regional competence of Boston. A further advantage of Boston is that Harvard and MIT are globally known universities which attract talent from all over the world.

Considering the link between knowledge institutions and cities, science parks and universities are not as interconnected as for instance other global institutions such as museums or national libraries. One expert from Paris stated:

> Let's give some examples, ... Bibliothèque de France is a world institution with very strong functionality to knowledge. So yes, I would put it at the same level as the British Library or library Grand Basque... The Louvre, the museum, I would put it the same, it's really a global institution, which spreads over information and more than information – exhibitions, like national gallery would do it in London. ... I think it's the same kind, but then if you start with universities, ... I don't think the universities are yet really linked enough to the city. There is a link between universities, but it's not yet what I would call knowledge intensive institution. It's not like ... in the Netherlands, for example, where the university is part of the city and share information with the people. No, I don't think we are that far. I don't want to say that there's nothing done, but we [in Paris] are not at the age of sharing knowledge between university and the city. (PA 2)

Thus, becoming a knowledge hub is related above all to the presence and quality of universities, and then science parks and university clusters can be built around those institutions. Therefore, in the following I will focus the discussion on the city's performance as sites of institutions of higher education. In addition, the trend of building clusters will be highlighted.

Knowledge hubs and education

In Munich, for example, the history of the two universities Ludwig-Maximilians-Universität (LMU) and Technische Universität München (TUM) goes back to the time of Bavarian Kings (MU 1), LMU having been established in 1472 and TUM in 1868 (LMU, n.d.; TUM, n.d.). Both are hubs on the national but not on the international level (MU 2). After the Second World War, the cooperation of the German military has pushed the success of the scientific location Munich as a knowledge hub (MU 1). Cooperation with the military is also part of the roots of Silicon Valley (Dembosky, 2013). Other cities cannot look back on such a long history because they are comparatively young, such as cities in Canada (TO 4).

Others have developed in recent years from a fishing village to a knowledge city, e.g. Dubai and Shenzhen. Dubai has imported the knowledge to educate their society: "There are 26 international universities in Dubai. Nearly all of the international universities are international branch campuses, which provide an accredited degree from the home institution. These institutions are from ten countries (Australia, UK, USA, India, Russia, Iran, Pakistan, France, Ireland, and Lebanon)" (Dubai private education landscape, 2014, p. 15). Dubai counts a total of 57 institutions of higher education. Shenzhen is probably a unique example of a city that has grown rapidly due to governmental plans (SH 1) – from a population of 2.39 million in 1995 up to 10.75 in 2015 (Statista.com, n.d.). The city neighbors Hong Kong and is currently set to grow up to be just as competitive. To reach this goal, Shenzhen has opened special economic zones in the 1980s with incentives for free market activities. According to a documentary on the city's rapid development, Shenzhen is called the "Silicon Valley of Hardware" (WIRED, 2016). The advantage of starting a tech startup business in Shenzhen is that it is quicker and cheaper to do research and to prototype products. Thus, the city is the perfect location to develop new products, which could be underlined by the number of patents granted. However, patents are investigated as an indicator of innovation and will be discussed in chapter H4 Creative city.

A true success story of becoming a knowledge hub is Singapore (Khveshchanka, Mainka, & Peters, 2011). The city's economic development began as a port hub for the shipping industry. But governmental plans have focused on the development of a knowledge society, today driven by ICT (IDA, n.d.-a). Since the people in Singapore were less educated in the beginning, they required the expertise of expats (SG 12). This has also been the case for Dubai, which has become a world city based on the country's oil resources. Today, Dubai is focusing on importing knowledge and the knowledge economy to succeed in future (DU 3).

The degree of education is acknowledged as an indicator of the knowledge society (UNDP, 2014). However, merely counting the number of people with higher education degrees and comparing them across cities and nations would not reflect the quality of the educational system. Globally the growth rates of higher education enrolment are continually increasing (British Council, 2012). One concrete example for this is São Paulo, where a special student credit that has made universities more accessible for lower income groups (SP 1). Nevertheless, the education quality remains very poor, especially at many private universities (SP 1), with public universities and more expensive private universities acknowledged as having a higher quality.

The main problem in São Paulo is that the best universities are the public universities (SP 1). To apply for public university, students have to pass an entry exam which contains questions about general knowledge. But most public schools do not prepare the students adequately to succeed in this exam, meaning that students of wealthy families who are able to pay for private education are privileged. Therefore, a few years ago the entry examination was adjusted to offer students from public schools and ethnic minorities special conditions to pass the test. In addition, many offers exist to acquire special qualifications after school instead of attending university. Thus, a lot of people educate themselves "after work" to reach a qualification (SP 1). Programs such as the German "dual curriculum" model, in which students alternate between practical training at a company and more theoretical university studies, do not exist.

Rankings of higher education institutes
Income inequality leads to unequal access to higher education (Bond Hill, 2015). Therefore, cities in which the living costs are too high will exclude students and creatives (BO 3, MU 1). Accordingly, the Quacquarelli Symonds (n.d.-b) ranking of the "QS Best Student City 2016" acknowledges further criteria in their ranking like the "Global Livability Ranking," the "Mercer Cost of Living rank," or the "Cost of a Big Mac." In this ranking, Paris is the leading student city mainly because of the high number of universities (18 universities that are ranked in the QS top university ranking):

> ... [W]hile Paris does have a reputation for being an expensive place to live, relatively low tuition fees mean that for students, it actually represents a more affordable destination when compared to many other popular student cities – though of course enjoying all the delights of life in this iconic European capital is certainly likely to demand quite a substantial student budget. (Quacquarelli Symonds, n.d.-a).

Of course, an informational world city should be a place that attracts talents, as this also includes students that will become knowledge workers in future. But being an attractive student city, as measured by the position in the "QS Best Student Cites" ranking, actually shows a negative correlation both with the number of students as a percentage of the city's population and with the number of international students as a percentage of all students (Table 6.2). In Paris, for example, only 18% of all students are from abroad. Within the QS ranking, London has the highest percentage of international students (42%). Language barriers may cause the rather small percentage in Paris in comparison with London. Interestingly,

Table 6.2: Pearson correlation of the cities ranking position within the QS Best Student City and the number of students as % of total population and with the number of international students as % of total students. Source: Dubai Statistics Center (2015), Goethe Universität Frankfurt am Main (n.d.), Quacquarelli Symonds (n.d.-a).

PEARSON CORRELATION	QS Best Student City rank
number of students as % of the total population	−0.40
number of international students as % of total students	−0.61

Dubai, not measured by QS, shows a high amount of international students (62%), coming, for example, from India (30%), Egypt (8%), Nigeria (6%), or Pakistan (6%). Hence, Dubai is not (or not yet) attractive for students from western countries, since only in a few cases the local institutions of higher education meet the standards of western educational systems (Kosior, Barth, Gremm, Mainka, & Stock, 2015; Schröder, Kuta, & Haidar, 2010).

To visualize the distribution of students as percentages of the population and the percentage of students from abroad, all cities that are investigated in the present work are listed in descending order from left to right according to their position in the QS ranking in Figure 6.12 (Dubai, Shenzhen, and Frankfurt were added manually as they are not a part of the QS ranking).

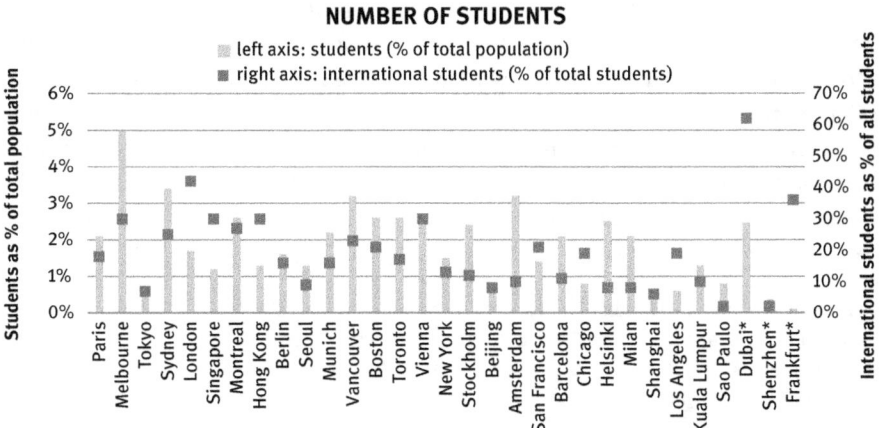

Figure 6.12: Number of students as a percentage of total population per city (left axis) and number of international students as a percentage of total students (right axis). Cities are ordered in descending order according to their rank as "QS Best Student City." *Not available in "QS Best Student City". Data source: Dubai Statistics Center (2015), Goethe Universität Frankfurt am Main (n.d.), Quacquarelli Symonds (n.d.-a).

In addition to the affordability of a student city, the quality of the university is important. Diverse rankings exist that compare universities on a global scale (Academic Ranking of World Universities, n.d.) and are given below:
- Academic Ranking of World Universities (Shanghai Ranking Consultancy, (SRC)
- Center for World University Rankings (CWUR)
- THE World University Rankings (Times Higher Education)
- QS World University Rankings (Quacquarelli Symonds)
- Performance Ranking of Scientific Papers for World Universities (Higher Education Evaluation and Accreditation Council of Taiwan)
- Ranking Web of World Universities (Cybermetrics Lab (CCHS), a unit of the Spanish National Research Council (CSIC))
- CHE-Excellence Ranking (Center for Higher Education)
- UTD Top 100 Business School Research Rankings (The UT Dallas' School of Management)

Indicators for these rankings are, for example, the number of publications, citations, patents and additionally awards received, e.g. the number of alumni and staff winning Nobel Prizes as an indicator of the quality of education. The rank is based on a score which is calculated for each ranking according to the underlying indicators. For instance, comparing the scores of the CWUR, SRC, THE, and QS rankings, the overall university scores show a positive Pearson correlation (Table 6.3). The QS ranking shows the lowest correlation with other rankings. In general, universities that reach a high score in one ranking will also reach high scores in the other rankings, as is evident in the visualization of the ranking's correlation in Figure 6.13. Interestingly, there are rather few universities that are able to reach very high scores (higher than 60). In addition, a high number of institutions only reach scores between 40 and 60 in total. Accordingly, those could be interpreted as the standard and universities with scores above 60 as elite.

Table 6.3: Pearson correlation of university scores in the rankings CWUR, SRC, THE, and QS for the fiscal year 2015 for 90 universities. Data source: CWUR, SRC, THE downloaded from O'Neill (2016), and QS from Quacquarelli Symonds (n.d.-b).

PEARSON CORRELATION	SRC	THE	QS
CWUR	0.93	0.93	0.68
SRC		0.83	0.67
THE			0.83

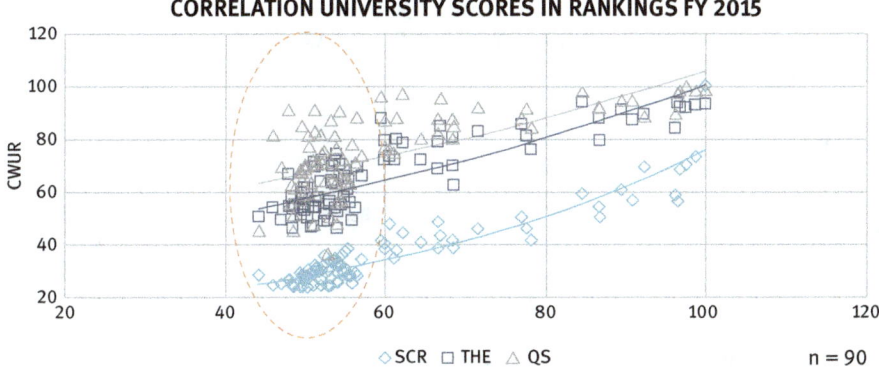

Figure 6.13: Correlation of university scores in the rankings CWUR, SRC, THE, and QS for the fiscal year 2015 for 90 universities. Data source: CWUR, SRC, THE downloaded from O'Neill (2016), and QS from Quacquarelli Symonds (n.d.-b).

However, the raw scores in any ranking rarely feature in a city's or university's self-marketing, with the rank being a much more powerful draw – for instance, "we are the best-ranked university in the world" is far more accessible than "we reached 100 points in university ranking XY." Using the example of the cities selected for investigation, their performance in higher education will be investigated based on the best QS ranked university for each of the QS Best Student Cities (only 29 out the 31 cities are available for the year 2015). Figure 6.14 visualizes the ranking position (CWUR, SRC, and QS) for the universities investigated. The size of the bubble (ranking position in QS) and position (x = SRC position, y = CWUR position) reflects that a city's rank cannot be summarized by one ranking alone. The different methodologies of the rankings result in slight variances of a university's position in global comparison. This becomes very clear if we take a look at the ranking positions of University College London (Table 6.4).

In the QS top university ranking, University College London ranks among the top ten universities, but not in the others. Looking at the average performance over all rankings, Boston and San Francisco are tied for first place, with Chicago on second and New York on third place (out of 29 cities). It is striking that the first

Table 6.4: Position of University College London in three global university rankings for the fiscal year 2015.

Institution	SRC	CWUR	QS
University College London	18	27	7

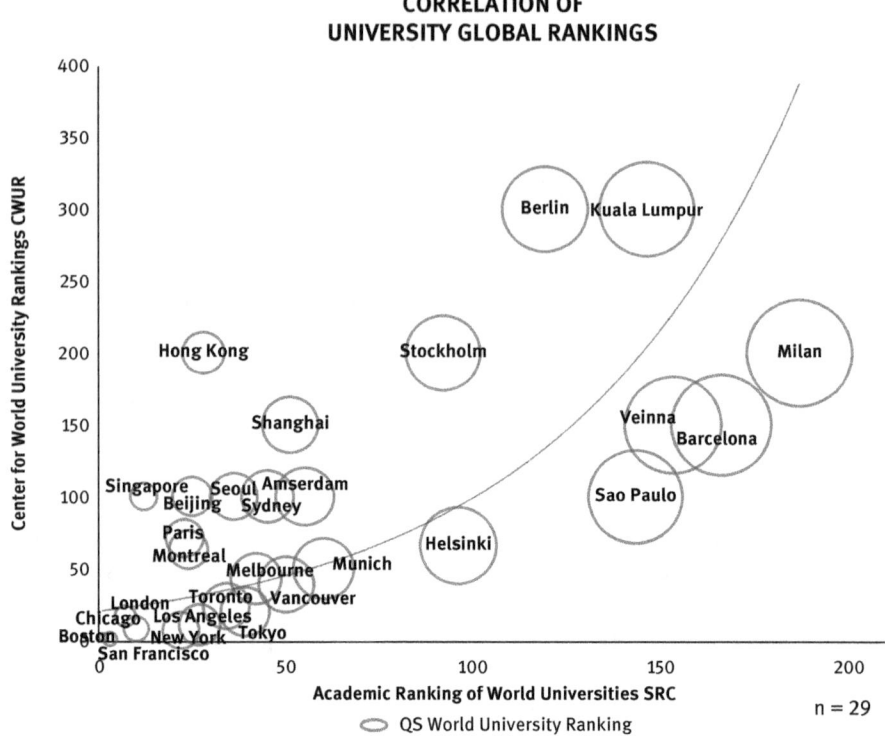

Figure 6.14: Correlation of the ranking positions within the three rankings: CWUR (x-axis), SRC (y-axis), and QS (bubble size). Universities = 29.

positions are dominated by US cities. The first European city follows on position four (London). However, as this calculation reflects only one institution per city, it cannot adequately reflect the landscape of higher education in informational world cities – all institution should be considered in a comparison between cities. Furthermore, it is important to choose criteria that reflect the quality of education and of the university in total. This is for instance attempted within the methodology of CWUR, in which the number of awards and Nobel prizes as an indicator is used. And as a final caveat it should be noted that "... achieving a higher ranking does not necessarily correlate with providing better education and research opportunities" (Holmes, 2016).

Impact of universities Ranking positions and the number of national and international students do not automatically translate into attracting the talents to stay:

> Harvard and MIT physically draw people from all over the world ... The trick is to hold on to them here – to hold some percentage of them here to staying here to create industries for the future. Boston has all these things but for various reasons we don't hold on to people very well-the cost of living, the cost of office space. The whole US have a serious immigration problem of knowledge workers. Students from all over the world come to the US. The colleges and the state invest a lot into those people, but when they are finished with their study and they want to have a job, they are not welcome. (BO 3)

According to the expert, Boston does attract students but is not able to get them to settle down. This is possibly true for the Boston Area, but following the investigation by Roberts and Eesley (2009, p. 3) "[m]ore than 38 percent of the software, biotech and electronics companies founded by MIT graduates are located in Massachusetts, while less than 10 percent of arriving MIT freshmen are from the state. More than half of the companies started by MIT's foreign-student alumni are located in the U.S., creating their primary employment and economic impacts here." Finally, the region is acknowledged as a global hub for R&D and tech start-ups, which is linked first to the personal network concentrated in Cambridge and second to the strong financial community with the ability to raise venture capital (Von Zedtwitz & Heimann, 2006).

This success is also acknowledged for other regions. The world's most well-known is Silicon Valley, which attracts talents from all over the world. Their cluster is not concentrated like the example of Cambridge in Boston, but distributed across the whole San Francisco Bay Area. Historically, this knowledge hub has evolved around Stanford University, which developed the "Stanford Industrial Park" (n.d.). Other regions have adapted the idea and demonstrated their intentions by establishing clusters which are also called "Silicon Anything," e.g. Silicon Alley in New York or Silicon Roundabout in London (Pancholi, Yigitcanlar, & Guaralda, 2015). Whether these initiatives will be as successful as the original remains to be seen.

Minguillo, Tijssen, and Thelwall (2015), have investigated publication activities of science and business parks in the UK for the time span from 1975 to 2010. According to their findings, science parks and research parks have a higher impact in fostering cooperation and scientific production than science and innovation centers, technology parks, incubators and other parks.

> Finally, regarding the industry-academia interaction, the public research base developed in the country represents a more relevant source of knowledge and technology than those located abroad and in particular within the same region due to the fact that on-park firms tend to collaborate with partners beyond their local region. The reason for this could be the lack of relevant and top quality universities nearby (Laursen et al. 2011). ... Furthermore, only the regions with great agglomerations have access to many international links. (Minguillo et al., 2015, p. 721)

Science clusters are therefore in need of collaboration with universities. According to an expert from Vancouver, this is the success story behind their local cluster formation:

> Here at the Center for Digital Media, we do a lot of collaborations and it's kind of unique in that sense, it's owned by four universities and ... our curriculum is a two-thirds collaboration with industries... I think there's a fair bit of that going on here in Vancouver, that universities are all quite engaged with working with their communities. I guess another example would be all of the work that's done at UBC and connecting to medical research companies or BCIT working with aerospace industry and another sort of electronics kind of companies, and my university, Simon Fraser, working closely with, well, just one example but Boeing and other aerospace companies on user interfaces and visualization... (VA 4)

When it comes to urban planning activities throughout the various cities, many new projects are under construction which focus on the cooperation of universities, R&D, and the free market. Some examples under construction are:
- WESTERN Sydney's Innovation Corridor, which will establish new "smart" jobs and homes (Paterson, 2015).
- "Kista Science City" in the neighborhood of Stockholm, which is already the largest ICT cluster in Europe. The hub is still in its development with the goal of becoming a real city (ST 4).
- "Aspern", a new suburb of Vienna, in which a former airport will become the location for a new science park in Vienna (a library is planned, as well) (VI 1).
- Paris has now introduced a new mega-university named "Paris-Saclay", with the ambition of becoming as excellent as Harvard, MIT, Oxford and Cambridge (Thoening, 2015). It is called the French Silicon Valley and will bring together 19 institutions of higher education and a business cluster in the outskirts of Paris.

Conclusion science parks and university clusters
According to one expert from Vancouver (VA 1), scientific activities and universities are "the heart of an informational city." The question concerning the importance of science parks and university cluster has led the interviewees to discuss the importance of universities, education and the quality thereof, as universities are without doubt the important infrastructure to educate future scientists and engineers. Comparing the universities' performance according to already established global rankings (e.g. CWUR, SRC, THE, and QS), only a few universities reach very high scores and the majority remain in the midfield (Figure 6.15). Appendix IV "Best ranked university in city" lists all investigated universities. However, due to the different methodologies applied in each ranking, the absolute position can differ substantially. This is demonstrated, for example, by the ranking of University

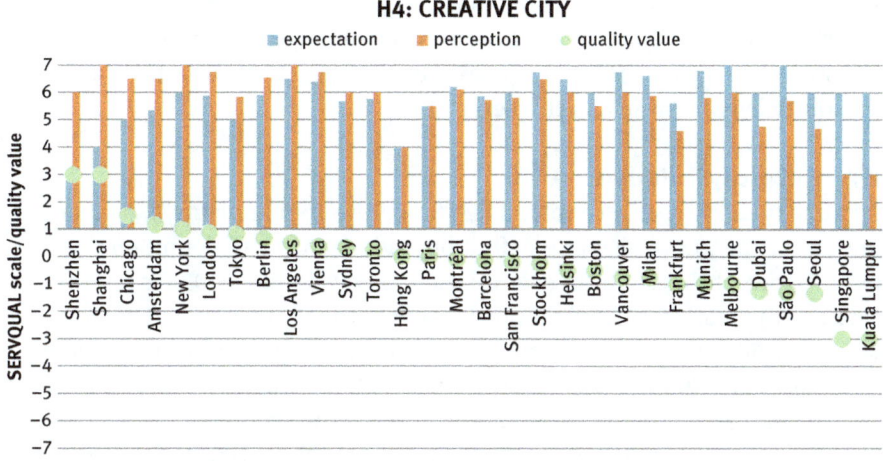

Figure 6.15: Quantitative interview results according to SERVQUAL (quality value = perception − expectation) for H4 (An informational world city needs to be a creative city).

College London, which is ranked 7th in the QS and 27th in the CWUR global index. To determine the city's environment as student-friendly, additional factors like the livability and cost of living are acknowledged as further indicators. The winning cities do not match the global university rankings, as the best student city in 2015 was Paris but the best universities according to QS are both located in Boston.

Accordingly, it is acknowledged by the interviewees and by the literature that a friendly environment for R&D and entrepreneurship are of advantage. The goal of a student city should therefore be to convince the students to stay in the region after they graduate. If we look at the examples of Cambridge in the Boston Area and of Silicon Valley in the San Francisco Bay Area, the agglomeration of universities and knowledge-intensive companies in clusters and science parks demonstrates the success of physical concentration. Other cities all around the world are adapting this idea and starting to build their own hubs of knowledge institutions. Having in mind that the world's best universities are located within the region of Boston and San Francisco – Harvard, MIT, and Stanford University are ranked at the first three positions in the SRC, CWUR and QS ranking – it will be interesting to see whether the knowledge clusters in other cities will be able to reach the same success on a global scale as these both cities do.

Finally, the cooperation between the free market and universities is considered to be a driver of the future economy, best demonstrated by the countless examples of science cities, "Silicon Anything"s and innovation incubator developments. Most of them are not isolated bastions of science, but integrate space for living and meeting into the planning to encourage the effects of networking and clusters as they are found in Boston and Cambridge. This importance of physical space for face-to-face interaction will be investigated in more detail as part of the fourth hypothesis in this chapter.

H4 Creative city indicators

As already discussed in chapter 3.1.2 ("Cognitive infrastructure"), a clear separation of the knowledge city definition and creative city definition does not exist. Hence, the creative class, as it is defined by Florida (2003), is represented through the workforce of the knowledge-intensive economy and the creative economy. He counts among them, for instance, occupations such as scientists and engineers, university professors, non-fiction writers, editors, think-tank researchers, and analysts, additionally to the traditional ones such as artists, writers, and performers. In all knowledge-intensive occupations, creativity grows in importance. At the same time, the importance of the "creative industries" like design, media, culture, and arts within cities' economic development plans grows as well. The economy and environment of a creative city should also attract talents. As an underlying indicator, the tolerance of the local society is acknowledged as important in European cities (Hansen, Asheim, & Vang, 2009). Furthermore, in US cities technology completes the conglomeration of Florida's 3Ts hypothesis that a creative city is determined by technology, talent, and tolerance (Florida, 2003). Thus, an open community of collaboration and getting to know that network in so-called "labs" extend the presence of "creative milieus" to so-called "milieus of innovation." Face-to-face interaction remains a crucial factor of innovation. To ascertain the validity of Florida's hypothesis of the creative class, the interview partners were asked if an informational world city needs to be a creative city. Therefore, to identify the importance of creativity within an informational world city the following hypothesis will be investigated:

H4   An informational world city needs to be a creative city.

Summing up the SERVQUAL evaluation, the interviewed experts on average have a high expectation of the importance of being a creative city (Figure 6.15). During the interviews 39 experts mentioned explicitly that it is important, 10 that it is at least of advantage and only 3 that it is not important to be a creative city. Interestingly, the interviewed experts in Shenzhen mentioned that they perceive the city as being creative but nevertheless do not see the necessity of an informational world city to be a creative city. This twist of understanding can be reasoned by the "traditional" way of how a creative city is acknowledged. For example, most people would say that Paris is a creative city because of all its cafés where artists and writers congregate, as well as the numerous museums, galleries, and performing arts venues in the city. But in the twenty-first century creativity is not only writing, painting, and performing, it is also about innovation and technology-based development. Hence, "Shenzhen is a field of experimentation that goes hand in hand with creativity and innovation. It is a testing ground for new ideas, computer programmers, and architecture" (SHE 1).

Following Landry (2011b) it is not the same to be creative and to be innovative. Merely counting the number of patents does not reflect the creativity of a region, but its innovation output, whereas downstream innovations and imaginative behavior are considered to be creative. As a matter of course, diverse indices and frameworks of how to measure the creative city have already been developed. A research group of the ARC Centre of Excellence for Creative Industries and Innovation (CCI) has investigated and compared the recently published indices to develop a composite index, namely the CCI Creative City Index, which captures the core definitions of a creative city (Hartley, Potts, & MacDonald, 2012). Accordingly, the indicators are grouped as follows:

1. Culture, Recreation & Tourism
2. Creative Output & Employment
3. Cultural Capital & Participation
4. Venues, Resources & Facilities
5. Livability & Amenities
6. Transportation & Accessibility
7. Globalization, Networks & Exchange
8. Openness, Tolerance & Diversity
9. Human Capital, Talent & Education
10. Social Capital, Engagement & Support
11. Government & Regulations
12. Business Activity & Economy
13. Entrepreneurship
14. Innovation & R&D
15. Technology & ICT
16. Environment & Ecology

The indicators above show that there is a mixture of world city indicators, e.g. "7. Globalization, Networks & Exchange," as well as political willingness "11. Government & Regulations," which are often mentioned in combination with other creative city aspects. Both will be discussed later. In the following, I will concentrate on indicators of creativity according to its definition as "creative economy" and the "creative people." In addition, the importance of space for face-to-face interaction will be discussed separately in the context of the fifth hypothesis in the next subchapter.

Creative economy
With regard to the question of being a creative city, some interviewees started to count the "creative sectors" that are of relevance for the city. Many mentioned traditional sectors like art (13 experts), culture and design (each by nine experts), museums (eight experts), fashion (seven experts), the film industry and music (each by six experts), and finally galleries, architecture, and tourism (each by five experts). Further rare mentions include media, printing, writing, performing arts, ballet, orchestra, opera and theater. With reference to Florida's 3Ts, technology has been mentioned by 14 interviewees, advertising and gaming industry by five, animation by three, and conferences and boot camps by two experts. Additionally, green city initiatives and research have been mentioned as creative, too. To bring together creatives with citizens, festivals have been indicated as important for a creative city (four times). Whether the festivals are, for instance, related to performing arts such as the International Circus Festival or related to engineers and programmers like the Ubisoft Street Festival is irrelevant (see chapter 5.2.7 Creative milieu).

Both the traditional understanding as well as technology and talents are part of the creative city. In addition, the growing importance of innovation has been described as the natural result of the financial crisis in 2008/2009. Innovation and startups are rising due to the proximity of many production industries in the investigated cities. Hence, "[h]unger makes the people creative. We are on the entrepreneurship way because there are no other jobs" (BA 3). Similar statements were given by two other interviewees. Another expert mentioned:

> I think that in future the question will be: What can I do and create with the information at hand? This requires creativity. And if you understand the creative city as the sum of all art, graphic design, music, and culture, then it is of course an advantage, because it creates a more beautiful habitat. It creates attention for further projects and has impact on society. (BE 6, own translation)

Finally, the term creativity has been widened from culture and arts towards technology and innovation. This can also be observed by the subcategorization of the creative sector in the Creative Cities Index CCI (Hartley et al., 2012):
1. Music and performing arts
2. Film, TV, and radio
3. Advertising and marketing
4. Software and interactive content
5. Publishing
6. Architecture, design and visual arts.

According to the CCI investigation, London has the biggest creative economy, calculated by total revenue and number of jobs, compared with Brisbane, Melbourne,

Bremen, Berlin, and Cardiff. The investigation of revenues and amount of employment is a traditional measuring method of an economic sector (Hall, 2000). In the overall CCI index, London is the most creative city, followed by Berlin and Melbourne. Further indices on a global scale have been developed by Florida, Mellander, and King (2015). These findings are on a national level and have Australia, the United States, and New Zeeland as the top three nations.

Places for music and performing arts as well as for visual arts are acknowledged as having a positive effect of attracting the creative class. According to one expert, New York is a creative city because it has galleries and is a hub for design and advertising. This is what is missing in Silicon Valley (NY 1). In contrast, a creative city like Paris has recently struggled with its galleries, as the artists are moving to other places (PA 2). To get an idea of the breadth of cultural amenities, the World Cities Cultural Report 2015 (BOP Consulting, 2015) has investigated 21 out of the 31 cities that are the focus of the present work. Figure 6.16 shows the number and share of cultural and creative facilities in these cities. Paris is, as assumed, the city with the highest number of galleries. Even the interviewees stressed that the future of Paris as continuing global creative hub is uncertain. Nevertheless, the city also offers more theaters and museums than any other. The second-highest number of institutions is found in London, and the third-highest number is found in New York. Interestingly, the report does not identify galleries or theaters in Shenzhen. However, one famous gallery in Shenzhen is the

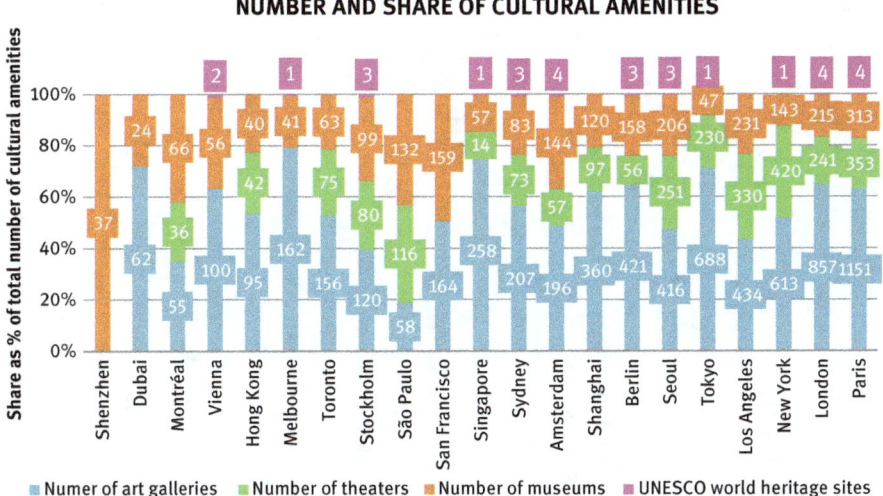

Figure 6.16: Cultural amenities of informational world cities in numbers and shares. Cities ordered from left to right in ascending order according to the total sum of cultural amenities. Data source: BOP Consulting (2015).

OCT ART & Design Gallery which was opened in 2008 in a renovated warehouse. Summing up, the number of cultural amenities varies a great deal between the cities. Furthermore, to identify the correct numbers is not easy because there is a high fluctuation of opening and closing galleries (PA 2).

Calculating the number of cultural amenities as a percentage of the total population, Paris is not the top of the list (Figure 6.17). Interestingly, San Francisco with its 323 cultural amenities and a population of roughly 800,000 has the highest number per capita. These numbers only apply to the city and not to the surrounding region, which has been criticized for its limited offer of cultural attractions (NY 1). In contrast, the number per capita for Paris is based on the metropolitan region (population = 12,005,077 and sum of cultural amenities = 1,821). However, the number of citizens does not correlate with the amount of offered cultural facilities. The number of tourists is therefore used as an indicator by the World Cities Culture Report 2015 (BOP Consulting, 2015), since cultural activities are a main attraction. Nevertheless, comparing the data of this report, no correlation between the number of international tourists with the number of cultural amenities or with the number of average art gallery visits could be found. Neither could a relation of the GDP and the number of international tourists. Finally, cultural amenities are not useful as a basis for the calculation to identify the most attractive city. The data provided by BOB Consulting do not allow reliable conclusions in this matter.

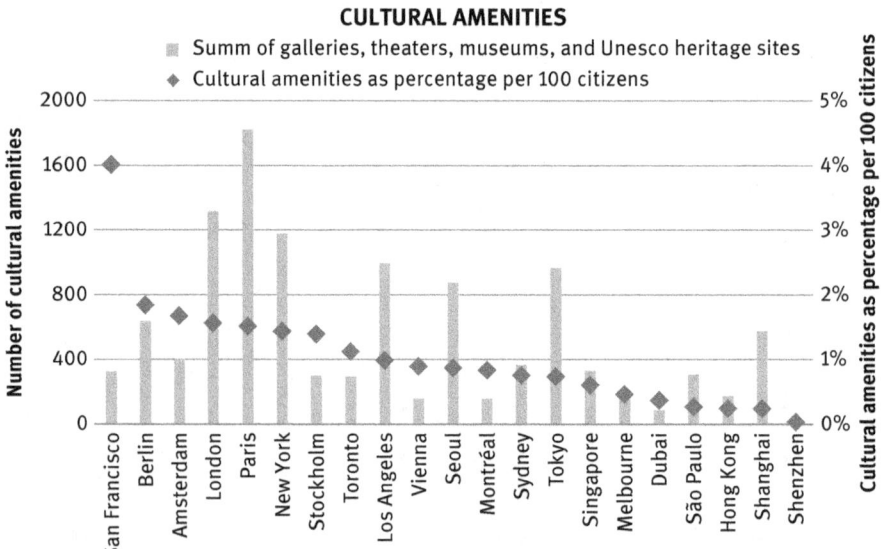

Figure 6.17: Cultural amenities as a percentage per 100 citizens (right axis) and the total sum of cultural amenities for each city (left axis). Data source: BOP Consulting (2015).

A further index is provided by to the MasterCard Global Destination Cities Index. According to this index, the most visited cities in the world are London, Bangkok, and Paris (Hedrick-Wong & Choong, 2014). London and Paris offer the most popular museums, i.e. the Louvre in Paris and the British Museum in London. Furthermore, it is not proven how the creative class, which also includes the knowledge workforce, interacts with cultural institutions or even if they use or visit them at all (Comunian, 2011). Cultural sectors like art and performing art are state-protected and supported. Therefore they are not as market-driven as others and are mostly stable during economic downswings, but benefit from upswing phases (Dapp & Ehmer, 2011). Thus cultural amenities are acknowledged as part of the creative infrastructure and are often included in city development plans (Comunian, 2011). Opera houses, theaters, and galleries serve as iconic symbols of the city and in addition, convey a cosmopolitan lifestyle feeling to the creative class. Examples are, for instance, the Walt Disney Concert Hall in Los Angeles, the Shanghai Grand Theater, the Sydney Opera House, or the Louvre in Paris (Figure 6.18).

Figure 6.18: Top-left: Walt Disney Concert Hall in Los Angeles. Bottom-left: Sydney Opera House. Top-right: Shanghai Grand Theater. Bottom-right: Louvre in Paris. Photos: Carsten Brinker (top and bottom-left), Wolfgang G. Stock (top-right), Agnes Mainka (bottom-right).

Furthermore, the creative industry impacts the digital economy (EY, 2015a), as digital and physical goods are sold on the internet, which means that books, music, games, and videos, as well as online advertising are part of the creative sector.

Today, the majority of creative and cultural content is still consumed offline. Nevertheless, an investigation by EY (2015a) has revealed an overall growth in the EU for online sales of Amazon (+44%), subscribers of Netflix (+61%), and the amount of collected money by entrepreneurs on Kickstarter (+81%) in the time span from 2011 to 2014. Furthermore, the advanced technology and increasing broadband connections have widened the spectrum of devices from PC and consoles to smartphones, tablets and smart TVs for gaming (Newzoo, 2016) and further consumption.

Combining the creative class, the digital economy entrepreneurship, and tech startups is part of the creative economy. The discussion about entrepreneurship in informational cities as part of the ICT sector can be found in chapter H1 ("Hub of companies with information market activities"). As already described, being a city with a high amount of creatives and entrepreneurs does not correlate with GDP growth (Murugadas et al., 2015). One interviewee has tried to figure out this relationship as follows:

> [In Vancouver there is a] constant build up and collapse of companies in this region. [F]or example, Pixar set up a studio here and three years later closed it... You can tick off thousands of companies that go through this cycle in this location. The only explanation that I can offer for Vancouver, this comes out of another cluster that we studied, that we did on biotechnology where exactly the same thing happens. [T]he University of British Columbia is very good at doing basic research. So, entrepreneurs start by taking a piece of basic research and developing it... Now, that applies more to locally grown companies that the entrepreneur built the company up to a point where it can then be sold to a much larger, usually, multi-national [company]. It still doesn't explain Pixar... and so on... There is a marginal explanation of what they are really doing is there are lots of independent, particularly, digital animation has us around here. And in effect, Pixar comes along, buys one, puts a nameplate on the door... Particular, as they start trying to apply large corporate management systems on them, everybody has walked out the door. But... there are some parts... that you still don't really understand. (VA 1)

The case of Vancouver shows an interesting example of entrepreneurship, and being entrepreneur-friendly is not the same as being market-friendly for large corporations.

Innovation and startups are driven by the creative class. To figure out the innovation potential of a city, the number of granted patents within a selected time period is used as calculation base (Florida et al., 2015). Looking at the 31 cities investigated, the numbers vary greatly (see Appendix V). The database used for this investigation is the Derwent World Patent Index (the method of patent investigation is described in chapter 4.6).

Accordingly, Seoul is the city with the highest amount of patents in total numbers but in relation to the number of citizens, the most innovative city is, as expected, San Francisco (Figure 6.19). Per annum, the most patents have been

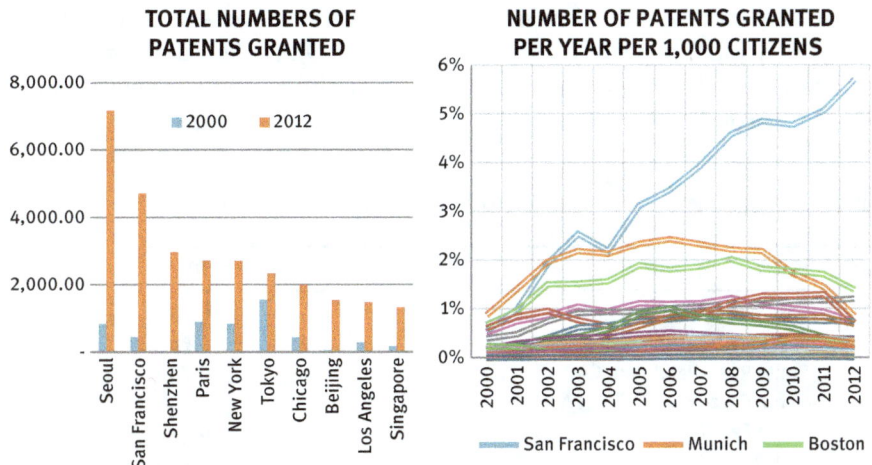

Figure 6.19: Left: Total numbers of granted patents in 2000 and 2012. Top ten informational world cities listed in descending order according to the number of patents granted in 2012 from left to right. Data source: Derwent World Patent Index (retrieved in 2013). Right: Number of patents granted as percentage per 1,000 citizens for 31 informational world cities. San Francisco, Munich and Boston are the top three of the list. Data source: Derwent World Patent Index (retrieved in 2013).

granted in 2006. Afterwards, a downswing can be recognized in the number of patents for the majority of the cities except for San Francisco. Munich has the second most granted patents for the time period under investigation, but in 2010 Boston has passed Munich in the number of granted patents per 1,000 citizens. The underlying data is only evaluated on the city level and does not include the region or surrounding area as, for example, the San Francisco Bay Area. However, in a composite index that includes amongst others indicators like R&D, education, and high-tech companies next to the number of patents, South Korea is the most innovative nation on a global scale (Jamrisko & Lu, 2016). The US follows at rank six.

Creative people

With respect to the 3Ts (tolerance, talent, and technology) one interviewee asked: "What is the creative city? People with their smartphones?" (MO 2). He is not wrong, since many of the smart and tech startups are building businesses by reusing data in a new way which may (and often does) eventually result in smartphone applications. To do this, it is of advantage to be in a "milieu of innovation" with other creatives and like-minded people. Hence, openness and tolerance have been identified as indicators of the creative class with a positive impact on the economic growth (Florida & Gates, 2001). According to Florida (2003), tolerance, also positively influences the flow of talents. Talents are not a static number, they are variable and therefore should be attracted by a city. Tolerance is an attractor for both US and European cities (Hansen et al., 2009). Looking at the most tolerant cities identified by Florida for the US, the majority, however, are not acknowledged as a world city (see Table 6.5).

Table 6.5: Top ten metro areas of the Tolerance Index. Data source: Florida (2012).

Rank	Metro	Tolerance Index Score (0–1)
1.	San Diego-Carlsbad-San Marcos, CA	0.751
2.	Napa, CA	0.747
3.	Santa Rosa-Petaluma, CA	0.739
4.	Santa Cruz-Watsonville, CA	0.738
5.	Santa Fe, NM	0.726
6.	Ithaca, NY	0.723
7.	Oxnard-Thousand Oaks-Ventura, CA	0.708
8.	Cape Coral-Fort Myers, FL	0.702
9.	Boulder, CO	0.701
10.	Ann Arbor, MI	0.693

Indicators of a tolerant population are the openness to people independent of their gender, race, nationality, sexual orientation, or "geekiness" (Florida, 2012, p. ix). The Global Creativity Index uses the indicators of openness to ethnic and racial minorities and openness to gay and lesbian people (Florida et al., 2015). These indicators are available on the national level and are based on poll surveys. The global index is constituted of the three Ts and ranks Australia, United States and New Zeeland on the first three positions. A regional index of tolerance has been developed by The Daily Beast (2011) on the state level for the US. In addition to race and sexual orientation, further indices are included, such as the number of hate crimes, discrimination complaints, and laws that support and protect legal rights for religious minorities and same-sex couples. The "winners" of this investigation are Wisconsin, Maryland, and Illinois.

On the city level, only a small number of indices have been developed and they are not on a global scale. One index that is similar to the indicators used by The Daily Beast is the LGBT-Friendly Cities comparison (LGBT = lesbian, gay, bisexual, transgender) (Miller, 2015). According to its results, San Francisco is the most LGBT-Friendly city. Out of the 31 cities under investigation, only Los Angeles ranks in the top 10 (9th). Being a gay-friendly city can also be recognized by offering gay neighborhoods and gay parades to celebrate this tolerance (see Figure 6.20). Nevertheless, such neighborhoods do not represent the whole gay community within metropolitan areas (Kelly, Carpiano, Easterbrook, & Parsons, 2014).

Figure 6.20: Gay neighborhoods. Left: Los Angeles – West Hollywood. Middle and Right: Toronto Church Street. Photos: Agnes Mainka.

The tolerance indicators of a society are used to identify the openness. The more tolerant a society is, the more talents feel welcome, no matter their religion, skin color, or sexual orientation, ultimately resulting in an increased flow of talent. This openness is more prevalent in western-oriented countries e.g. North America, Europe and Australia (Florida et al., 2015). Nations in Asia and Africa show a lesser tolerance with respect to religious and sexual orientation. But even in western countries, some differences can be made out, as talents in the US are more likely to flow between places and creatives in Europe are more likely to choose destinations near their family or friends (Murugadas et al., 2015). All in all, tolerance is one indicator that can attract talents. Further indicators of an attractive city are, amongst others, the livability, cost of living, and creative networks (Wright, 2016). As mentioned previously, the creative class does not only consist of the software developer or researcher with moderate or even high income. Among the creatives are also the painters and performers who are mostly not that wealthy. In a creative city ideally, both are present. Two interview partners from Paris and from Berlin stated that they do not believe that their city will keep the status of a creative city as it is known today.

Berlin is acknowledged as "poor but sexy." The city has in the past struggled with establishing authority, especially in East Berlin. After the reunification, many buildings or even whole street sections remained empty and no one was aware of the owner. This boosted the creative redesign of these places which led to a vibrant "creative milieu." But this explosion of open space was due to a unique, once-in-a-lifetime event in history and therefore it is not clear if Berlin will be able to maintain its role and reputation as a creative city in the future (BE 6).

Regarding Paris:

> Paris is a paradox because it has a large amount of creativity. Of course, ... there is a lot of artists, a lot of companies that do art or design in different stages, in these fields – architecture, fashion, cinema etc. So, in this aspect, it is a very creative city. It has the potential of a

creative city – the human and the intellectual and the financial potential of a creative city. But then the paradox is that Paris is a very conservative city. The danger of Paris is that it becomes a museum and it is in some part becoming a museum – lots of museums, beautiful city, in fact, a small city in comparison ... to New York, London, [or] Berlin. ... [T]he people that live in Paris are pretty rich people, lots of foreigners, and lots of people coming... So, those people are very conservative, they want no change. They don't want to be like this and this is the paradox. They have a huge potential and sometimes a big amount of conservatism. That maybe explains the fact that Paris has lost somehow this leadership in creativity in the last 50 years. (PA 2)

Theses qualms mentioned by the interviewees should not mean that art and culture are becoming less important, as "quality of life is an important location factor and one aspect of quality of life is culture" (MU 1, own translation). One main reason for the uncertain future development is that citizens are confronted with changing circumstances:

Before World War II, Paris was obviously the capital of art ... Today it's not anymore. French artists are much underestimated in the market and the so-called good artists or well-known artists ... go away. They go to London, ... New York... [or] Berlin. Lots of my artist friends live in Berlin because it's more creative, ... more innovative galleries, ... there are more crazy exhibitions... [and] space for artists. You very much cannot find here, because [Paris] can't expand. (PA 2)

The rising density and the high cost of living have been mentioned by five interview partners as detractors for creatives and young people, e.g. in London, Boston, and Munich. In other cities such as Vienna and Barcelona, the experts are more optimistic and consider the creativity in the city to be growing. Cultural behavior like tacit norms to avoid risks is mentioned by one interviewee from São Paulo: "the people are afraid of risks, e.g. afraid cycling to work or so and all kind of risks hinder Brazilians to be creative" (PA 3)

To attract talents in spite of high costs of living, jobs in the creative economy and educational institutions of art, performance, fashion and others are needed. To support innovative projects, the city of Munich, for instance, gives awards to locals in diverse categories such as art, design, architecture, or comics. Other interviewees mentioned that more needs to be done with respect to keeping the creatives, as well, and not only the technology talents. Today, we can see that in diverse regions, technology and creativity is merged. For instance, the Google Cultural Institute in Paris or the growing gaming and animation industry (Newzoo, 2016).

As has been shown, the creative class is an umbrella term that is used for very diverse groups of people with very diverse needs and ambitions. On the one hand, the creative class comprises creative people working in the culture and art sector and on the other hand, creative people working in innovation industries such as tech startups (Cohendet, Grandadam, & Simon, 2011). Both are creative in their own way, but the big difference lies in their income, resulting in different needs in their everyday life. Nevertheless, both groups are attracted by cities, especially by a creative or innovative milieu.

Conclusion: Informational world cities need to be creative cities
According to the experts interviewed, the creative city is widely recognized as the sum of economic sectors that could be counted as creative. Most experts started to list the "creative" institutions and organizations that are present in the respective city. Interestingly, the cultural and art-related institutions were mentioned roughly as often as technology- and innovation-related developments. The investigation of cultural amenities as well as the analysis of granted patents have revealed the already well-known hubs of the creative economy. Among the 31 cities, Paris is the iconic city for culture in total numbers and San Francisco leading the list based on the per capita calculation. Nevertheless, both have to struggle with critics, e.g. even though San Francisco as a whole has many cultural amenities, Silicon Valley is not known to offer many of them (NY1), and Paris is acknowledged as having become too conservative and might thus become a museum in itself (PA2). When calculating the most innovative city by the number of granted patents, Seoul leads, but San Francisco is the hub of technology and innovation if the numbers are calculated as a percentage of the population (in this investigation, the city was used as a basis and not the whole San Francisco Bay Area). According to Landry (2011a), it is not necessary to be at the top of the list, but it is of importance to find a niche in which the city is outperforming most other cities. It is also still undetermined what relation holds between the knowledge workforce and cultural amenities (Comunian, 2011), and many interview partners have argued that they did not have the time to visit cultural attractions. This might be reflected in the fact that no relationship between cultural amenities and tourism or the GDP could be established. Nevertheless, cultural amenities that are iconic symbols, like the Louvre in Paris or the Opera House in Sydney, imbue the city with a metropolitan lifestyle. It is not the quantity of cultural amenities that makes the difference, but their quality.

Furthermore, a creative city is attractive due to the creative people living there. This has been investigated by Florida (2014). Referring to his creativity index, creative places are determined by the three Ts: technology, talent, and tolerance. In his definition, the openness towards minorities attracts further talents. Florida is using the gay index as the main resource to investigate the tolerance of a region. Additional indices have been developed to measure the tolerance of a specific region or city. Those are based on crime rates against immigrants and minorities, or by the existence of laws to protect the rights of minorities or women. Nevertheless, in the majority of these rankings, world cities are not at the top. Rather, mid-size cities are the leaders of "attractive city" rankings for creatives, because they are less expensive, offer more space and are safer. Especially space is seen as a booster of creativity as mentioned, for example, in the case of Berlin's rise of creativity which was driven by open space after the reunification. And the lack of space is acknowledged as the problem of Paris which has struggled not to become a museum in itself.

H5 Physical Space for face-to-face interaction
Even in a creative city, not the whole city can be considered creative. Most common are so-called "creative milieus" or creative quarters near the city center (Collis, Felton, & Graham, 2010). Talents, industries or galleries tend to cluster in a physical place that fosters interaction and innovation. Conglomerations of like-minded people pop up in diverse city labs to co-create or to co-learn something (Carrillo, Yigitcanlar, García, & Lönnqvist, 2014). These clustering effects are also due to entrepreneurs and the startup scene, sharing coworking spaces or so-called incubators. Thus, they are in proximity with other people with diverse educational backgrounds and occupations, which encourages a "milieu of innovation" (Camagni, 1995; Hitters & Richards, 2002). With regard to space for face-to-face interaction in creative cities, the following hypothesis will be investigated:

H5 Physical space for face-to-face interaction is important for an Informational world city.

According to this hypothesis, 53 of the experts interviewed explicitly mentioned that it is very important to have physical space for face-to-face interaction. The SERVQUAL evaluation also reveals this opinion (Figure 6.21). The average expectation for all cities has a score of +6.2. Additionally, it demonstrates that in most cities there is a lack of availability of such spaces, which is why the average quality value is negative (-0.9). However, people are going to sit more and more in front of their monitors and do need these places (FR 1).

> Knowledge does not emerge only from reading. Knowledge does not emerge from intelligence. Knowledge emerges specifically through exchange, through inspiration, through reflection and for this you need fellow human beings. Sure, there are people that are able to do all that by themselves, but most people need external inspiration to reflect certain circumstances in a new way. And for them, this interaction is very important. It is necessary to step out of the box. (MU 1, own translation)

Hence, "[t]o do a project only online is hard. To meet face-to-face and to discuss things is very important" (AM 6).

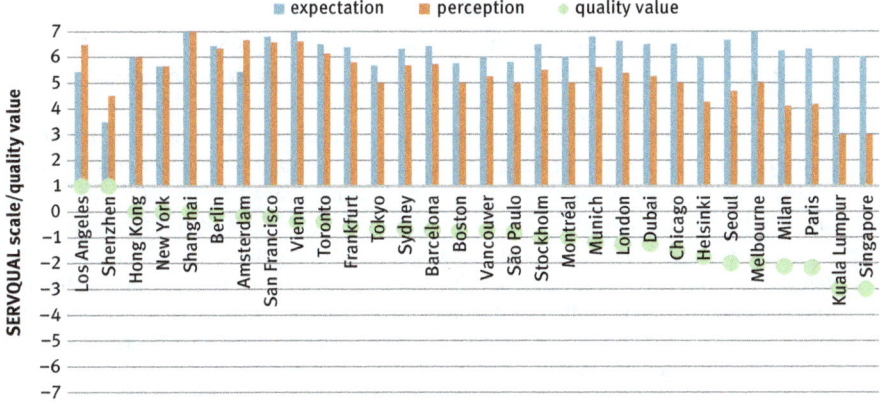

Figure 6.21: Quantitative interview results according to SERVQUAL (quality value = perception – expectation) for H5 (Physical space for face-to-face interaction is important for an Informational world city).

Only two experts mentioned that they did not see the necessity of such places, and others have voiced some doubts about their importance. For instance, one expert from Hong Kong (HK 2) has stated that "[t]here is no real reason for the Silicon Valley clustering. Technically, if we are all connected and it's all about the flow of information you can live everywhere... but people will cluster anyway." In addition, physical places for face-to-face communication should not be established only for creatives and knowledge workers, as they should meet with other people with diverse backgrounds (BE 6). Interestingly, most of the interviewees have associated those places with knowledge workers and entrepreneurs only, discussing the availability of coworking spaces or other places of knowledge sharing like universities rather than creative clusters or creative milieus in their city. The discussions were mostly dedicated to the culture of communication and the availability and importance of an architecture that should enhance the interaction between creatives and knowledge workers. Both aspects will be investigated in the following.

Architecture of face-to-face interaction

Interviewed experts from London, Milan, and Paris mentioned that there was a lack of places for face-to-face communication in the city. For 13 experts, cafés, restaurants, and bars are those places. In addition, several interviewees also mentioned informal spaces like museums, galleries, and parks. In London, for instance, people working in the finance sector meet at weekends in pubs to relax and for conversations at Canary Wharf (LO 3). In Munich, beer gardens are well-known as places to meet each other. Those places should also offer a Wi-Fi internet connection (MU 1). Thus, the architecture of the city is important to offer space for an informal meeting at diverse places with heterogeneous people (Rantisi & Leslie, 2010). One expert from Boston (BO 3) explains the impact of a city's architecture as follows:

> It's the way like their office buildings are structured, the way how street culture works, coffee culture and that sort of thing. If you go to California it feels completely different. Harvard Square is a kind of European feeling. It's lots of small shops and cafés, restaurants that have seats along the sidewalk, book stores etc. Candle Square is where Microsoft Research, Google Research, and the Computer Science Complex at MIT is – very cold! You walk blocks, nothing but glass windows from the front of offices buildings facing you [(Figure 6.22)]. No place to sit. And then they try to plant a café every now and then but it just doesn't feel the same way at all. We have some disadvantages for knowledge workers where they can meet face-to-face by our architecture and weather.

Figure 6.22: Boston – Cambridge office buildings with glass front. Photos: Carsten Brinker.

Thus, if your city is not located in a warm climate zone then the most common choice is to meet at a commercially used place. For one interviewee from Frankfurt, this has become a common standard, but there are other possibilities, as well:

> So I would say there is a lack of free indoor meeting spaces where you can bring your own thermos of coffee or whatever. I think that most people just expect these days, when they have a meeting that they have to buy something. I mean, I always do. I don't even think

about a free meeting space. But in Fremantle, where I come from, in the Perth region, Fremantle has a place called 'The meeting place.' It's free. It's a community facility where you wanna have a meeting with a group of people, then you can go to 'The meeting place.' You don't have to pay. You can use the kitchen. (FR 3)

Formal places have also been established for face-to-face meetings. For instance, one expert from Los Angeles (LA 2) has stated that

> [i]n LA there are increasingly very interesting spaces where people do meet to collaborate. Renting small offices. There is an open office concept. Not necessarily people working for the same company but where they dump together in the break room and that is very popular in the software industry. So when there are people who doing consulting they can bump into each other have companionship and exchange ideas. So that's become pretty popular throughout the city.

Such coworking spaces enhance the opportunity to meet people spontaneously and get in contact by accident. According to one interviewee from San Francisco (SF 2), libraries have been just that for many years already:

> People who are more in the digital technology than me are often talking about how important it is to have these human interactions... Workspaces are more and more being designed that people working around a big table or they working collaboratively in a coworking space. I think the library is for some people a surprising place which is fulfilling their needs and they are excited when they see that they can come here and work here and see other people and they think that they discover it before anybody else. It's funny, too, because of the coworking, it feels like some of our branches, they have been doing this for years. The people come in and take out their desktop and punk it on the table and then they put out the big monitor and put it there and you just look: 'Aha?' (SF 2)

Unconventional working spaces have already been introduced in tech firms, for example in California. "When you visit Google or Facebook. It's all fluent and open and there is food everywhere. This is what we want. The west coast has this much more than the east coast has" (BO 3). Consequently, people tend to meet at places where they have space to (co-)work and to meet like-minded individuals no matter if it is at a working space or in the open kitchen of their company. As one of the early adopters, Berlin has been mentioned for its popular coworking spaces by several experts. One expert from Berlin (BE 7) has noted a rapid growth of such spaces. In 2012 about 50 such spaces existed, rising to more than 100 a mere two years later. The success of coworking spaces goes hand in hand with changing business models, which are becoming more and more open. According to Chesbrough (2003), innovation models in the economic market have changed from closed to open innovation processes. Hence, the process of research, development, and selling of a product do not follow just the principles of doing everything alone and with one's own capacities anymore. Further, a flow between stakeholders, technology, and capacity opens the way for new markets and enhances the innovation process. For example, face-to-face communication in an organization can be described as a knowledge transformation process (Nonaka & Takeuchi, 1995). Tacit knowledge plays a significant role and is best transferred through physical social relation. Coworking spaces enable both principles, open innovation and tacit knowledge transformation, bringing people, mostly entrepreneurs or freelancers, together who could contribute to the innovation process of each other's product or service. For example, a programmer might be working on a smartphone app but have no idea how to design a beautiful user interface or even how to sell the final product. By accident, this programmer might meet a designer or a marketing person at the café of the local coworking space and get some insights into how these things work. And in turn, the programmer might be able to solve some WordPress bugs for them, creating a win-win situation for all parties. Hence, coworking spaces "are defined as localized spaces where independent professionals work, sharing resources and are open to share their knowledge with the rest of the community" (Capdevila, 2014, para. 7). In the beginning, this trend arose due to the changing labor market of knowledge workers. They are able to work and live anywhere where they have access to the digital infrastructure (Beyers & Lindahl, 1996). The financial crisis of 2008/2009 acted as a catalyst, since many office buildings became empty. To enliven these buildings, shared spaces were used as a new renting model. For instance, "[t]he first [coworking space] CWS in Barcelona was launched in 2007 and currently, more than a hundred spaces in the city define themselves using the term 'coworking'" (Capdevila, 2014, para. 17). This development is related to the emerging spaces of the creative industries. Creatives reuse empty industrial spaces as galleries or performing spaces due to their relatively cheap availability (Collis et al., 2010).

In practice, not all coworking spaces are innovative. Capdevila (2014) undertook a qualitative investigation of coworking spaces in Barcelona and divided them into three groups: "no innovative communities," "innovative communities," and "highly innovative communities." The typology of this investigation is fuzzy, but at least it demonstrates that unconventional handling is understood as innovative. For instance, "[t]he CWS managers avoid sponsorship understood as getting free products or putting a visible logo at the entrance. Instead, they propose challenges and innovative approaches to reach win-win agreements" (Capdevila, 2014, para. 59).

Looking at the 31 cities investigated in the present work, spaces of this kind are available in all of them. Of course, the number of spaces varies and a definite list of all coworking locations does not exist. In Figure 6.23, the number of coworking spaces found for each city are listed in descending order. The numbers were arrived through a Google search on November 18, 2016, using the search terms "[city name] coworking space." There are some websites and blog posts that are trying to help people

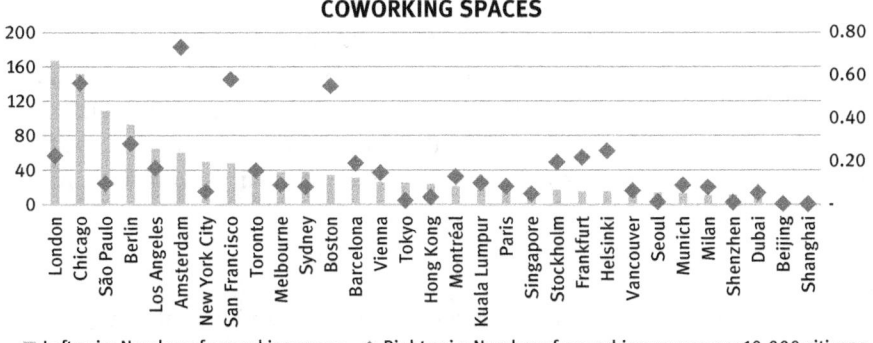

Figure 6.23: Total number of coworking spaces per city and the number of spaces in relation to the population.

to find those places and are constantly updating their list of coworking spaces. For instance, a blog post about coworking space in Tokyo (tokyocheapo.com/business/drop-in-coworking-spaces-tokyo), the coworking wiki (wiki.coworking.org), or the matching platform Coworker (www.coworker.com) do offer overviews about available coworking spaces for some cities. These lists were used in conjunction with the internet search to identify the number of coworking spaces. The numbers should not be taken as absolute, because coworking spaces are being set up and shut down at a high frequency. The full list of identified spaces and sources is available in Appendix VI (Coworking spaces). Interestingly, Berlin, which is the city that has been mentioned by the most interviewed experts as the pioneer of coworking spaces, is not at the top of the list. This could be due to the varying definitions of coworking spaces. Some spaces are just shared work spaces or offices with short time rental options. Other spaces are acknowledged as cafés with a good Wi-Fi connection and many plugs for Laptops. Nevertheless, the city with the most identified coworking spaces in relation to the number of citizens is Amsterdam. The city is rather small and offers 60 coworking spaces while having a population of smaller than one million. In Beijing or Shanghai, it is rather difficult to find coworking spaces. There are some first attempts to offer innovative spaces for entrepreneurs and freelancers, where sharing of resources and knowledge can happen, but in relation to the size of the cities only very few exist. In addition, the language barrier might constitute an obstacle to find such coworking spaces via a Google search. Interestingly, one blog post about coworking spaces in Amsterdam lists the public library as a place for people looking for a desk. The importance of public libraries as physical space for face-to-face communication will be investigated in a later subchapter of the cognitive infrastructure of informational world cities.

Not every interviewee had experience with coworking spaces or even knew that those spaces existed in the city. Many said that this might be due to their professions and community, which does not intersect with the startup and freelance scene. Similarly, the culture of communication and work practice can exclude some citizens from this community because they are not aware of those spaces. Therefore, I will discuss cultural diversities in communication according to the 31 cities in the next section.

Culture
Something that everybody will find in the city are cafés and bars at which people can meet. In Amsterdam for instance, the cafés and restaurants are trying to attract knowledge workers, e.g. by offering Wi-Fi (AM 2). One interviewee from Berlin (BE 6) mentioned that their newspapers are full of cartoons that are making fun of all the people with their laptops sitting in cafés. For him, this is indeed an indicator that the knowledge workforce is meeting there. In São Paulo, the people meet mostly for lunch or dinner. Lunch is more informal, hence the people working in the same neighborhood might meet each other at lunch restaurants by chance. This offers opportunities to exchange with people that work in another office (SP 4). In Barcelona, one expert (BA 6) stated that it is important to offer free Wi-Fi at public spaces because open spaces are the meeting places. This is indeed possible in a city like Barcelona which has a Mediterranean climate. Finally, getting into a community is more important in some regions. In Los Angeles one interviewee has reported how important personal contacts are to becoming successful:

> It is the story of legends. It's the people here in the entertainment industry. They will go to a club to hear a new performer. Then there are sitting two very famous people from the entertainment industry look at this new performer. That person is unknown. They decide they are going to mentor that person and that person becomes very famous. So that face-to-face. Getting to know somebody here. That is what makes the industry. (LA 1)

Personal contacts and meeting the community are very important for other businesses as well: "In Asian cities like Hong Kong, Singapore, or Beijing it is difficult to find those places. It is kind of a 'speakeasy' concept, e.g. in the Hutong area in Beijing. In Beijing, places are not public. It is very hard to find those places; you need to be in the community" (SHE 1). This is also the key in Tokyo. To build a business in Japan, much time must be invested in personal contacts (TOK 2). It is necessary to create confidence through both office meetings and getting together for dinner or at a bar before any business happens. This is similar to Dubai, where businessmen meet at bars to get in contact with each other and conduct their business. To meet at conference rooms considered old-fashioned (DU 2).

Summing up, the community is the key to identifying the best spots to meet with other people to conduct business or create creative things. For instance, Frankfurt is a city that is not generally acknowledged as being very creative. One expert mentioned that this is due to the community which is not so strong in Frankfurt (F3). The people are moving in and out and there is no feeling of belonging to the city. But to do creative or innovative things, communities and networking are very important:

> You can't have that kind of technology revolution that we are having here [in Chicago] without a social component. People don't talk about that very much but it's hard work to organize the meetups, even things like getting people to go. Getting people to mix. Because these communities often times don't like for instance urban planners and architects and technologists. I do a lot of that. Mixing people, inviting them to meet at a restaurant or something like that. (CH 2)

Especially in the tech scene, collaboration is important, be it face-to-face or digital. Code and experiences are shared to support each other. One example from Boston reveals the importance of the open source mentality and sharing within this community:

> [T]he local group had broken up the flu in January here. The mayor was like everybody gets a flu shot. 'Here is the location where you can get for.' And it was just a list. 'Ok, what we can go with that?' This group was like: 'That's not very user-friendly. I can't say what's the closest location to me'. Chicago gets a similar thing. So the local group got the Chicago code. We set up on a server here... Within like 40 hours we had working app with maps. Having that kind of structure in that community that collaborative ability made providing a good solution. So much faster than it would be without that infrastructure. Without that community, it's like 'forget to build that ourselves'. (BO 2)

Hence, this community is productive due to local coworking and global digital exchange. The important factor is that they are sharing their knowledge with open source principles to help each other. This is a kind of co-learning that is also described with regard to upcoming social learning or hack labs in many cities (Carrillo et al., 2014). One of the most well-known communities is the Open Knowledge Foundation, heading OK Labs in cities all over the world. They have built a community of people that are using open data and open source code to develop new tools to solve local problems and to share their results with the global community similar to the case in Chicago and Boston, working globally on a digital level and still meeting face-to-face in each city lab.

One expert from Toronto reported about an interesting phenomenon of communities in the music industry:

> Some of my post docs did some work looking at the music industry... And comparing those industries, how the kind of music and musicians and how they work in Toronto vs. how they

> work in Halifax. And one of the things that was a really interesting one, is that musicians who are working in Halifax actually do more stuff. They have a broader scope. Because there are fewer of them. So if you are a bass player. You might play with a jazz trio one day, you may play with a heavy metal band a couple days later and you might be working on a documentary film the other. That is because there are not so many bass players in Halifax. In Toronto, when you play with a jazz trio, you only play with jazz trios. You will never ask about or even know about the other options. Because Toronto is large enough and actually silos more. If it gets bigger people tend to focus on the one thing they do and ... the music industry ... is a good proxy for a lot of the kind of creative industries. What you see is that people tend to silo up and they are competing more against each other and they don't tell each other about opportunities. Whereas in a smaller market like Halifax the people have to collaborate more than to compete. In a city like Toronto, you have this siloing. People who are working in IT, doing stuff and meeting other people from IT, but they are not really meeting with designers if you would think about industrial design or whatever. The larger places become the harder those things become ... But I also think, because Toronto is so large it has the siloing even if you have more spaces that are kind like that it is hard for them to be effective. (TOR 1)

Many of the interviewed experts related the fifth hypothesis to knowledge workers, the tech scene, and coworking spaces. According to Saskia Sassen (NY 5), next to coworking spaces, there is also a trend towards small performance spaces in the art and entertainment industry:

> What we see is the return of the small performance space as the hottest space. So when you go to the Bowery which is a totally degraded place. If you go at certain times at a day, mostly late at night. It is full of people. You know, they dance, they do sound poetry, all kinds. It's not just to dance or to rave. It's also the performance of arts of theatrical etc. This area has really grown... Once that full Madison Square they probably came from the suburbs. The typical hip New Yorker is not going there. They are going to small performance space. So it is a sense of also yourself representing, it's a form of creative zone than sitting in an audience applauding the best-selling musician of blabla... But the performance festival in New York is the hottest festival and it's all counter it's not in the center. (NY 5)

Conclusion: Physical space for face-to-face interaction

With regard to hypothesis H5 (*Physical space for face-to-face interaction is important for an Informational world city*), the results definitely confirm that face-to-face interaction is a main requirement. Informational cities should not be cold places like Candle Square in Boston, where Microsoft Research, Google Research, and the Computer Science Complex at MIT have built their glass complexes. A culture of meeting and meeting places should be established. For this, the infrastructure of a city is important: Wi-Fi in cafés or open spaces enables meeting places to also function as working spaces, allowing people to work or get in touch with others at informal places. Examples for this are Berlin, where many people connect to the Wi-Fi of cafés to do their work, or São Paulo, where the special lunch restaurants offer the opportunity to get in touch with people working in other offices.

Further, the labor market has changed in the recent years. There are many people that are working as entrepreneurs or freelancers and are able to work at any place with internet access. Nevertheless, they still often like to use shared offices or coworking spaces. Each city that was investigated offers coworking spaces. They vary in their innovation profile but as a minimum always offer a place that opens up the possibility to get in touch with other people. Globally, tech communities and coworking spaces follow a very open culture. Networks and workspaces are easy to find online through a simple Google search, but becoming involved in the community may require much more effort, e.g. in the music industry in LA or with businessmen in Asia. This is more of a 'speakeasy' principle or a VIP circle that includes new people by invitation only. Additionally, the size of a city can result in a siloing effect. Only selected people are able to get in touch with each other and are likely to work together.

Finally, the examples show that places tend to become smaller, with a more familiar climate. This is true for innovative coworking spaces that explicitly invite the people not only to share hardware but also their knowledge and experience. In the creative sector small performance spaces are locally organized and in the case of New York, attract the hip New Yorkers more than the mainstream bestseller musicals do. It is, however, not clear whether the informational world city accelerates this movement of smaller and more familiar meeting places or not. The experts interviewed mentioned that these places are not known if you are not into that community. Hence, the SERVQUAL evaluation found a lack of such places, even though a simple Google search showed that interested persons can easily find them. The interviewees also mentioned that an infrastructure like Wi-Fi connectivity and coffee house culture are important. Some have mentioned the example of public libraries which can serve as a meeting and working space. Additionally, the library offers access to a huge content infrastructure. To which extent the public library is acknowledged as a provider of content infrastructure and as space for face-to-face communication will be investigated in the two following hypotheses.

H6 Fully developed content infrastructure provided through digital libraries
Libraries in informational world cities inhabit a special role as a content provider, be it through digital access or as physical space for face-to-face communication. Hence, to foster lifelong learning access to digital and physical libraries is crucial and libraries are acknowledged as part of the cognitive infrastructure of a knowledge society (Stock, 2011) in addition to the knowledge output which was

investigated in a previous section. The use and access of digital libraries represent the consumption of knowledge resources (Stehr, 2003). Furthermore, digital libraries do not only meet the needs of the knowledge society but are information service providers for companies beyond that. To review the importance and state of the art of digital libraries in the 31 cities, the following hypothesis will be investigated:

H6   A fully developed content infrastructure, e.g. supported by digital libraries, is a characteristic feature of an informational world city.

During the interviews, many experts have stated that it is not important to have digital content support from any particular institution located in the city (19 experts), but it is important to have access no matter where it comes from. The common argument was that through the ICT infrastructure access is provided to all of the information in the World Wide Web. Interestingly, the SERVQUAL evaluation does reveal that there is a lack of such institutions according to the average perception of the experts interviewed (Figure 6.24), since for most of the cities, a negative quality value can be observed (average 1.34). Accordingly, 17 experts stated that there is a need for improvement and that their public library is not the ideal content provider yet. Aside from the content and digital access, the ability of the population to use this content was discussed. In connection with the location aspect, ten experts emphasized the need of open data, open source and open access to city and regional sources. To digitize regional data would be a support that could be encouraged by public libraries (three experts). Finally, the high cost of licensed

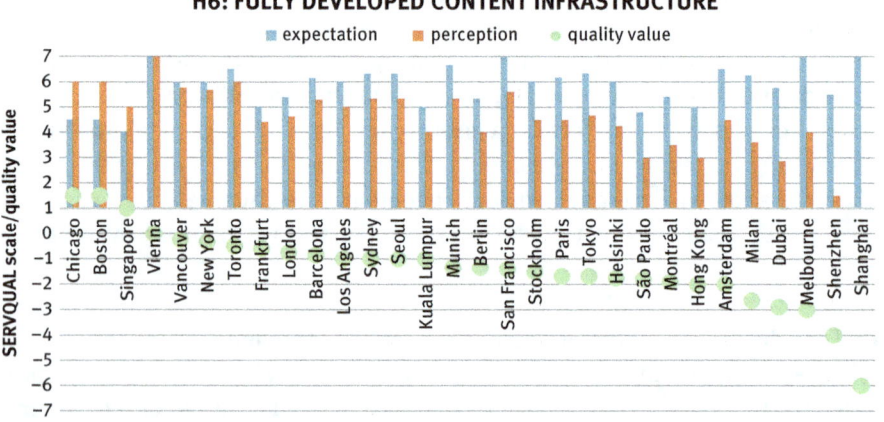

Figure 6.24: Quantitative interview results according to SERVQUAL (quality value = perception − expectation) for H6 (A fully developed content infrastructure, e.g. supported by digital libraries, is a characteristic feature of an informational world city).

material will hinder many projects (four experts), and a network of libraries would probably be the only option to offer the access to the public (one expert). Summing up, the interviewees on the one hand discussed the access to full content through any provider and on the other hand through the public library. In the following, I will discuss the arguments by the experts interviewed and briefly present the digital library services of the 31 cities investigated. The role of the public library as physical spaces will be discussed afterwards in the next subchapter.

Access to full content database
A digital library can be understood as content collections on behalf of any community or as an institution or service provided by librarians (Borgman, 1999; Levy, 2000). They are more than metadatabases, as they provide full content of diverse media in a machine-readable and processable format (Mainka, Hartmann, et al., 2013). To offer such a service for the public, as in the case of Reykjavik, which offers access to all scientific publications published in specialized journals on Elsevier's Science Direct for the whole population of Iceland (van de Stadt & Thorsteinsdóttir, 2007), brings with it high costs for the public institution. Considering this, it can be said that "libraries are very important content providers" and that "[d]ue to copyright and other licenses, no other institution will be able to offer such content" (BE 1, own translation).

In Singapore, for example, the library is distributing the costs among the national network of libraries (Chellapandi, Wun Han, & Chiew Boon, 2010; R. S. Sharma, Lim, & Boon, 2009). There is no branch that has to apply for databases on its own. Of course, Singapore is a special case since the city is a nation state and benefits greatly from this governmental constitution. On the city level, we can find a number of other networks of libraries that cooperate and share resources, e.g. in Amsterdam and Paris (AM 7, PA 2). In contrast, public libraries in London mostly have to bear such costs on their own because they are not organized as a network across the whole city, but merely they are just connected within their borough (LO 4). In Munich, the importance of sharing resources is acknowledged and the library is therefore connected to a diverse set of networks across the nation and Europe (MU1). One interviewee from Berlin (BE 7) has emphasized that in Germany university and regional libraries are accessible for the whole population and not only for members of the university. Thus, full content access to many databases does not have to be provided by public libraries to allow the public to access such information. A similar openness is, among others, also encountered in Canada, while the access is very limited in some other cities. For instance, in Dubai, the university libraries provide an enhanced access to information and databases with full text (DU 1), but these are not accessible to the public (DU 2, DU 3).

It is certainly questionable whether it is important who is providing the full content. One expert from Toronto (TOR 1) argues as follows:

> I think you can be a fully kind of informational city by having content infrastructure but it could be private. Is that the best? No. And I am all for open data and all kind of the other things. But… you can qualify if you have the content infrastructure and have that kind of information and it might be in private hands. So, the openness of it is, I think, of help but I don't think the openness is absolutely required.

The completely opposite, however, is currently happening in France. Paris has undertaken a lot of projects to offer eBooks and online content to the public. According to one expert from Paris (PA 2), the French people love to rebel against the rest of the world, by, for instance, fighting against Google and starting their own, similar projects. While, in the end, Google is likely to win this battle of content providers, the French government nevertheless tries to offer their own digital libraries with full content. Currently the public library in Paris provides more online services than full content services (PA 3). In Montréal, also a French-speaking region, a similar development can be recognized: "There are many plans but there are too many barriers especially on the licensing of materials that make it difficult to offer full content" (MO 5, own translation). In Montréal, there is an ongoing project to publish eBooks by authors from Quebec written in French language, but these authors are afraid of having to give up their copyrights, which makes it difficult for the project to grow (MO 6). This project originates in the Quebec region, where there is a general preference for pushing local authors and the local language (MO 4, MO 5).

In addition, censorship of information is a problem in some regions, for instance in China. One expert from Shenzhen (SHE 2) has described the digital content market as follows:

> It can be very frustrating to do research because a lot of sites are blocked or limited. It is not just about the absolute censorship; it is also about knowledge competitors. A lot of sites are slowed down so that it comes not worth using them… There was a fight between China and Google. China had a selfish interest in holding off Google because they wanted to help the home grounding companies. If you want to get global information that can be very hard. You cannot access 'New York Times', 'Bloomberg'… You can see that articles exist but you cannot read the full text. [At the] illegal market you can get books that don't actually exist. It is easy for print shops to exist. You send everything to be printed to print shops because it is very cheap. They then have the drawings, competition rendering etc. They collect this treasure of information, sell it on DVD or as a book on the black market.

Furthermore, offering online full content services is always tied to high costs and licensing questions. As one interviewee describes, "[w]hen you have access to those databases it is amazing what you can do. I have access to a database [at the office]. When you gave my name in it, you will get a fifty-page report of everything that I have done… They would never give that out for free" (LA 1). In addition, the usefulness of access to full content databases for the public needs to be considered. Los Angeles' public library is offering selected services for their users which they think they meet their needs: "We offer many databases. We don't have to say offering Bloomberg real time but we do offer internet access and I think that within the libraries we offer instructions. We offer instructions on using different information sources… We have some specialized subscription databases that offer a… real selection of full-text databases. They can work better for the general public" (LA 2). Digital libraries are not used by the whole population (BA 3), as, for example, archives of Ph.D. theses or other content is not mainstream content and therefore only interesting for particular individuals working in specialized fields: "Only experts know the existence of them because they have no use for general public" (BA 3).

"The thing what you need is good broadband access. It doesn't matter where the digital library is located. For instance, the digital public library of America dp.la, Boston is the hub of the development of the dp.la. A digital library is not in any place" (BO 3). As can be seen in this quote, the ICT infrastructure is acknowledged to be highly important, since it offers access to all the global knowledge and not only to the local knowledge. The local knowledge is primarily important for cultural issues (MO 1). The location of the library can be of interest when it comes to its own digitized documents. For instance, the public library in San Francisco is making available historical geographical data of the San Francisco region. But the access to this data is open for all people with an internet connection (SF 1). Another example comes from Canada, where hockey is a very prominent leisure activity. This is a regional phenomenon and to open a database on best ice rinks is expected to be beneficial for the community:

> There is kind of APIs that you can plug yourself on the city data… There are ice rings in the city there you can go and play hockey during the winter. And the quality of the ice is compared by the city… There is a guy that created a website to basically show on a google map the quality of the ice in all of the parks of the city. So I mean this is a kind of need application of that. (MO 2)

The libraries in Barcelona, for instance, offer access to open data such as maps, but access to music or books is not fully developed (BA 3). The problem of slow progress in digitization and open data in some locations could be due to the skills and experience of the decision makers. One expert from Amsterdam (AM 6) argues that "people who are responsible for this are not really digital-minded." Free flow of information and national policies that enable this are important to advance the development (SG 4). One pioneer in offering open data and full content on city-related content is Vienna. The urban planning department offers a metadata and full-text retrievable database with all content and data published by the city (wien.at, VI 1). According to one librarian from Montréal (MO 6), libraries should strive to act as a mediator between data and the society:

> I am a librarian and I think librarians or archivists work in their library or in their active center, but not work with people. They need to be more engaged in the society to bring together explicit knowledge with tested knowledge. They need to be a more intermediary resource for the society, especially to use data. Open data need to be mediated, curate or mediated by some people, maybe by [information scientists], but also by librarians and archivists, because ordinary people don't know to use data. (MO 6)

Libraries should help to bridge the gap between non-information literate and information literate people in the knowledge society (MO 6). The topics open data and free flow of information will be discussed separately in subchapter 6.2.

Merely offering databases and online content does therefore not mean that the whole population is able to use the data or even access it. According to the interviewed experts, public libraries should offer instructions on how to use digital content, give assistance in information retrieval, and as a minimum offer digital equipment and Wi-Fi connectivity at their branches. The availability of technological devices and internet connection is related to the physical library which will be discussed further in the next subchapter. However, a fully developed content infrastructure is not the be-all and end-all of the twenty-first century as demonstrated by the example of Seoul. In Seoul (SE 1) one expert mentioned that South Korea plans to digitize all school materials and books by 2015. Accordingly, all educational content would be available online with additional multimedia hyperlinks and features for self-regulated learning (Kim & Jung, 2010). While this was the original idea behind this governmental plan, due to recent findings it has been adjusted towards using online and paper materials in parallel. Having access to materials anywhere and anytime does not automatically enhance the quality of learning. According to an article in the Washington Post, "[a]bout one in 12 students between ages 5 and 9... is addicted to the Internet, meaning they become anxious or depressed if they go without access" (Harlan, 2012). Thus, the digital transformation can also cause negative health issues in addition to the exclusion of parts of the population:

> I think digital libraries are subservience to people knowing how to access information and read. Just because is it a digital library it doesn't mean that you know how to use a library. So the first question is: Who is training people to use libraries? How do you access digital information? Why do you believe that everything is on Wikipedia? ... There are lots of libraries in LA. We have a huge construction of libraries and I think lots of people use them. (LA 3)

Digital library investigation of 31 informational cities

In a previous project on public libraries in informational cities (Mainka, Hartmann, et al., 2013), the 31 cities were investigated with regard to their digital and physical services. In this investigation, online services of public libraries have been identified and the availability of these services was tested. The underlying indicators are listed in Table 6.6.

Table 6.6: Indicators of the digital library. Source: Mainka, Hartmann, et al. (2013).

Group	Indicator
Web-OPAC	Web-Online public access catalog (OPAC)
	Web-OPAC in English
e-documents	e-journals
	e-books
	digital images
	audio books
	music
	e-magazines
	videos
	newspapers
	bibliographic databases
	other e-resources
databases with access to full papers	databases with access to full papers
guides	video guide
	podcast guide
	seminars
	text documents
	FAQ
	other guides
international access	website in English
digital reference services	e-mail
	chat / instant messaging
	SMS
	web form
	Skype
social media	blogs
	Facebook
	Twitter
	Flickr
	YouTube
	other social media
apps	apps
own digitizations	own digitized documents / collections

Among the digital services of a public library, full-content service is just one aspect. However, to identify the availability of full-content information through the public library service, the areas of e-documents (e-journals, e-books, digital images, audio books, music, e-magazines, videos, newspapers, bibliographic databases, other e-resources) and the library's own digitized documents are of interest. These materials can be made accessible through the web. To make these documents searchable, a Web-Online Public Access Catalogue (OPAC) is needed. Additionally, the availability of an OPAC in English extends the access of the materials to users with limited language skills (as long as the user is able to read and write in English).

30 out of the 31 cities offer an available digital library (retrieved in December 2012). For Dubai, only a website including an English translation with some information about the library can be accessed. Nevertheless, the identified 30 digital libraries support all a Web-OPAC in their respective national language and 22 of them are also available in English (Figure 6.25). The website of Vienna's public library is additionally available in Sign Language for users that are not able to read (VI 3). The access to e-documents varies between the types of documents. The most common type is e-books, followed by e-journals and bibliographic databases. Videos are available only in 14 libraries and digital images in 13 libraries.

In addition to the availability of e-resources, the ability to use them is an important aspect, as mentioned by several experts. Therefore, the types of instruction (FAQ, text documents, seminars, and video guides) for using the digital library services were scrutinized. The results demonstrate that such instructions and guides are not a common standard for all libraries (Figure 6.25). FAQs are the least common instruction guides (provided by four libraries). The most common types of instruction are text documents (ten libraries) and only seven libraries offer two different types of instruction. It should be noted that it does not reveal anything about the quality of these instructions, i.e. whether they are good or bad, it merely demonstrates that user guides on how to use the digital library are not yet a standard.

According to the statements of the interviewed experts, there is no common development of digital libraries in the twenty-first century. One expert from São Paulo (SP 3) estimates that the digital library in his city is still in a twenty-first century development stage. For him, the digital library landscape in Brazil in general is like a desert. As in other cities, full content access is, for the most part, only available for researcher and students in technical or university libraries, which means that only a certain group of the population has a good access to knowledge and information in São Paulo. In Milan all experts also mentioned that the digital library service is not sufficient, characterizing it as more traditional

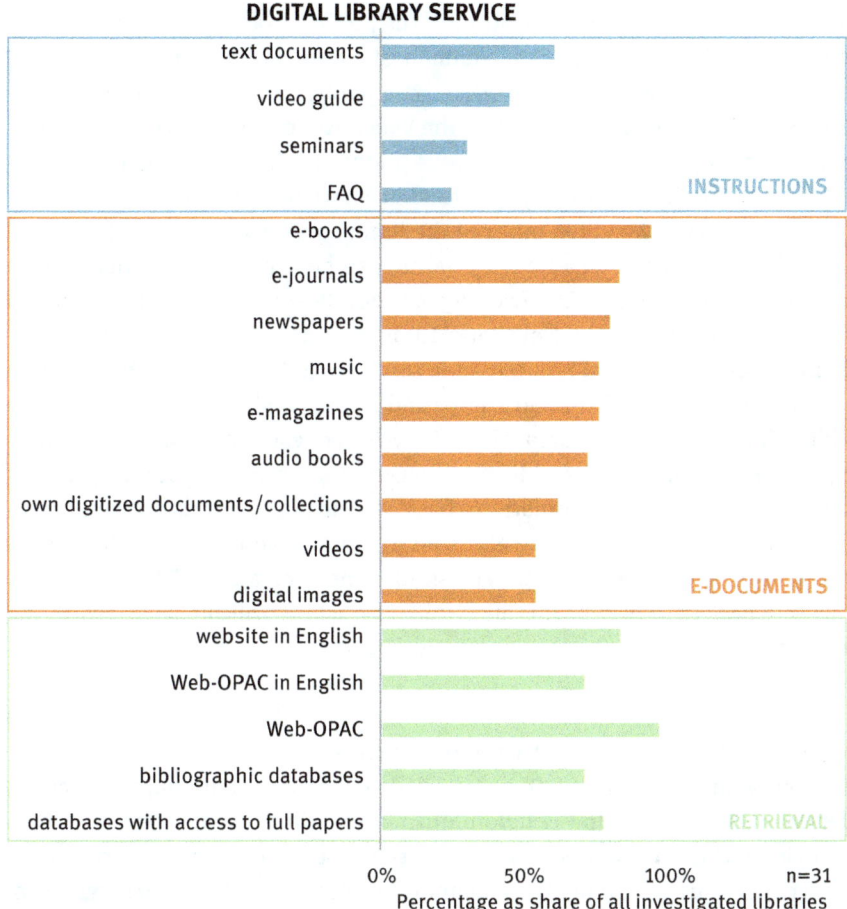

Figure 6.25: Digital library service provided by Informational world cities' public libraries. Data source: Mainka, Hartmann, et al. (2013).

and not offering much online access to full content. Again, however, the university libraries provide superior access for researchers and students.

Still, the findings demonstrate a digital transformation in many cities, even if this transformation could be enhanced and accelerated further as, for instance, mentioned by experts in Stockholm (ST 2, ST 4). In New York, the public library is just starting to open the access to more digitized content: "The NY public library system has an amazing e-book infrastructure but it is like in a pilot phase" (NY 2). To access relevant research content in London, physical presence in the library is required, as the digital access is very limited (LO 5), but there is a move to enhance digital content (LO 6). A problem acknowledged by one interviewee in Barcelona is that there is a lack of standardization: "[The public libraries] offer many digital repositories and archives but no common standard or project is developed" (BA 1). Only in one case an interviewee from a public library said that the library has good funding. This was the case for the Vancouver public library, which offers a great deal of digital content: "I'd say we're pretty good in compared, for example to libraries in Britain... We're very well-funded, so we're able to supply" (VA 3).

According to the weighted index of the library survey of Mainka, Hartmann, et al. (2013), this advantage of good funding can be verified by the high scores that were reached (Figure 6.26). Looking exclusively at the support of the digital library, the public libraries of New York, Toronto, and San Francisco are the ones with the widest range of online-accessible services. At the bottom of this ranking, the public libraries of Tokyo, São Paulo, and Dubai can be found. According to one interviewee, one possible reason for Tokyo's bad result might be that libraries and Japanese society in general love to treasure their printed materials, regardless or even in spite of the population's habit to read entire novels on their smartphones (TO 3). Most Japanese are also said to prefer sending a fax instead of an email, revealing their resolute stickiness to printed material (TO 5). However, printed materials are part of the physical space which will be discussed in the following subchapter H7 (Library as physical space).

Conclusion: Fully developed content infrastructure
Following Dehua and Beijun (2012), knowledge in an informational world city is understood as a service that is ubiquitously available through digital networks and devices. These networks can be represented as digital libraries providing content collections on behalf of any community or as an institution or service provided by librarians (Borgman, 1999; Levy, 2000). Summarizing the cities investigation no common development of digital libraries in the twenty-first century can be discerned. Furthermore, expert opinions diverge as to whether such fully developed content services with access to advanced databases, full text, and specialized content is even needed. Some have argued that it would be sufficient to have private providers for such services or to only offer limited full-text service for the public. As a counter-example, the French-speaking communities in Paris and Montréal strongly advocate local efforts to offer digital libraries with the intention to preserve their culture and language. According to the index of library services, the digital libraries provided by the public libraries of New York, Toronto, and San Francisco are the most advanced with regard to the diversity of supported services. To offer e-books and e-journals has become somewhat of a standard service in recent years, which is why more than 70% of the digital libraries investigated offer them (see Figure 6.25). A feature of local digital libraries supported

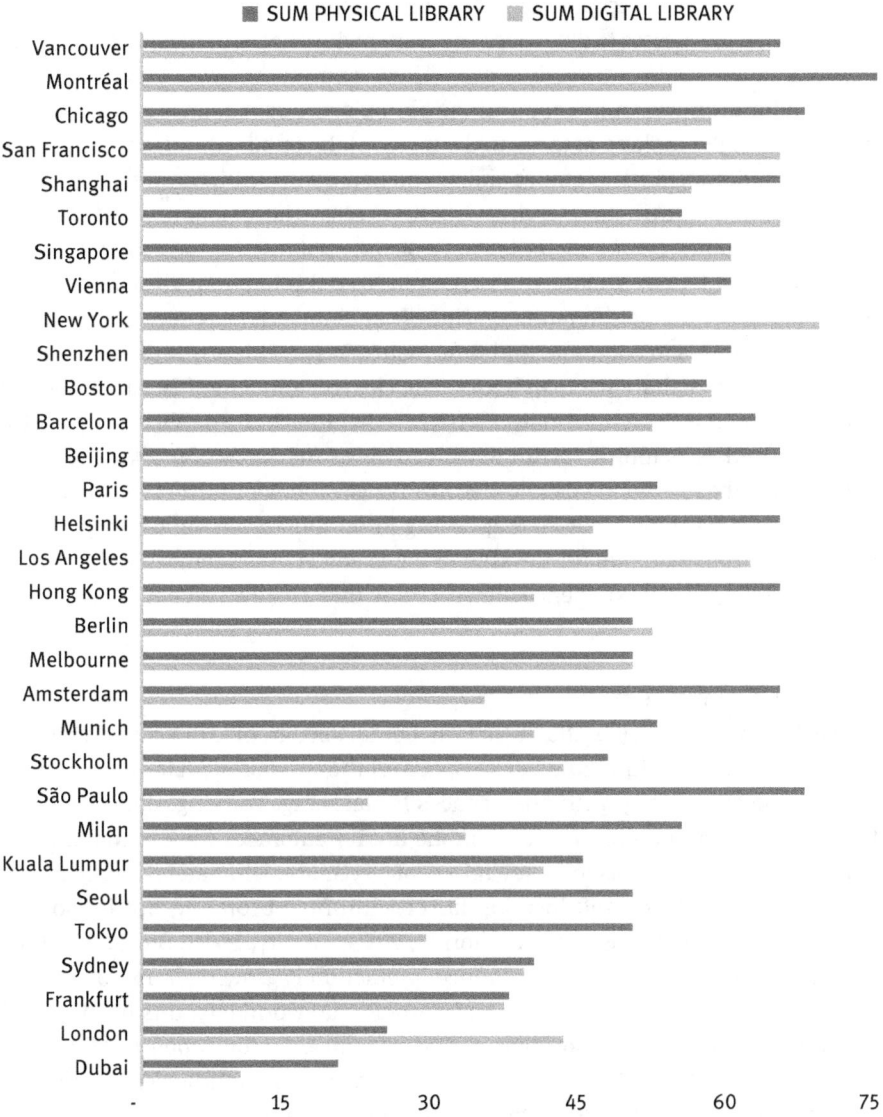

Figure 6.26: Ranking of informational world cities' public library service represented in descending order according to total scores reached in digital and physical library service. Data source: Mainka, Hartmann, et al. (2013).

by public libraries is the dedication to its role as a repository of open data, for instance, to open the access to digitized local content.

Furthermore, access to databases and content does not automatically mean that the population is able to use this service. Thus, many interviewed experts stressed the need for information literacy and the library's role as mediator of information literacy skills. But looking at the actual support of the public libraries investigated, there is a noticeable lack of guides and instructions, with seminars and FAQs being the least common instruction services. A better access to information would enhance the library as a place not only for books in shelves but also for the community (BE 6). During the interviews, many experts emphasized this importance of the physical library as a place for the community: "The role of libraries is important for people to meet, to get access to computer and web connection ..., cultural activities in some libraries. It's more than just loan books. The city [government of Helsinki] tried to close a library but the Finns protested" (HE 4). In Munich the public library is also acknowledged as a cultural institution which meets the needs of the whole society and not only those of young people who are using technology (MU 1). "Actually, one of the interesting things that are happening in our library [in Vancouver] is a big move to... becoming a hub for the kind of the maker environment with 3D printers and where people can experience things. I think that's really the next..." (VA 4).

H7 Library as physical space
Space within the library is changing from being a big archive of printed material towards increasingly functioning as a space for the community. Library buildings support the community with open spaces for learning, working, and collaboration. In addition, the presence of a public library enhances the economic value of a city and may serve as local revitalizer of city space. Ultimately, public libraries are considered as a soft location factor within the economy, but for society they serve as basic infrastructure (Florida, 2003; Landry, 2008; Stock, 2011). As already discussed, space for face-to-face interaction is growing in importance in an increasingly digitized world. In an informational world city, this need should not only be served by private places like cafés or bars, but open public space should also be offered by the city. So while information and books are increasingly available through digital libraries, the physical library assumes a new role. Whether the physical space of a library has adjusted its role in line with the needs of the knowledge society of the twenty-first century will be investigated with the help of the following hypothesis:

H7 Libraries are important in an informational world city as a physical place for face-to-face communication and interaction.

Public libraries are understood both as a public space and as a cultural institution. During the interviews not all interviewed experts were familiar with the idea of a public library as meeting space. In total, 24 experts mentioned that visitors ought to be quiet in the library, or that people just go there if they like to lend a book (two experts). The SERVQUAL evaluation reveals that on average per city many experts do not have a high expectation of public libraries serving as physical space for face-to-face meetings (Figure 6.27). For instance, the average expectation scores for Melbourne, Hong Kong, Tokyo, and Shenzhen were three or lower. In 23 cities the perception of the real use of the public library as collaborative space is lower than or at most as high as expected. For 17 cities the quality value is negative, which indicates that libraries could improve their space with regard to serve as a place for face-to-face meetings. Furthermore, the experts stated during the interviews that in some cities the library as a physical place is only important for a certain group of the population, for example only for children and students (four experts), which is dictated by the opening hours largely overlapping with the most common work hours (LA 1). For them, the library is important as a space for working and learning (five experts). In China, the situation is a little different because "the bookstores are more important than the libraries. People are using book shops like a library. They sit there and read the books and spend the whole day there" (SHE 1). In Shenzhen, for instance, the opening hours of the library also overlap with the common work hours, which

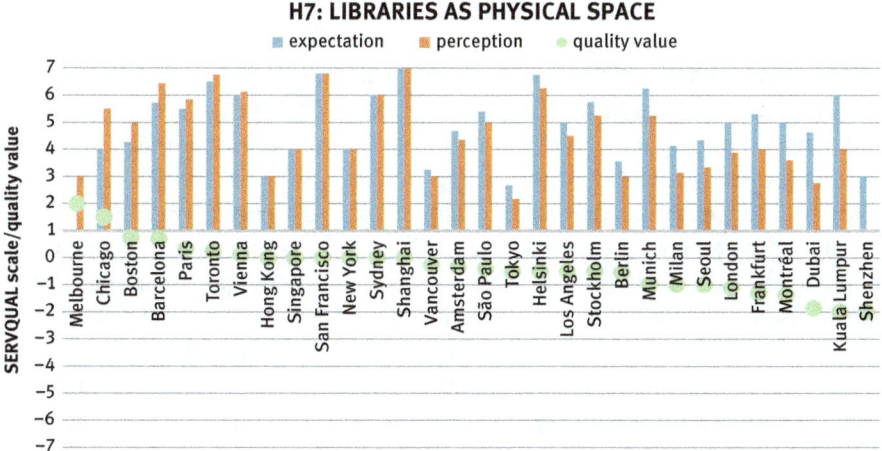

Figure 6.27: Quantitative interview results according to SERVQUAL (quality value = perception − expectation) for H7 (Libraries are important in an informational world city as a physical place for face-to-face communication and interaction).

makes bookstores automatically more attractive for the public. Libraries may also serve as a place for social inclusion (four experts). For instance, people that do not have the access to technology and internet, but also people who do not know how to use ICT, need the library to access information (four experts). Thus, libraries can be understood as a bridge between the digital and physical world. Three experts argued that the physical library has lost its role as an information provider, which is why all collections and data should be available online (six experts). One expert also stated that "[l]ibraries will change to an office for librarians who work for digitalization processes" (MI 1).

On the other hand, 44 experts mentioned explicitly that it is important for a physical library to serve as space for face-to-face communication. It has always played a role as an event location (14 experts) for book readings, readers clubs, events for children, and much more. In addition, a library should offer space for collaboration (eight experts), as some have started to do with maker spaces, which offer 3D printers and other hands-on equipment. Libraries have become a community place (four experts) but there is still space for improvement (six experts). However, as public institutions they often have to struggle with funding. Ultimately, a definition of the role of a physical library in the twenty-first century is needed (four experts). And a total of 22 experts argued that there are other places than the public libraries which offer space for face-to-face interaction. Therefore, in the following, I am going to discuss the arguments and examples given for a physical library's role by the experts interviewed. In addition, the results of an investigation of spaces within the physical libraries of the 31 cities will be discussed.

The physical library's role in the twenty-first century
The current state of the public library and its role as a physical place reveals major differences throughout the world. In informational world cities, we can find libraries that are used traditionally as learning and working space, but also as innovative and creative centers that foster collaboration. As a public institution, public libraries have an educational mission (MU 4), and are thus aimed at the whole population. Traditional libraries offer vast collections and quiet working spaces, mostly used by people who are reading, doing research, or learning. In Beijing, for example, very crowded city with high costs of living, the space within the National Library of China is highly frequented. Even though the space within the library is very crowded, users are working quietly (Figure 6.28, right picture). Other libraries offer traditional working space and shelves with self-service materials as well as space for meeting and collaboration. Figure 6.28 shows on the left the Stockholm Stadsbibliotek, which offers a mixed space for borrowing materials, working, using PCs, asking a librarian and a staging events like talks or music performances.

Figure 6.28: Physical space of public libraries. Left: Stockholm Stadsbibliotek. Right: National Library of China in Beijing. Photos: Agnes Mainka.

> As much as I am a huge fan of libraries and books. I think the standard model of the library was never about a workspace. Librarians tell you to be quiet and stuff like this. Not made for brainstorming, not made for collaboration. It's a place made to come to research and relax. Books are still critical for libraries. Libraries have a job to modernize and adapt to a culture that wants to be more collaborative and that... But I haven't seen libraries really do that a lot. They try to fit the digital stuff into the old framework of quiet research and kind of things like that. (BO 2)

As stated above, the traditional library serves the needs of the whole population and includes people who otherwise would not have access to the information provided:

> The physical space of the Boston public library is a great library and it's always played a role as the place where the lower social classes and recent immigrants could educate themselves and could come and leave the ghettos and begin to see a different world. The cultural experience of sitting with lots of other people and reading books is something that kids couldn't have sitting at home in front of their screen. That's still important. (BO 3)

Additionally, people who cannot afford their own internet access can make use of the free Wi-Fi at the library or the technology offered by it (LO 4). Public libraries serve as an information provider, and this can be done by physical collections or through the access to PCs and Wi-Fi at all branches. For instance, in New York, it is "a huge effort in the city to take those areas where people don't have access to digital content and give them access to [it]" (NY 2). In LA, "Wi-Fi is available even [when] the library is closed e.g. Sundays" (LA 3). Even after opening hours people sit outside of the library building, using the still-available Wi-Fi (LA 3).

Self-education and life-long learning are two aspects of a library that make them as important as they are for the knowledge society (Stock, 2011). One expert from São Paulo (SP 4) stated that education is the key for libraries to become more attractive:

> I think we lack good basic education, fundamental for kids. This is something that's lacking here, so that's why... only certain degree... go to the libraries, ... theaters, ... [and] movies. These are expensive things here in Brazil and books also are very expensive ... And people usually don't have the habit to read ... They have a habit of watching TV... but that's it ... No, it's something that has to be put more effort. (SP 4)

In São Paulo and in throughout Brazil the government is planning to establish a library next to each school which should support as many books as the school teaches children (SP 1). At this stage, this is merely a plan and the future will show if this comes true, but for children and the ordinary population, such institutions are important (SP 1). Books and reading are important for education, as, in general, the population in Brazil does not have the habit to go to the library to meet someone (SP 2), preferring to go to the shopping mall.

The issue that has arisen today in many cities is that a lot of people can work quietly from any place they like to. They do not have to go to the library building. This may cause many buildings to not be visited very frequently by the citizens. In cities where living space is very expensive, the library can offer space for students to study and work (PA 5). However, most libraries in high-cost Paris, for instance, remain largely empty. It seems as nobody is aware of their existence, with the exception of the *Bibliothèque Sainte-Geneviève*, which is more a kind of a meeting place and attracts a lot of people (PA 5). To fill the physical library with life, the space will have to be used in other manners. One expert from Boston (BO 3) reported that he is giving his classes at the library: "We used the library as place for classroom because nobody was using the library for studying or meetings."

In contrast, in other cities, the number of people using the physical library is very high:

> I don't know how many people have registered in a library. It is like a club. In short, it is the biggest club in the city after the football Barcelona. Almost every one of us goes there. Libraries are always full in Barcelona. Libraries also play a big role for integration. People that came from foreign countries. One of the first points of contact are the libraries. They make relations. Maybe they find internet points. It is also easy to follow

programs like language [education]... Is maybe one of the vehicles to make that model of neighborhoods... I think that everyone knows where the public library is... I think we have a good experience. (BA 6)

In North America, public libraries also support immigrants with language skills or help citizens with writing job applications, as several experts stated. Their role as a sole information provider is therefore extended with further public services. Additional services are, for instance, lectures and seminars for illiterate and non-information literate persons. In addition, "libraries are one of the bridges between the digital and physical city" (BA 1). Information and content, not only in the case of scientific information, need to be discussed and shared with other people (BA 1). Hence, public and other libraries are important places for knowledge exchange. In Frankfurt and in Munich the libraries have programs that inform the citizens, for instance, about social and political developments (MU 4, FR 2). Furthermore, libraries can serve as a hub for information and news: "The library in Beverly Hills is a gathering place. Hurricane Katrina in New Orleans, they found out that the libraries were the places where people went to get the information. Mean when you really get all the support systems the library was the center place to go" (LA 1).

As in many other cities, in Berlin the government plans to cut the budget for physical library spaces. In San Francisco (SF 1) one expert reported that people "freak out" whenever any branch is going to be closed. "They get very upset. They really want their little local spot to meet people and talk about things. And they don't just freak out. They give money. And we have one library for every two miles" (SF 1). The library is for the public and therefore the San Francisco library is working closely with their customers: "We are currently in the process... We revisit our open hours depends on what the public wants from us. We would change our hours in respond to that and we are expanding our hours in many locations and responds to public feedback. They want us open more" (SF 1).

It is important to define the role of the physical library and, in addition, to communicate this role to society (BE 6). Today, in Berlin the public libraries are used as meeting space only by the elderly population, who read their newspaper and meet others of their age (BE 9). In Vienna, most of the public library branches are very small. In the past, they used to be established as part of almost every social housing project (VI 2). Nevertheless, the central library is larger and the chief librarian, Markus Feigl, has started to redefine the role of this library into a community place, one not only for the elderly. His vision was to reduce the number of printed material hosted in shelves in the library to offer more space for the people to meet and to exchange knowledge (VI 2). This is a shift from the traditional library towards a modern interpretation of a library's physical role.

But this needs to be communicated carefully. He stated that the media reported on his idea as if he wanted to throw away the books and even compared his development plans with Nazi book burnings. This is because "books are also a cultural good and you cannot implement your vision of a community space library that offers more space for people than for books immediately. This hurts a lot of people who feel very much attached to books and they will protest against this development" (VI 2, own translation).

In other cities, the library space has already been redesigned for different kinds of activities and users. In Helsinki, for instance, there is a library in the neighborhood Espoo which offers many games for young people and which is a noisy community place, but also still offers space for quiet learning. As the library is open until 10 p.m., the library is not only available for students and children (HE 1). The university libraries, for instance in Berlin, are used by students to get in touch with other students and to flirt, foremost (BE 3, BE 7). To make the library space more attractive for the public in the Eastern part of London they have been renamed "IDEA STORE." Still, "libraries are not a natural place to go, not the first place to meet" (LO 3). In Barcelona, public libraries are "spaces to do performance and other activities in parallel and ... offering cultural activities" (BA 6). One expert from Chicago (CH 2) stated that library space "is not just face-to-face communication. I think its face-to-face collaboration. They offer the most important public space in the city, more important than parks, more important than the public way outside."

Libraries have started to implement "hands on" seminars and projects for their users: "[In Chicago] we run classes there, we do arts programming there, we actually have startups that are getting free office space in our libraries and then they have to give back to the library. Run seminars and things like that. That's a great setup actually. Free rent plus these really talented individuals have to give something to the community" (CH 2). The maker space is where these seminars are given at the Chicago public library (Figure 6.29). The Toronto public library also has some business development and incubation taking place in its halls (TOR 2). They also offer a printing machine to allow the public to create own books (Figure 6.29). But the question is: Should this really be the role of a library? In Boston, for instance, these kinds of activities are more related to locations near the universities in Cambridge:

> We have some amazing libraries but... you might see that more in coffee shops. You might see this in Cambridge Innovative Center and its right next to MIT. And the whole building is full of startups, adventure capital firms, little research divisions of companies, and it must have a hundred conference rooms. The walls are painted with whiteboard paint. There are kitchens, and food, and coffee and you can just go there and work and collaborate and get a room whenever you need one. It's producing that kind of environment. It's really fast internet connection. The whole environment is set up for that kind of digital economy and collaboration and working together. And just like the innovation district at this side of

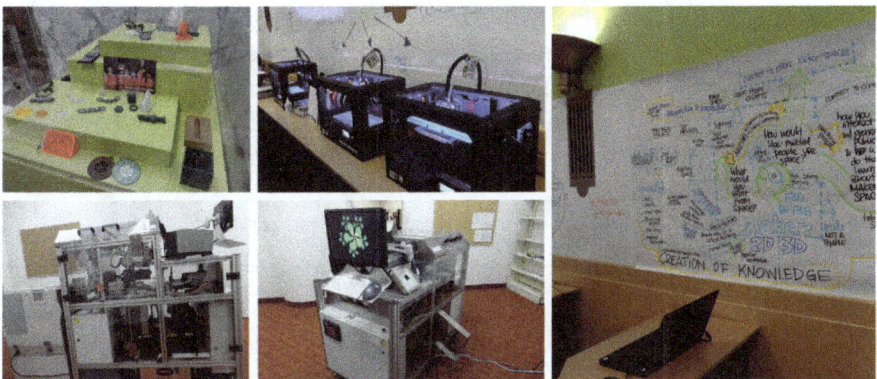

Figure 6.29: Maker spaces. Top both and right: Maker space at the Chicago Public library. Photo: Carsten Brinker. Bottom left both: Toronto public library "Espresso Book Machine." Photo: Agnes Mainka.

> Boston is building those types of modern spaces that the class of engineers, researchers, creatives want. This kind of buildings is what they want instead of libraries. (BO 2)

There are much more examples of the physical use of public libraries other than being a quiet reading and learning space as this development is not only recognized in public libraries, but also in university libraries like the Mc Gill medicine library, as one interviewee from Montréal states (MO3), saying there is now more space for the students than for the books. Accordingly, one expert reported how their physical public library in Vancouver (VA 5) has been used recently:

> We do have folks that come here and engage in connecting learning opportunities. They are here for the space. They are here for the quiet. They use their laptop. We do have a project on the goal. We will be creating an "inspiration lab" or digital "media lab." That will also be flexible to allow for different types of entrepreneurial activity within the creative sector. For example, if there is a local resident with an idea and want to pitch it to someone remotely then we will have the room. We do have meeting rooms that people can use for different types of activity. I think increasingly people are seeing the library as not just a space. More people are recognizing the potential of the library as public space for beyond study and beyond kind of recreational uses. There is more potential for different types of things like entrepreneurial.

Physical library service

As the discussion above reflects, there are many ways how the physical space of a library can be used. Mainka, Hartmann, et al. (2013) have investigated the spaces and services offered at physical public libraries of the 31 informational world cities. The indicator catalog of the physical library is presented in Table 6.7. In the following I will concentrate on the available spaces in the various libraries (learning, meeting, working, and children's spaces). As discussed by several experts, physical libraries play a crucial role as mediator between the digital and physical world. Therefore, the aspects of Wi-Fi availability and seminars on information literacy will be presented in addition.

Table 6.7: Indicators of the physical library. Source: Mainka, Hartmann, et al. (2013).

Group	Indicator
spaces	learning spaces meeting spaces working spaces children's spaces
use of technology	RFID interlibrary loans (borrow anywhere and return anywhere) Wi-Fi
architectural landmark	architectural landmark
the attraction of spaces	drinks / food the attraction of spaces
information literacy	seminars on information literacy

In Figure 6.30 the different kinds of spaces that are offered in the public libraries of the 31 investigated cities are represented. However, only the central public library or the biggest library near the city center was used as a reference of this investigation. According to the findings, the most common spaces available are

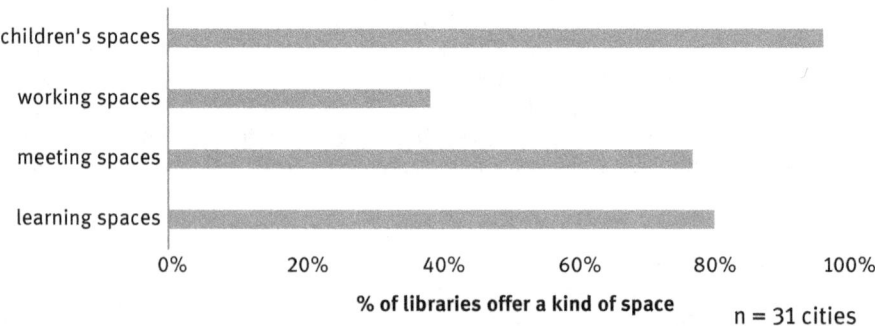

Figure 6.30: Spaces in physical libraries of 31 informational world cities. Data source: Mainka, Hartmann, et al. (2013).

for children. Except for the Shanghai Library, which is a public library joined with the Institute of Scientific and Technical Information Shanghai, which offers additional special industry information research, all public libraries offer children's spaces. Less common are working spaces that offer the opportunity to collaborate with others.

It should be noted that there are several other kinds of spaces in libraries not investigated here. For instance, the public library of Amsterdam also offers a theater, a video screening room, a music space, small single learning spaces, conference rooms, and a live radio broadcasting space (see Figure 6.31 for some examples). Spaces that are open for people who want to create something are generally called "maker space" or "innovation lab." As in the previously discussed examples of Chicago and Toronto, most common are 3D printers, but a book printing machine, media digitization systems, and other hands-on equipment and software like digital design are examples of how people can become an active creator in a library. At the San Francisco public library, for instance, it is planned to build a youth space that will include a recording studio (SF 1). Out of the 31 public libraries, 40% are already offering some kind of maker space (Born, 2015).

Figure 6.31: Spaces in the Amsterdam public library. Photos: Agnes Mainka.

In addition, libraries play a crucial role as mediator between the digital and the physical world. Offering Wi-Fi at library buildings enhances the connectivity to information for all of society. Additionally, information literacy courses for people who are not familiar with technology and information retrieval are needed. According to the results of the study, 90% of the investigated public libraries offer a Wi-Fi connection. Some are free and others are for registered users only, but only in very few cases the users had to pay for Wi-Fi access. Seminars on information literacy are a little less common, but far from rare: In total, 71% of the libraries offer information literacy seminars. However, one librarian from the Toronto public library (TOR 4) reported that not everyone is availing themselves of such courses, instead simply asking other library patrons, giving and receiving help from another.

Libraries do not only act as a social inclusion space. The buildings and the spaces are in many cases explicitly designed to attract the citizens. In previous studies, it already has been discussed that libraries are important "place makers" (Skot-Hansen et al., 2013), with an attractive library able to enrich a neighborhood. The public library of Vienna is one example that tries to do that. As the chief librarian (VI 2) reported, the new building is situated next to "red light" establishments, which attract the attention of the police very often. Nevertheless, the library is frequently used by the public, the building being directly connected to the public transportation network. One negative aspect of the architecture is, however, the entrance, which is not on the bottom level and therefore makes it inaccessible for people in wheelchairs if the elevator is broken. Nevertheless, in most cases (91%) central public libraries in informational world cities are attractive spaces that could be acknowledged as "place makers" (Mainka, Hartmann, et al., 2013). Some examples are presented in Figure 6.32.

Figure 6.32: Public libraries as "place makers." Top left to right: Amsterdam Openbare, Vancouver Public Library, Stockholm Stadsbibliotek. Bottom left to right: São Paulo Public Library, Paris Public Library, New York Public Library. Photos: Carsten Brinker (São Paulo, New York), Agnes Mainka (others).

According to the weighted index of public libraries in informational world cities, the libraries in Montréal, Chicago, and São Paulo are ranked at the top when looking purely at the physical library service (Figure 6.26). However, most libraries perform very well in this ranking, offering a variety of services within the physical and the digital space. The only exception is Dubai. In 2012 the old town library was more of a newspaper reading room for immigrants instead of being used by the whole public (Figure 6.33). And the digital library was not available online, except for a short description of the library in English. According to the information on the governmental website of Dubai, the website (www.dubaiculture.gov.ae) has been updated, but no additional features, e.g. access to the digital library, are available (as of December 2, 2016). Most of the library's activities and programs are intended for children and, in addition, a traditional library service with books in shelves and quiet reading areas has been established. Hence, Dubai, like other cities in the Gulf area, has undergone a vast transformation in the last years, with many foreign investors and cultural influences arriving in the city (Kosior et al., 2015). It will be interesting to follow Dubai's transformation due to its increasing number of immigrants and investors from foreign countries.

Figure 6.33: Dubai physical public library.
Photo: Agnes Mainka

Conclusion: Physical libraries are spaces for face-to-face communication
In summary, libraries have made a transition in the recent years to meet the needs of the knowledge society in the twenty-first century. An increasing access to online resources is being offered by the digital library, whereas space for the public is growing in importance in the physical library. There is a trend of reusing the traditional library space in diverse manners. Of course, space for quiet working is still offered, but other needs are also being met. Thus, the library is a space for the public to get access to information and to technology. Examples are the Wi-Fi access and information literacy seminars in public libraries, both of which offer access to information for those who would otherwise not be able to afford it. Public libraries are also mediators between the physical and digital world: It is not only in seminars that people can learn how to use information online but also just by being in the library, with patrons also helping each other, as stated by one librarian from the Toronto public library (TOR 4).

That public libraries could serve as public space for face-to-face interaction is not a universal interpretation by all interviewed experts. Hence, a lot of coworking spaces and incubators have opened their doors in recent years at which this use of space is more likely to happen (BO 2). However, coworking spaces are not public spaces. Entrepreneurship and startups are very common in cities of the knowledge society, and public libraries can build a bridge between this community and the public, as in the case of the Chicago public library. Here, entrepreneurs receive free office space for their startup businesses, but have to give something back to the community in return, for instance by giving seminars on how to design objects for 3D printing (CH 2). As public institutions, libraries assume an educational role and should offer the amenities for life-long learning. Seminars that are run by smart individuals out of the community enable such new opportunities: The library space is used in a new manner, individuals who like to share their knowledge get a platform to do that, and finally, the library can give something back to those individuals. For instance, in the case of the public library of Chicago, a win-win-win situation evolved – for the library, which gets filled with life, the public, which get access to sophisticated seminars, and for individuals, who receive free office space for their startup.

Repurposing the public library as open public space for interaction and collaboration needs to be implemented with caution and should not be established in a top-down development. There is a need to communicate the vision carefully and to avoid creating the impression that the library is going to be closed or will not be able to serve the information needs of the public. Depending on the community, these needs can differ and therefore a library should not only open the space, it should foremost open a dialogue between the community and the library and identify what the citizens want. This has happened, for instance, during the development phase of the new public library in Helsinki, which will open its doors in 2018. This building is going to be a "place maker", and with its special architectural design it will fill the new build neighborhood (next to the central station) with life. In most cases, the public library's architecture is a place maker in an informational world city (Mainka, Hartmann, et al., 2013).

In summary, for most of the hypotheses investigated so far (H1–H7), San Francisco has ranked highly. The city is, of course, a very interesting phenomenon which evolved around ICT innovation. The San Francisco Bay Area is home to the elite university of Stanford, has a high density of ICT-related corporations, and a thriving entrepreneurship landscape. Furthermore, through the public private partnership of San Francisco and Google, the city is offering free public Wi-Fi access. The public library is also very well-developed, as demonstrated by the 4th rank in the comparison between the 31 informational world cities. Special attention has been given in nearly all discussed hypotheses to the increasing importance of entrepreneurs and tech startups. In the following chapters, I will now turn towards the political will and cityness and their impact on informational world cities.

6.2 Political will

The development of a city can be influenced by various and diverse factors. In this chapter, the political willingness of a city to become an informational world city will be investigated. In addition, the growing digitization of governmental services will be discussed, particularly whether the investigated cities are characterized by e-government, e-participation, and e-democracy. With this digitization process, increased attention is being paid to the availability of open data and a free flow of information. The political will and its transformation based on an enhanced ICT infrastructure for the 31 cities will be investigated with regard to the following hypotheses:

H8   Political willingness is important to establish an informational world city, especially with regard to knowledge economy activities.

H9   An informational world city is characterized by e-governance (inclusive e-government, e-participation, e-democracy).

H10 A free flow of all kinds of information (including mass media information) is an important characteristic of an informational world city.

H8 Political willingness
The political willingness of a city or region is most visible through agendas or master plans that define the goals of the future development, even though it can only be investigated in retrospect whether they have been accomplished. According to Yigitcanlar (2010), cities that have the willingness to become a knowledge, smart, or informational city focus, for instance, on financial support and strong investments, on agencies to promote knowledge-based urban development, on an international, multicultural city character, metropolitan Web portals, value creation for citizens, creation of urban innovative engines, assurance of knowledge society rights, low-cost access to advanced communication networks, research excellence, and robust public library networks. As there are very different examples of successful cities, e.g. New York, which has grown historically and was mostly driven by its economic development, or Singapore, which has been led by a strong national vision, it is open to debate which role the government plays in recent knowledge-based urban developments. Thus, the following hypothesis will be discussed:

H8 Political willingness is important to establish an informational world city, especially with regard to knowledge economy activities.

Among to the interviewed experts, 46 explicitly mentioned that the political willingness is very important to establish an informational world city. Eight experts stated that this development is only possible with a strong political willingness. "If you do it without political willingness you will lose. You want to have support" (CH 2). Another six experts consider the political willingness at least helpful. Following the SERVQUAL evaluation (Figure 6.34), the experts from Shanghai, Hong Kong, Chicago, Boston, Shenzhen, Toronto, Amsterdam, and Stockholm have a higher perception of the political willingness than they would expect. This assumes that in the cities in which the expectation is high, the governmental influence to become an informational world city is acknowledged as positive. In Shanghai and Hong Kong the average expectation is less than 3, which implies that the government had not been much involved in the development towards an informational world city. In New York, Vienna, and Singapore the perception of the political willingness is high. It is scored 6 and above. In these cities, the quality value is zero. Thus, the experts agree that a strong political will can establish a successful informational city.

The perception of the importance of political willingness depends on the confidence in the political system (eight experts). Thus, "if you are measuring the

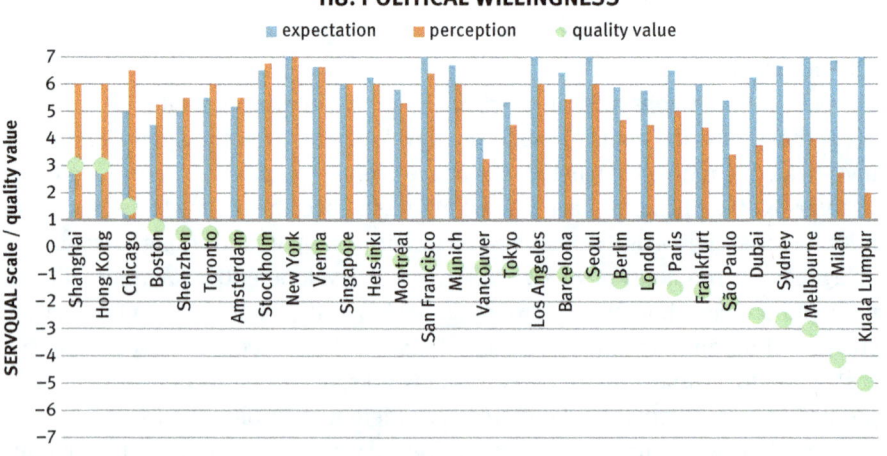

Figure 6.34: Quantitative interview results according to SERVQUAL (quality value = perception − expectation) for H8 (Political willingness is important to establish an informational world city, especially with regard to knowledge economy activities).

words of the politicians then of course [it is important], but when you measure the action then not" (BA 3). "[P]olitics can either help or hurt. But it is not necessary. A lot, in terms of being an informational city, goes back to the industry and what industry is doing. The government can really help that process or could really hurt that process" (TO 1). Accordingly, six experts mentioned that the political willingness is not important at all and that the development towards an informational world city is caused by the economy (eight experts). "The willingness alone will not bring innovation. There are always people out of the society that bring the innovation. E.g. in the San Francisco Bay Area, Google and Apple have not been planned by the government" (FR 4). On this view, the development of the knowledge society is a natural evolution (BA 8), as many things happen without political intervention. For example, the Web 2.0 is not an invention of any municipality. It came from the economic market and grew into its own system in which people meet to participate (VI 1). To empower the economic market, the government may introduce tax incentives for companies of the knowledge economy to enhance the city's attractiveness to them (three experts). In total, the average quality value according to the importance of political willingness for 18 cities is negative. This implies that in most cities the political willingness could be stronger. For Vancouver and Tokyo, the expectation and perception are very low compared to the other cities. This implies that the interviewed experts do not agree that political willingness would be important, as it currently is not in their city.

Summing up the interviews, money and funding are the keys for political power and impact (14 experts). The government is foremost responsible for the implementation of the needed infrastructures and spending on public welfare (three experts). As education is fundamental to become an informational world city, adequate funding of education and universities must be ensured (ten experts). Additionally, it is of advantage if the city is open to highly educated people, open to smart city initiatives, and also open to citizen participation. Finally, open government initiatives like open public data also increase the openness and transparency (five experts). In the following I will discuss the arguments given by the interviewees in more detail by discussing the political impact, funding possibilities, infrastructure issues, and finally the evolution of bottom-up and top-down developments.

Political impact
In identifying the political willingness of a city it is not easy to define whose willingness is of importance, since the political willingness of a city can be different from the willingness of the region. For instance, "[t]he Chicago region is heavy fragmented. So we have 300 units of different local governments in the Chicago metropolitan area. But the connection or the willingness to plan together, work

together regionally, to create a vision, the prudential regulation signal is poor" (CH 1). In contrast, one expert from Helsinki (HE 3) stated that the political willingness is not only on the city level, but that the whole region is on the way to establishing itself as an informational world city/region. Determination by the region or nation can be an advantage, but also a disadvantage, as one expert from Paris (PA 6) emphasized that the city government has no power. In Paris, the development of the knowledge landscape is governed by the ministry of the whole region and not by the city's mayor. Further, "the advantage of London is that it is the central city of UK and government is centralized here. Nevertheless, the development and grow of London is not able to be controlled by politics. They can give incentives into the right direction" (LO 1).

Sometimes, the political will is focused solely on one topic, as in the example of Vancouver:

> The political will is mainly focused on 'the greenest city,' and all that sort of thing. We did have a digital strategy committee... but it hasn't really had much impact and I think... they're too focused on [the greenest city] to give proper attention to the knowledge economy, digital creative city, informational city stuff. They have put some effort into it but not nearly as much as they could have. (VA 4)

In New York, the city has made a vast step forward due to its mayor Bloomberg, who introduced an information sharing system called 311 and open data (NY 2). Cities in the US depend largely on their mayors and their work. Boston has also had a powerful development that is generally attributed to the mayor's willingness:

> We have an incredibly understanding mayor, who pushes hard on this front. He really wants this innovation. The innovation district is really his effort... He is not digitally literate himself. He just understands how important it is and really pushes and supports this... He is a really good leader and you got to have that. It doesn't happen without that. (BO 1)

> We have buildings for the digital innovation economy. They start having grants and things like that to give companies some incentives to be there. And that took a strong political will to make that happen. This rural transform in that area added jobs and technology etc. I think that stuff can happen anyway if the infrastructure is there like at Boston University. But the political stuff can encourage it that it grows. (BO 2)

This importance is also acknowledged in other cities. For instance, in Munich, a panel has been established to work on labor market and economic development to push the knowledge economy (MU 4). In Seoul, a similar body was renamed to "ministry of knowledge" to represent the importance of knowledge within the economy (SE 1).

Funding

The political willingness is represented by its impact on the city's development, which is why funding and investment play a crucial role. Depending on the governmental structure, funding comes from different units and is not necessarily decided by the city itself (ME 2). "In Montréal, for instance, we have a special situation. The region of Québec would like to become an own nation and due to this, the government of Québec is investing a lot into the culture and other things that should enhance the development towards a knowledge society" (MO 3, own translation). In general, public spending decision have a large impact. In Los Angeles, for instance, a vote about the funding of the public library revealed that 67% of the population are "pro" libraries. Thus, the government had to keep investing in the library system due to the citizen's democratic decision.

Decisions on governmental level are commonly hierarchical. In Canada, for instance, there is the federal, the provincial, and the city level.

> The problem is that the capacity of getting money is all at the federal level… So where I am going on that it's basically that political willingness yes, but it has to come from more than one level. The city wants to become a knowledge city it needs to be either very rich or to get the help of the federal government or the provincial government. We have got Ubisoft which is the French game creator company. They are benefitting from quite a lot of subsidies from the government… they don't have to pay taxes and some stuff. So we do see that there is a political willingness to create a city like that by having companies in the field of information in general. (MO 2)

Incentives seem to be a common procedure to attract companies of the knowledge or information economy. San Francisco, for instance, is home to the Twitter company headquarters. According to one expert (SF 2), they have to pay fewer taxes and in return they stay in the city and hire local people. One expert from Shenzhen is arguing that "the political willingness is about finance. Every single district is fighting against each other to attract companies to come to their district" (SHE 1). Hence, China has introduced "Special Economic Zones" (SEZ) in order to attract national and foreign companies, and laws and regulations like "Private Property Rights Protection," tax incentives, and a land use policy have been implemented (Wang, 2013). Among other cities in China, Shanghai and Hong Kong have set up such SEZs as well. The UAE have adopted a similar procedure and opened "Free Economic Zones" to attract foreign investments, e.g. at Dubai Airport Free Zone, Dubai Media City, Dubai Internet City, etc. (UAE Embassy in Ottawa, 2016).

Infrastructure

A city that is only driven by market regulations is not social (Robinson, 2016). Cost of living, housing, and rents for offices are a problem which may or may not be the result of governmental policy decisions (BO 3). For instance, to offset the high costs of living in Helsinki, much funding is needed for students to access education. Nevertheless, "the Finnish people are very proud of their very good educated children" (HE 1). One interviewee from Vancouver (VA 1) emphasized that the governmental development is acknowledged as very slow.

> For instance, they have made a good job 20 or 15 years ago to implement the needed ICT infrastructure. But 5 years ago they talked about to open public Wi-Fi spots but in the end they did not. It was too expensive for the city. Thus, the big telecommunication provider implements the Wi-Fi hotspots and you are just able to use it if you have a contract from this provider. (VA 1)

In such cases, the infrastructure exists, but is only accessible for those who are able to pay for it.

> So the question is rather more basic and that is how do you even have enough money to take care of the infrastructures and services that are in place. Fire, roads, schools, police, water etc. So not alone information technology in addition to that... So people who can pay for the service will get it. But it's not a public service because we don't have any money for that kind of things anymore... [For instance,] LA is a working class city... So I just came back from Palo Alto, Silicon Valley, and the contrast in wealth is huge. There is a really rich white influent area. LA is not. And a lot of services are provided by private companies. (LA 3)

The key to such wealth might be education as stated by several experts. For instance, "Hong Kong was a manufacturing city before and now it is international and teaching in English" (HK 3). The universities, as places where smart people are educated, are very important (MO 1), e.g. Silicon Valley did not evolve due to any particular political willingness, but rather due to the Berkeley and Stanford universities. In Boston, the universities and knowledge people helped to establish the knowledge and ICT industry, as well (BO 2). Thus, universities, science parks, and research institutes are important to become a knowledge society and to attract them reflects a city's willingness (MU 1, MU 3).

That education is important has been acknowledged, but the problem of an adequate funding still looms large, in particular in São Paulo:

> Education is acknowledged as the major problem in São Paulo and in the whole of Brazil. In recent years, the educational sector has made big efforts. Thus, we have enhanced the alphabetization rate, the access to universities and so on. Education will be the most important topic in the future as well. The future will show if their efforts will help to manage this problem. (SP 2, own translation)

"What is missing is to identify how this university knowledge can be transformed into economic goods and development" (SP 5). However, according to another expert the politicians in São Paulo do not have the long-term view to resolve the big problems (SP 4).

As previously mentioned, the municipality or the city's mayor does not always have the power to significantly impact the development. In Vancouver, for instance, the universities are under the control of the federal government and not of the city (VA 1). What can be supported by local political willingness is, for instance, attracting further research institutions to the city (FR 3). In Frankfurt, for example, a research hub is going to be established as the world's biggest research and development center in the field of logistics and mobility. However, there is a difference in the finance model of scientific activities. In Germany, compared to Switzerland or the US, a tremendous amount is financed by the government and only a small amount by private organizations and companies (MU 1). To keep the position of Munich as a hub for scientific activity in the future, a more diverse model in financing should be pursued (MU 1). London, in particular, has the advantage of a very high concentration of banks and finance firms (LO 2), but according to another expert from London (LO 6), there is still a need for more funding, e.g. of projects on knowledge exchange.

Top-down or bottom-up?
In some cases, a development has happened without any political incentives. In Berlin, for instance, many things were able to happen not because the politicians have taken specific measures to encourage development, but rather because they did not do anything against it – the startup scene in Berlin, for example, was a bottom-up evolution (BA 1), even though startups are acknowledged as job creators that need to be supported. According to one expert from Berlin (BA 1), the problem in Germany is that entrepreneurs are frequently working with venture capital, thus making no profit and not having to pay taxes. When these startups end up becoming profitable, they are sold to foreign investors, e.g. from the US or are transformed into a public limited company. In consequence, they do not give anything back to society. However, in some cases by ignoring certain circumstances, the government can help a development to flourish. In Shenzhen, for example, there is a high amount of forgery- and hacking-based industry with which the government does not want to be associated with (SHE 2). Nevertheless, this industry is a fast and easy venture for entrepreneurs that like to experiment and build prototypes. And this is what has made Shenzhen to a "Silicon Valley of Hardware" and attracts many entrepreneurs (WIRED, 2016).

In contrast, there are also initiatives by governments to push entrepreneurship and citizen's engagement (top-down).

> It is interesting because the mayor's officers do strange initiatives... They having these 'the entrepreneur residents project'. They having this teams of two or three entrepreneurs who work to solve city problems that are mainly data and infrastructure related and that kind

of initiative. They hoping that by sponsoring that through the mayor's office that they are going to be seen.... They are going to city departments like us [the public library]... and say: What are your big problems? Where are the places where you have troubles meeting the demand of your clients? And then will trying a respond for it like making apps for you or doing data analysis. (SF 2)

In Barcelona, this kind of engagement is encouraged: "We have a lot of activities which are done by the local government to engage citizens. A lot of things. Special programs for entrepreneurs like incubators for those people. We have the compact city model. Which means to mix industry and citizens at one cluster" (BA 1).

One expert also said that they "need civil engagement to make the political willingness arrive" (MO 6). To have citizen participation is an important aspect in enhancing the city's development. In line with this, another expert from Barcelona (BA 8) stated that many projects are driven by democratic processes: "This city has... to make a lot of steps to arrive at decisions with the civil society in urbanism, in mobility and in all the projects" (BA 8). In Barcelona, a protocol has been introduced that specifies that each of the ten districts must include citizen participation in development projects for the city and then find a common decision for the whole city. "It's not an easy way to find a solution. I think we can improve the process but it is a good way" (BA 8).

Nevertheless, in the twenty-first century it is important to have a digital strategy for the city government and for all institutions, including the public library (VA 1). All the plans need to be made transparent to locals. Hence, for instance, in Vancouver, those documents are available to the public online (VA 3). While local benefits, public engagement, and the protection of civil and social rights should be the goals within this political willingness, "[t]he vast majority of investments in it are being made simply in the interests of profitable returns. Our political leaders are not shaping the markets in which those investments are made, or influencing public sector procurement practices, in order to create broader social, economic and environmental outcomes" (Robinson, 2016, para. 35). According to Robinson (2016), the municipality is not the right entity to identify own ideas or be creative. The creativity is already in society and develops in many small start-ups. Political leaders need to invest in these existing ideas and transform them into "top-down" developments to achieve real impact in their city. This can be done by the government through bridging the institutional levels of the economy, university, and community.

Conclusion: Political will
The discussion on the political willingness has reflected that there is no "right" answer: In some cases, the interviewed experts have not seen the necessity of

political intervention. In London, for instances, the interviewees are not convinced that politics will have a large impact on the knowledge economy (LO 1). However, in far more cases, the experts agree that a political willingness is important or at least helpful. Furthermore, the question arises whose political willingness is of interest when it comes to a city's development towards becoming an informational world city, as it is often the case that different regulations and authorities apply e.g. for the educational system and attracting of universities. From the economic perspective, different approaches have been attempted to attract companies, e.g. by the introduction of "Special Economic Zones" or "Free Economic Zones" in China and the UAE, or by tax incentives for companies like in Montréal and San Francisco.

The most common argument mentioned by the experts interviewed is that the political willingness in an informational world city should at least ensure an advanced educational system. In addition, attracting universities and science parks is acknowledged as a foundation for a further development. The best example of a vast development within the knowledge economy built on this is Silicon Valley, which has its roots at the Stanford University. The support of advanced infrastructure, as for example citywide public Wi-Fi, is acknowledged as a consumer good that may be offered by private companies. Finally, the cities' development has not only been driven by top-down initiatives from the government. Citizens' engagement and entrepreneurs may come up with creative ideas to solve urban problems. Accordingly, the government should push these developments and enhance them through an adoption into "top-down" programs. One example was given by San Francisco and its "entrepreneur residents project", which brings together startups and city institutions to identify and resolve problems through data analysis or creating new mobile app services.

There does not exist an index or measurement tool that could be used to identify and compare political willingness on a global scale, which is why real-world examples that can give a short insight on this complex problem have been discussed in this subchapter. However, it is very hard to identify whose political willingness determines the development of a city, and to which extent a mayor or municipality is able to impact it. Attempts to include the political willingness as a measure within a framework of knowledge cities' development have been introduced, for instance, by Ergazakis, Ergazakis, Metaxiotis, and Charalabidis (2009) or by the "Most Admired Knowledge City Award" (MAKCI), which is based on the knowledge-based urban development framework by Yigitcanlar, O'Connor, and Westerman (2008). Both approaches are not feasible for comparison on a global scale, since not every city provides a master plan on how to become a knowledge, informational, or smart city. Nevertheless, both approaches are emphasizing best practice examples of knowledge cities and how political willingness impacts the development of those cities. For instance, Melbourne has been investigated as a best practice example and was the winner of the MAKCI Award in 2010, 2013, and 2016:

> However there is good evidence from the Melbourne experience that education and R&D institutions, three tier government and communities are altogether supporting the emergence of Melbourne as a KC [knowledge city]. Global recognition of Melbourne as an emerging KC and processes that have been established in Melbourne provide some useful insights for policy makers of other cities in designing, developing or moving towards a KC. (Yigitcanlar et al., 2008)

H9 E-governance
Political willingness in an informational world city is also represented in its implementation of e-government services, e.g. "metropolitan Web portals" (Yigitcanlar, 2010). Enabled by an advanced ICT infrastructure, digitized services may help to solve urban problems. They may also be used in diverse ways for citizen-to-government, business-to-government, and even for government-to-government interaction and transactions. With the 24/7 access to information through the internet, a demand for such a service offered by the government with regard to public information has emerged (United Nations, 2016). The idea of e-government is not new, but the implementation of adequate services is still a work in progress. Additionally, further possibilities have been opened up the online availability, such as e-participation and the integration of diverse stakeholders in decision-making processes to enhance e-democracy. Combining online services, enhanced possibilities of participation, and the integration of diverse stakeholders in the decision-making process is labeled as e-governance (Harrison, Burke, Cook, Cresswell, & Hrdinová, 2011). Of course, the maturity of this development may differ due to various reasons. To identify the current state and factors that hinder the development of e-governance, the experts were asked for their opinion on the following hypothesis:

H9 An informational world city is characterized by e-governance (including e-government, e-participation, e-democracy).

Looking at the SERVQUAL evaluation, a negative quality value can be observed for most cities (Figure 6.35). This indicates that, for the experts in these cities, the perception of the city's own development is below the expected maturity in an informational world city. Overall, experts from 25 cities have stated explicitly that they have e-government and in eight cities that they additionally have e-participation. It is common to offer online information about the city and the

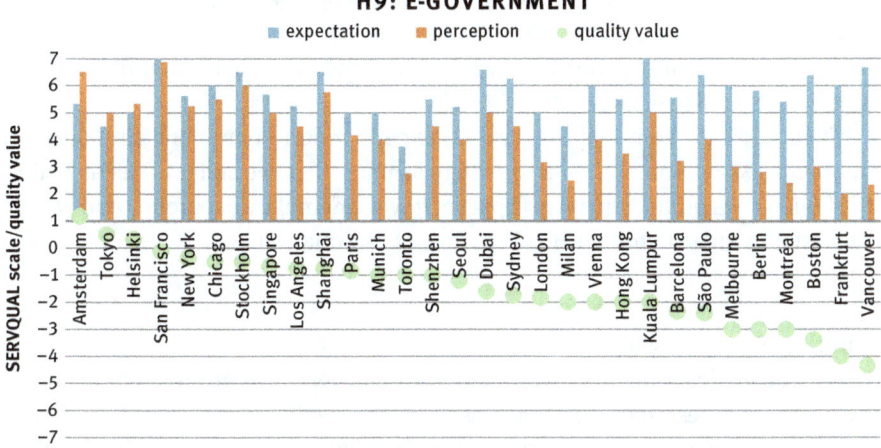

Figure 6.35: Quantitative interview results according to SERVQUAL (quality value = perception – expectation) for H9 (An informational world city is characterized by e-governance (including e-government, e-participation, e-democracy)).

government, but a platform that asks for citizen engagement is hard to find. In contrast, nine experts report that there is neither e-participation nor e-democracy in their respective city. At this stage, e-services cannot replace the traditional system of voting. The enhanced ICT infrastructure should rather be used on top of face-to-face discussion and personal participation (SP 5, MU 3). Ten experts also do not see the necessity of e-governance as a characteristic feature of an informational world city: "You can have an informational city without e-government. It's another kind of. You don't need citizen participation when you look at some Asian cities" (BA 1). According to another expert, without e-governance "[y]ou have characteristics of an informational city but e-governance is part of the outcome of the complete package" (TOR 1).

According to previous research on governmental websites (Fietkiewicz, Mainka, & Stock, 2016; Mainka, Fietkiewicz, Kosior, Pyka, & Stock, 2013), five pillars of e-governance can be determined: (1) information dissemination, (2) communication, (3) transaction, (4) interoperability (horizontal integration), and (5) participation. Based on the fifth pillar, participation, the integration of enhanced democracy can be established through a decision-making process that includes all stakeholders (Harrison et al., 2011; Palvia & Sharma, 2007). Following one expert from Munich (MU 1), e-services are a special need of people of the creative class. The creative class is used to being independent and autonomous. Thus, those people would like to have a voice in decision-making processes and they want to have as many e-services as possible to save time. Sharing the power in collaborating projects in general is acknowledged as e-governance or open government (Harrison et al., 2011). Accordingly, I will discuss in the following first the maturity of e-government and second the implementation of e-governance on the city level.

It should also be noted that several experts reported an increasing demand for opening governmental data to society as part of e-government. The availability of open data on the city level is part of governmental transparency and will be discussed in the subchapter H10 Free flow of information.

Maturity of e-government

As previously described, e-government and e-governance are not synonymous. E-government describes the services that are available through ICT (Palvia & Sharma, 2007), e.g. websites, social media, mobile applications, while e-governance refers to the collaboration aspect of managing the city. With the help of ICT, participation and decision-making processes can result in an open government (Harrison et al., 2011). That diverse degrees of maturity exist is visible, for instance, in the "United Nation E-Government Survey", which is based on national data (United Nations, 2016) and investigated 193 nations around the globe. In it, nearly half of the nations are acknowledged as having a very high (15%) or high level of e-government development (34%), with 35% having an average and 16% a low e-government development level. Most of the informational world cities investigated here are located in nations belonging to the very high development category. Exceptions are Shanghai, Beijing, and Shenzhen in China, São Paulo in Brazil, and Kuala Lumpur in Malaysia, which however still at least belong to the highly developed nations.

Next to the ICT infrastructure (based on the "Telecommunication Infrastructure Index") and human capital (based on the "Human Capital Index"), the "E-Government Development Index" includes the scope and quality of online services (United Nations, 2016). Accordingly, to investigate the maturity of e-government on the municipal level for the 31 informational world cities, previous research has investigated the city platforms available online(Fietkiewicz et al., 2016; Mainka, Fietkiewicz, et al., 2013), employing a survey on the five pillars of e-government as detailed in Table 6.8.

This five-pillar model is based on the five-stage theory of Hiller and Bélanger (Hiller & Bélanger, 2001). As the e-services available online are not introduced step by step, the terminology used changed from stages to individual pillars. For instance, a municipality does not have to install online transaction services before they can enable public participation. Both can happen individually and

Table 6.8: Five pillars of e-government based on the five-stage model according to Hiller and Bélanger (2001).

5 PILLARS OF E-GOVERNMENT		
	Information	Make information available online; Inform citizens through mobile applications, for instance about current news.
	Communication	Start a dialogue between the government and citizens (or businesses) through ICT; Use social media channels to reach the citizens where they are.
	Transaction	Pay taxes or apply for licenses online. Citizens are able to make verified transactions with personal e-IDs.
	Interoperability	Share information across organizational/administrative boundaries.
	Participation	Involve citizens and other stakeholders in decision-making processes.

independently. According to one interviewee (BE 1), the implementation of e-government services depends on the people working for the administration and their willingness to use and introduce new services. In the following, the pillars will be discussed according to the results identified for each city (Figure 6.36).

Pillar 1: Information To make governmental or city information available online through municipal websites is the first pillar of e-governance. According to the survey of Mainka, Fietkiewicz, et al. (2013), the municipal websites of New York, Helsinki, Berlin, Singapore, and Hong Kong offer most of the requested information online. Thus, one can get for instance information about actual press releases, information on health care or about politics. In addition, the availability of that information can be investigated by the usability of the websites. To provide access to the citizens accessibility standards are needed. According to

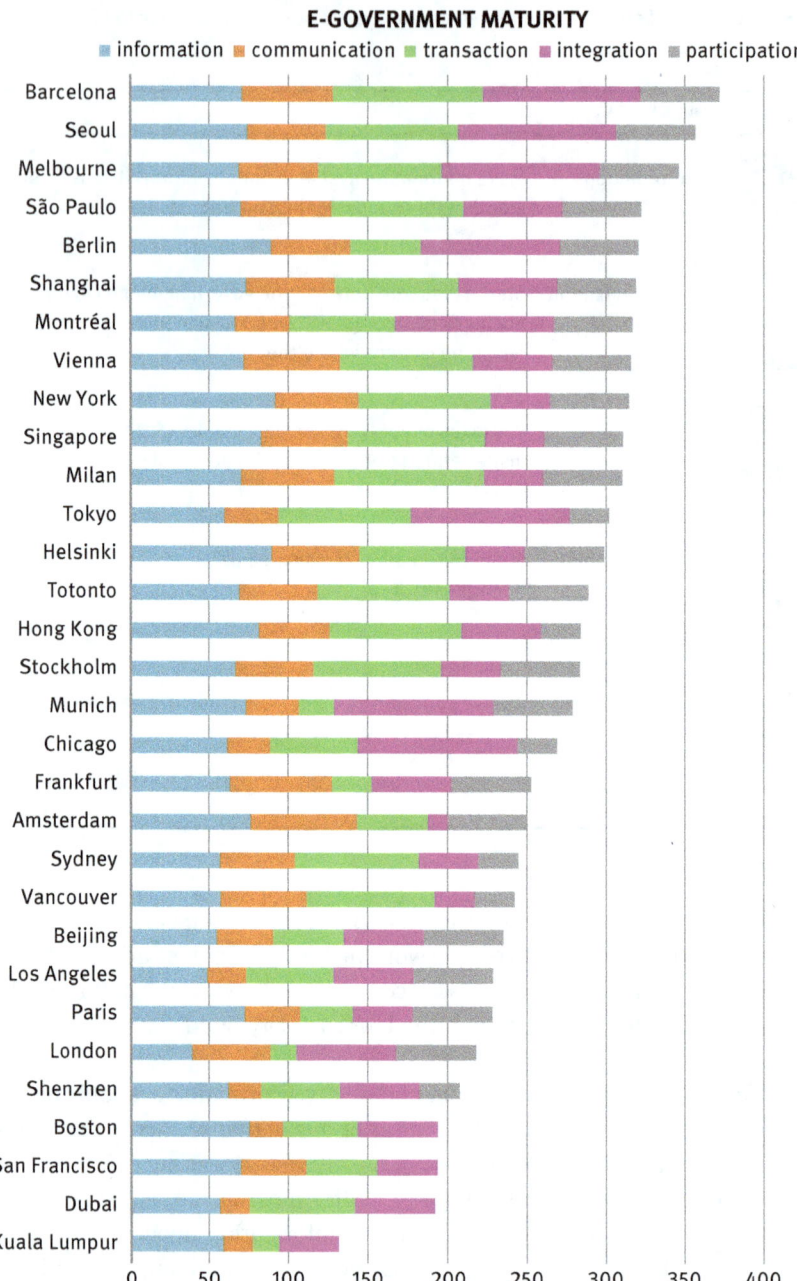

Figure 6.36: The maturity of municipal e-government for the 31 informational world cities. Data source: Mainka, Fietkiewicz, et al. (2013).

the usability of the municipal websites Vienna, Seoul, Shanghai, Stockholm, and Munich are the top rated municipalities.

During the interviews some experts started to explain what kind of online information they think is of importance. For instance, information like FAQs (SY 1) or the availability of future governmental plans (ST 1) have been stated as important. A further argument to offer online information is that the governmental services could easily be translated into the most widespread languages instead of training the staff to speak all of them (BE 1). This is also part of the survey on the first pillar (information), which checks for the translation of the municipal websites into the three most frequently spoken local languages and English. The devil is often in the details: Hong Kong, for example, offers much information online, but this should not be equated with e-governance. According to one expert (HK 1), the government in Hong Kong is decidedly not embracing digital services: "Everything is print on paper. In China, in general, PDF documents and e-mail correspondence does not count as an official record. You print out the e-mail and stamp it so that it counts." It is thus not clear for the public if the available information can be relied upon. In contrast, other cities embrace social media channels to spread current information to the public. For instance, Twitter is used as an information channel for news in Tokyo, such as warnings for earthquakes or smog (TOK 4).

Even when a city is offering a great deal of diverse information, this does not necessarily mean that the city is transparent (CH 1). To offer information online and to offer open data are two different things. Hence, a municipality can offer some data they like to see in the spotlight, and keep data which they would rather not be scrutinized from the public. Open data is a more comprehensive philosophy based on the public right of access to information and includes, for instance, data on public spending. Cities that are located in nations or regions that have established freedom of information laws are per se acknowledged as more transparent (Corrêa, Corrêa, & da Silva, 2014). According to Relly and Sabharwal (2009), the first nations to introduce laws on transparency were the Scandinavian nations: Sweden in 1766, Finland in 1951, and Denmark in 1964, with the United States following suit in 1966. These nations are the forerunners on the path to truly open government. However, freedom of information laws do not equal open government per se (World Justice Project, 2015).

Today, many cities, regions, and nations have introduced open data portals that offer free access to a variety of open urban government data (Mainka, Hartmann, Meschede, & Stock, 2015b). These data concern local issues in a machine-readable format and can therefore be reused, for instance in data visualizations or mobile applications. In combination with real-time data, such information turns into valuable gadgets in citizens' everyday lives (Mainka, Hartmann,

Meschede, et al., 2015a): "We have Apps like the parking App which informs me if I need a new park ticket or the bus app which informs me whether the bus is working..." (SF 1). Finally, mobile applications make governmental and city-related information ubiquitously available.

Pillar 2: Communication The second pillar refers to two-way communication. As the investigation of the 31 informational world cities revealed, it is in many cases possible to make an appointment with the local administration online, or to correspond through email, and comments and feedback are welcome. The most mature municipalities according to their two-way communication are Amsterdam, Frankfurt, Vienna, Milan, Barcelona and São Paulo (Mainka, Fietkiewicz, et al., 2013). According to one interviewee from Helsinki (HE 1), e-services are used to communicate with city administration: "People working in city departments are easily available through email or through their social media channels."

In particular, communication within the field of e-government includes social media in addition to phone calls, emails or other contact methods. Today, it is a common practice for governmental agencies to use Facebook, Twitter, or YouTube as a communication channel (Mergel, 2013). A further investigation on the social media use of the 31 informational world city's municipalities, identified that, on average, general government accounts are run on 5.9 different channels (Mainka, Hartmann, Stock, & Peters, 2015). General government accounts refer to social media profiles that represent the official municipality or official city website. In addition, further social media accounts are created for various agencies, institutions, or politicians. As presented in Figure 6.37, Barcelona is the city with highest number of different social media channels followed by Melbourne, Sydney, Munich, and Boston. Boston, for instance, is represented on different channels especially due to accounts that are created for governmental agencies, institutions, and individual politicians and not only by general government accounts.

The most commonly used platform is Twitter. In total, 24 out of the investigated cities are using this platform. The Chinese counterpart, called Weibo, is used by two cities, Hong Kong and Shenzhen. The second most frequently used platform is YouTube, followed by Facebook. Accordingly, the most published content by governmental accounts can be found on Twitter, which is due to its simple and short posting possibilities. Berlin, for instance, posts about 500 tweets per month. The government of Seoul is tweeting nearly as often as Berlin and additionally uses blogs as regular publishing method (40 posts per month). Furthermore, it is very common to use photo broadcasting platforms like Flickr and Instagram. Barcelona, for instance, posts roughly 70 pictures on Flickr and 30 on Instagram per month. The general idea in making use of social media platforms is to reach citizens where they are. Among the investigated platforms, the

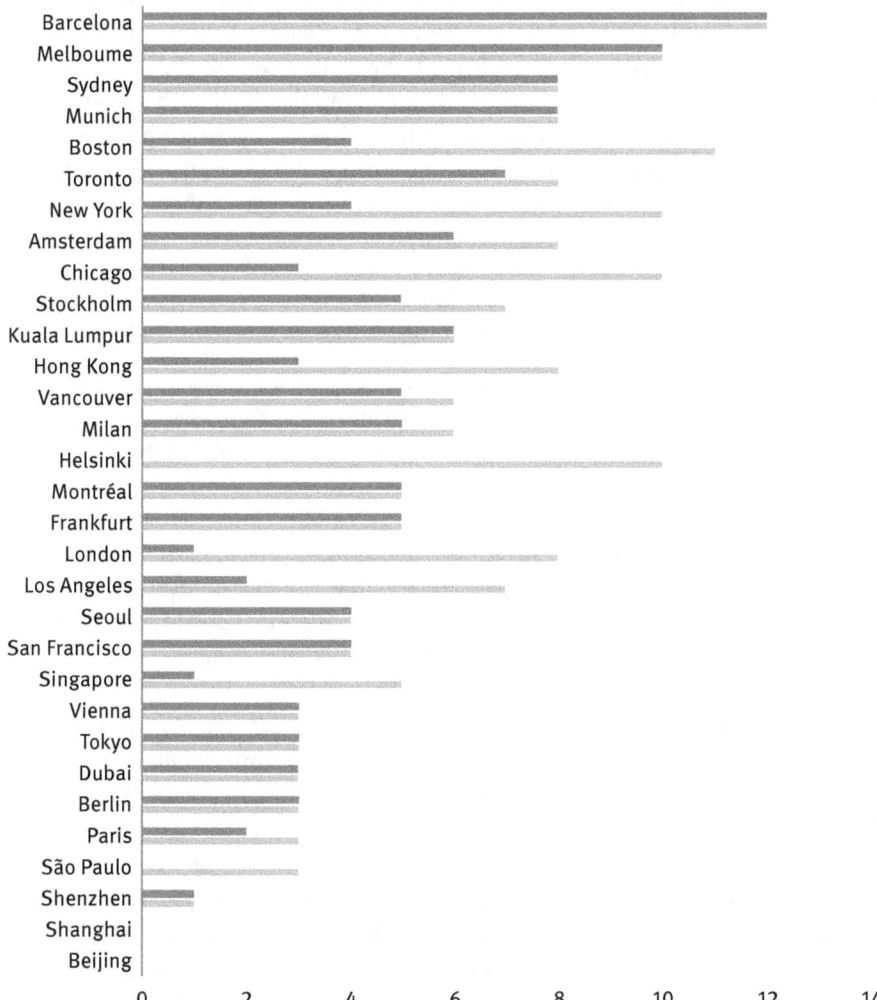

Figure 6.37: Social media platforms used for governmental purposes in informational world cities. Data source: Mainka, Hartmann, Stock, et al., (2015).

most followers have been reached at Facebook and Twitter/Weibo. Hong Kong and Shenzhen have reached follower numbers of about 600,000 and 300,000, respectively. Nevertheless, this is only 8.3% of the population in Hong Kong and 2.8% in Shenzhen.

Social media channels are only one of several communications possibilities with the city or city department. For instance, in Boston, a citizen relationship management (CRM) system called 311 has been introduced. This is a service where people can call, walk in personally, or communicate through a mobile app. This is a direct communication channel for the citizens: "There is a pothole, there is a streetlight out, my trash didn't get picked up, the sidewalks were broken, those kinds of infrastructure issues [are managed through 311]" (BO 2). This has changed citizen engagement substantially:

> Like ten years ago the people don't think like 'Oh, let's send a report while a streetlight is broken to the streetlight department' and think they will fix it. [Today,] you make a photo, mark the location and it goes right to public works. And when public works fix it you will get a notification back. Ten years ago that was not possible. For people to have that kind of interaction with government... becomes very important... I just go online to do this. I think it's really important to keep the government in a conversation on innovation. (BO 2)

Citizen relationship management systems are modelled after the customer relationship management systems (CRM) frequently used by companies to simplify and centralize their customer engagement activities. Understanding citizens also as customers is therefore one aspect of the changing philosophy towards an open government that is giving the society a voice in city development.

Pillar 3: Transaction This pillar refers to all financial and non-financial transactional e-government services. For instance, filling out a form online and wiring it directly to the appropriate administration without printing, or paying taxes online. According to the maturity evaluation of the 31 cities (Mainka, Fietkiewicz, et al., 2013), Milan and Barcelona are the most mature cities with regard to transactional e-services, whereas Kuala Lumpur and London are the least mature. "E-services are something that the population today expects. That you can communicate online or order postal voting ballots through an online form helps a lot" (BE 1, own translation).

Many municipalities have introduced online identities to offer transactional services. According to the interviewed experts, the population in Sweden and South Korea can manage nearly all transactions online with the help of their personal online ID (ST 4, SE 2). In Singapore, the population is able to use e-government services with their "Singapore Personal Access" (SingPass) which offers the services of about 60 agencies (SG 5). Online IDs have already been introduced in various countries to offer verified e-government services. Due to their digital identity cards, Denmark and the UK are highly ranked within the UN e-government development ranking in 2016 (United Nations, 2016).

On the city level, further services and online transaction possibilities have emerged through smartphone applications. A city that is offering many such mobile services is Dubai. The governmental mobile application can, for instance, forward fines for speeding (DU 3). However, the application also integrates very sensitive personal data. It is, for example, possible for husbands receive an information alert in real-time if their wife is attempting to leave the country (DU 2).

Mobile applications can be combined with other city services like paying for a parking ticket or bus ticket. But whenever ICT is used for sensitive personal data or payment the service needs to be secure and trustworthy (Srivastava, Chandra, & Theng, 2010). According to an investigation of mobile applications that are based on open urban government data, only 17 apps out of 471 offer mobile payment (Mainka, Hartmann, Meschede, et al., 2015a). Many of the investigated apps covers topics like public transportation or traffic, but most are mainly designed to deliver information in combination with geographical maps and GPS. In the US, for instance, mobile payment applications are a growing market, but the acceptance is only rising slowly and mostly among the younger generations (Board Of Governors of the Federal Reserve System, 2016).

Pillar 4: Interoperability The fourth pillar refers to information sharing across governmental and non-governmental organizations. Organizations need to define a common standard to work within a network (Gottschalk, 2009), but while the technical process of data sharing seems to be a manageable issue, the integration of standards for knowledge transfer on the semantic level is acknowledged as difficult (Cullen, 2010). Integration of governmental agencies is also introduced mostly in the backend, and not presented on governmental websites. Hence only municipalities that publish their governmental goals, including the use of ICT, could be investigated. One example is Singapore, which is publishing and informing their citizens on how they introduce ICT solutions on a municipal level with the help of its city development plans. One important aspect in this was to emphasize that the governmental agencies are no silos. Instead, the action plan has stressed the interagency collaboration (Ke & Wei, 2004).

On the investigated municipal websites, the interoperability has been investigated from the user perspective (Fietkiewicz et al., 2016). Websites that offer special entry pages for various user groups or include standardized forms can be summarized with the term "boundary objects." Such documents make it possible to share asynchronous knowledge among diverse user groups, e.g. through an introduction on how to use a document. However, municipal websites share information to all citizens and can enhance the accessibility of this information through boundary documents. Among the investigated municipal websites, only Tokyo's governmental websites offer a "how to use" guide, and 26 websites offer separate entry pages for different stakeholders, e.g. business, citizens and tourists. Nevertheless, this is only one method how the government can make information better accessible for different user groups, but does not tell us anything about the information sharing between the agencies within a municipality. According to one interviewee from Barcelona (BA 2), it is a big problem that different parts of the government do not work together very well. Instead, they often have to fight one another, for instance, when applying for a higher budget.

One good example of interoperability is the citizen relationship management system 311 in the US. This system has been introduced, for instance, in Boston and New York to manage government-to-citizen interaction (Hartmann, Mainka, & Stock, 2017). It is intended as an one-stop shop where citizens can access all governmental (non-emergency) services they need. It is available through a mobile application, the internet, via phone and in personal. Thus, 311 is able to be accessed by all user groups. Additionally, through 311 a diversity of services is available via one platform, requiring all governmental services to be connected to the network of 311.

Today, many online portals can be found that offer open governmental data containing non-personal public information. These exist on the international level, e.g. the European Union (data.europa.eu), national level, e.g. Germany (govdata.de), as well as on a municipal level, e.g. Munich (opengov-muenchen.de). The amount and quality of the data vary between these portals. In an investigation of the open data movement in Amsterdam, Barcelona, and Paris, for instance, it was found that the cities offered between 130 (Paris) and 424 (Amsterdam) datasets (Mainka, Hartmann, Meschede, et al., 2015b). The data is available in different formats, e.g. non-machine-readable as images and PDF documents and machine-readable as CSV, XML, or JSON. Generally, the data at those portals is gathered from various sources, mostly government agencies, and interoperability is needed to make them accessible within a common standard. In the cases of Paris (opendata.paris.fr) and Barcelona (opendata.bcn.cat), both platforms also host the data and therefore need to be able to integrate the diverse sources. Amsterdam (amsterdamopendata.nl), instead, is a meta-platform that is linking to all city-related data sources and thus has no integration at one place. Of course, for users or app developers it is easier to use one platform instead of meta-platforms as in Amsterdam, which is why access should be offered at one place with one common standard.

Pillar 5: Participation The last pillar refers to e-participation and focuses on a digitally enhanced democracy by including further stakeholders in decision-making processes. Online participation can be realized, for example, through online surveys, political discussion forums or online voting (Fietkiewicz et al., 2016). However, the involvement of non-governmental individuals in discussions on future city plans has so far been introduced, for the most part, only in case studies and not become a standardized practice yet (Susha & Grönlund, 2012). E-participation faces a number of new challenges (Sæbø, Rose, & Molka-Danielsen, 2010): On the one hand, the right stakeholders need to be identified and addressed during the development and decision process, and on the other hand, the political administration need to become more open towards a participatory cooperation.

In a previous study on the maturity of e-government (Mainka, Fietkiewicz, et al., 2013), e-participation was investigated according to the possibilities of citizens to become involved in online surveys, available forums for questions and discussions, online participation in community meetings and, finally, according to online voting possibilities. The most municipalities offer online questionnaires to involve citizens (23 cities), with discussion forums at which citizens can communicate and ask questions being the second most common measure (21 cities).

In Frankfurt, for example, a website called "Frankfurt gestalten" ('shape Frankfurt', frankfurt-gestalten.de), which is inviting all citizens to discuss city planning and other issues to create a better quality of living (FR 5). But to which extent these attempts impact the decision-making process is not clear to the experts interviewed. According to the website's own description, the focus is to make political information more transparent for citizens. Therefore, people can identify information about political decisions with reference to their location, for instance, the neighborhood they are living in or the city center where they are working. The platform only addresses citizens, and it is unclear whether politicians and administrations are aware of the discussions on the website.

Another expert has emphasized the use of a software which allows citizens to comment directly on a map (HE 4). The methodology used is called SoftGIS and relies upon geographical information developed at the Aalto University in Helsinki (Kyttä, Broberg, Tzoulas, & Snabb, 2013). It is based on a citizen survey that includes the location of the respondent participant to identify environmental issues for the urban planning process. In other cities, e.g. Stockholm or Boston, the city uses social media platforms to encourage citizen participation (BO 1, ST 4).

> You can do almost anything here digitally. There are tons of opportunities for online participation. From social media to surveys to forum systems. The city is pretty connected. It's a small city. It's a lot easier here than in a city like New York City. But the mayor is absolutely adamant about hearing the voice of everybody. Of all the constituents... [T]here is always room for improvement but we do it very well. (BO 1)

One expert from Helsinki (HE 2) is criticizing such platforms. According to him, a discussion platform has been established, but not many citizens have used it, and in many cases, the comments have been at a very low level. A further argument from Hong Kong is that the people would not have the time to participate in decision-making processes (HK 2). A similar statement was given by an interviewee from London (LO 1). He opines that the people do not care about politics, they just care about their personal issues. Today, a fast and easy way to participate are online petitions. The people create e-petitions for everything but whether they are helpful is another question (BA 2). In addition, it is debatable whether e-petitions have the same weight as paper petitions with signatures. If the citizens are the only user of those platforms and their voices never get heard by decision makers, this does not qualify as e-participation or e-democracy.

> I mean they still do projects without informing really. They inform the people once the project is designed and ready to go... There has been a project where they say they've done more... [T]hat is the project of 'Les Halles'... So they've said that they are going to do a large so-called participation process which has been, in my point of view, a caricature of a participation process. (PA 2)

In Paris, there is no e-participation or e-democracy (PA 6) and there is not enough transparency, for example on budget spending (PA 6): "We are in the middle age of participation in Paris... [But] I'm not sure there is even an awareness of the problem" (PA 2).

Nevertheless, it is becoming more common to involve citizens in the decision-making process (Janssen & Helbig, 2016). In São Paulo, for instance, one interviewee has stated that the city is actively promoting participation processes in city development plans, e.g. with posters in the metro stations which invite the citizens to get involved. E-services could add incentives to reach more people, but are unable to completely replace personal communication (SP 5). To move all interactions online makes the people anonymous. Some experts think that it would be dangerous to have e-democracy and e-voting, saying that it is important to take the time to vote in person (LO 7).

Implementation of e-governance

E-government services have enabled new possibilities to share information, to communicate, to make transactions, to cooperate and finally for citizens and businesses to participate in political processes. The utilization of these possibilities to enhance planning, innovation, and funding also on the municipal level is called "e-governance" (Palvia & Sharma, 2007), while acknowledging citizens and businesses in a city as important innovators in the decision-making process is labeled as "open government" (Harrison et al., 2011). This refers to the term "open innovation", which originates from economics and is used to define the free flow of knowledge and innovation between different stakeholders (Chesbrough, 2003). On the city level, these innovation flows can occur when the government accepts innovative ideas from outside its own agencies, e.g. from citizens and businesses. Together they may enable, for instance, new services based on open data or establish better conditions to start a business.

> It only helps in so far as your creative class looks at the government and says: 'They value the things we value...' But often times that creative class just looks if the government is not doing that. 'Who cares? We do our own thing!' And they do... Chicago was like that for a long time. Before mayor Manuel came in, we were not an e-government and yet we have a still thriving information economy. (CH 2)

Therefore, to establish e-governance or open governance the community needs to be active and engaged. According to some experts, this is only possible if the community trusts in the government and acknowledges the political willingness of their municipality. In Milan, for instance, one interviewee has stated that there is a lack of political willingness to implement e-governance (MI 7). Thus, the citizens do not see the possibility to participate. A further interviewee from Tokyo mentioned that the politicians in Japan have no good standing (TOK 3), meaning that young people are not interested in politics and election turnout is very low. In Barcelona, an equal problem exists, as one expert has stated: "The problem is that we don't have much experience in e-participation and e-government or democracy. The councilor has asked the people: 'Which project would you prefer to be realized?' The people's option was 'none' because they don't want this councilor and representatives at all" (BA 3). In general, there is a skepticism that e-services will lead to more democracy. Hence, "how many really have e-democracy in the world? This is very low" (BA 1).

Furthermore, the political willingness is also about integrating all people working for the administration. They should also get the feeling that the digitization process is able to help them in their everyday work and not merely makes everything more difficult. Currently, in Barcelona, there are a few forerunners that implement some good ideas, but there is no common standard on how to implement e-government services (BE 1). Thus, the availability of e-services depends almost entirely on each individual department's willingness to implement them. For instance, in Frankfurt the immigration administration has no e-services, whereas the city office supports many online applications (FR2, FR 3). Particularly in Germany, e-government seems to be a national development that is only adapted if regulated by law (BE 7).

However, e-governance goes one step further and asks for sharing power between all stakeholders in the city (government, citizens, organizations, businesses, etc.). In Boston, for instance, "[t]he political system is partly conservative and doesn't welcome a kind of distribution of power and loss of central power" (BO 3).

In Tokyo, e-governance is not possible because there is no horizontal integration (TOK 2). The agencies do not work together, making it difficult to invite additional parties to cooperate. And São Paulo is a city that seems to be very far behind and removed from this development, as one expert stated: "I don't think I will see it in my lifetime here... It will require a lot of changes in the mindset of the politicians, of the population ... This would be a dream" (SP 4). This change in the mindset can be observed, for instance, in some projects for developing public services such as building or rebuilding public libraries (Mainka et al., 2016). The public library of Helsinki, for example, has actively engaged the citizens in the design development of the new public library, calling this "open and participatory planning", with the hope that if the users of the library have been involved in its development, they have a feeling of ownership and stronger affiliation to the library.

To identify open government initiatives some new indices have been introduced, investigating the degree of participatory and democratic governance on a national level. One example is the "Sustainable Government Indicator" introduced by the Bertelsmann Stiftung (Bertelsmann Stiftung, 2016). It is based on the three pillars "Policy Performance," "Democracy," and "Governance" and evaluated with 67 indicators. The indicators of the policy performance are investigating the national conditions in a broader sense with respect to economic, social, and environmental policies. Aspects that are more related to open government are evaluated according to the quality of democracy and of governance. For instance, the right to vote, popular decision-making, and corruption prevention are investigated as part of the quality of democracy. Part of the quality of governance is the citizens' participatory competence according to citizens' knowledge about policies, people voicing their opinion to a public official, as well as the voter turnout. Another global investigation that includes e-participation indices is published by the "United Nations E-Government Survey": "The Survey evaluates e-participation, i.e., use of online services to engage citizens and non-citizens including through provision of online information (e-information), interaction with stakeholders (e-consultation) and engagement in decision-making processes (e-decision making)" (United Nations, 2016, p. 145). Both indices are based on a combined methodology considering quantitative as well as qualitative data. Top-ranked nations (according to the "Sustainable Government Indicator") are Sweden, Finland, and Denmark and (according to the "UN E-Government Index") the United Kingdom, Australia, and the Republic of Korea. However, on the city level, no comparable index has been published. Municipal openness, the transparency and the availability of open data have been in the focus of several researchers, e.g. in the "Open Cities Index 2016" (Canada's Open Data Exchange, n.d.). However, one expert cautioned that the implementation of e-democracy should always go hand in hand with advanced data security (SP 3).

Conclusion on e-governance
All in all, the investigated pillars of e-government (information dissemination, communication, transaction, interoperability, and participation) have shown that there are different approaches regarding how ICT is used to make governmental services accessible in a digital era. ICT opens the way for more transparency and adds possibilities of participation that were unimaginable ten years ago, e.g. making complaints or retrieving information through the web and mobile apps on a 24/7 basis. To communicate with their citizens, local authorities have adapted to use social media channels as Twitter, Facebook or YouTube. E-identities have also been introduced to verify correspondence and offer online transactions in many municipalities. However, sharing information across governmental and non-governmental parties requires common standards. One good example of governmental interoperability is the 311 citizen relationship management system introduced in some municipalities in the US, intended as a one-stop shop where citizens and businesses can access all relevant governmental information and services. It should be noted that e-participation has been discussed controversially by the interviewed experts. On the one hand, it is acknowledged as very fruitful, for example in the adoption of citizens' information in combination with location-based data. Thus, for example, in Helsinki, such data was able to be used for urban planning based on citizens' feedback. On the other hand, online platforms do not always have the expected impact and frequently attract people to comment on a low level or do not even reach them.

However, e-services are just the tool with which the government tries to reach citizens or transfer information. In order to implement e-governance or open government, the active engagement of all city stakeholders is required (government, citizens, businesses, organizations, etc.). As stated by many interviewees, the municipal government still has difficulties in becoming more open and sharing power. Thus, a real e-democracy is not established right now in any of the investigated cities. To compare the degree of e-governance, some indices exist on the national level, but not on the city level. However, the data evaluation is highly complex and is based on qualitative as well on quantitative data. On the city level, comparisons on transparency and open data can be found, which will be investigated in the following subchapter as an aspect of the free flow of information.

H10 Free flow of information
Following Yigitcanlar (2010), the "assurance of knowledge society rights" is a basic element of an informational city. One of these rights is the freedom of information. Without freedom of information, it is possible to be an informational

city but not a knowledge society (Lor & Britz, 2007). Affordable access to knowledge, the exchange of knowledge, critical thinking, and exchange of ideas are fundamental requirements of the knowledge society. Furthermore, freedom of information is essential for a transparent, open, and democratic government that helps to prevent corruption (La Rue, 2011). Independent information flows through the mass media as well as online and offline public data enable the open exchange of information and knowledge. The definition of the informational world city used in the present work includes the emergence of the knowledge society, which is why the dimensions and the existence of a free flow of information will be investigated with the following hypothesis:

H10   A free flow of all kinds of information (incl. mass media information) is an important characteristic of an informational world city.

As represented in Figure 6.38, the interviewed experts have a rather high expectation of the free flow of information in an informational world city. According to the result of the SERVQUAL evaluation, the average of all expectations is 6.4. A lower expectation can be observed by the experts from Vienna, Chicago, Los Angeles, Hong Kong, and Berlin. This lower rate is caused by the definition of "all information." Hence, for nine experts the protection of private information is

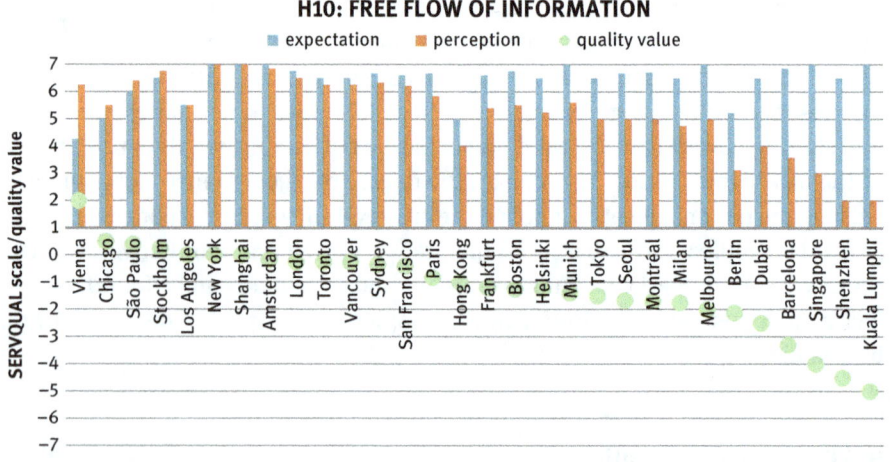

Figure 6.38: Quantitative interview results according to SERVQUAL (quality value = perception − expectation) for H10 (A free flow of all kinds of information (incl. mass media information) is an important characteristic of an informational world city).

essential which would be eroded if "all information" were free. There needs to be a balance between privacy and transparency (LO 8). In the following, free flow of information therefore refers to public data and the mass media.

Accessing information requires information literacy. For example, state censorship in China constitutes a formidable information barrier, with some information only available on the black market, according to one interviewed expert (HK 2). Thus, a free flow is not given in China. A further precondition to a free flow is the ability to access the internet (four experts), a field in which libraries can contribute through computer terminals or free public Wi-Fi (LO 6, LA 2). In the SERVQUAL evaluation, a positive quality value was found for the cities Vienna, Chicago, São Paulo, and Stockholm. São Paulo, for instance, has been described by the most experts as a city with a free flow of information, but the information broadcasting is very selective, e.g. what is shown in the top news (SP 5). Mass media inhabits a critical role for the free flow of information. A diverse media landscape of TV stations, newspapers, and magazines is expected to inform the society about current news and politics on national and global level (17 experts). Of course, the media industry also faces serious issues, such as limited competition leading to oligopolies and cartels (six experts), or the threat of governmental censorship (three experts). A trend can also be observed that mass media, like newspapers, are increasingly being read online (Chandra & Kaiser, 2015), in particular by the younger generations (VA 1). This medium opens new ways of communication and discussions in an online environment, but an overabundant flow of information could also result in an information overload (LO 4, TOK 3). Additionally, a negative aspect is the growing amount of hate speech that can be found online (HE 3), leading to the difficult problem of balancing freedom of speech with other valid interests. In Sweden, for example, there is a law against hate speech but also the right to freedom of expression (ST 1). The free flow of information is a result of both the freedom of expression (four experts) and the freedom of access to information (ST 1).

Within the context of public information, municipalities are increasingly opening their public datasets via online platforms to create transparency. Open data has therefore become an important aspect within the context of free flow of information (six experts). Of course, politicians need to keep sensitive information confidential (BA 8) but beyond that any information that increases transparency should be open to the public (three experts). This requires political willingness to open public data (BO 2). In some cases, this development is prevented due to the administration's fear of losing the control or of publishing erroneous data (BE 6). Four experts argue that access to public data should be free, so that it can create engagement and innovation. Overall, according to the interviewed experts the topics mass media and open data play a critical role within the free flow of information and will be further investigated in the following.

Mass media
In relation to the question of free flow of information, many interviewed experts have started to count how many broadcasting corporations, newspapers, or publishers are located in their city. "We have public television stations, ... commercial television, ... publishing, tremendous free street publication all over the place. When you are out of the streets there are glossy magazines, neighborhood newspapers [etc.]" (BO 4). The diversity of mass media, e.g. newspapers and TV stations, plays an important role in informing society (BO 3). In Montréal, for instance, three newspapers are published. Two are in a conglomerate and one is independent. It is important to have independent publishers (MO 2). One expert from Toronto (TOR 2) expected that in his city the free speech of media is higher than in any other place in the world. In São Paulo, as well free speech of media is given, but it seems to be somewhat selective media coverage (SP 5). The overall impression is that news mostly cover crime and traffic jams. In Shanghai, for instance, current news about politics are not broadcasted (SHA 1). While the global press is accessible in China, it is largely in English, and most Chinese people do not speak English (SHA 1).

The free flow of information within the mass media is evaluated by Reporters without Borders on the national level in the "World Press Freedom Index." This evaluation is based on an expert survey and the underlying framework and covers the topics: pluralism, media independence, media environment and self-censorship, legislative environment, transparency, infrastructure, and abuses (see 3.2, "Political will"). Table 6.9 presents the World Press Freedom Index 2016 for the respective nations of the cities investigated.

According to the overall investigation of the world press freedom, a decline was found for the years 2013 to 2016 (Reporters without Borders, 2016). The mass media is under pressure by "oligarchs" that are buying media outlets all over the world. Along with governmental pressure, this is increasing attempts to influence the media with regard to publications about e.g. religious ideologies. A further problematic issue is internet access block by the government and the destruction of printed publications that governments dislike. Enhanced "self-censorship" is acknowledged, further amplified through legislation in many nations that allows journalists to be published for crimes such as "insulting the head of state," "blasphemy," or "supporting terrorism."

The results show that only 12 of the 21 nations are found to have good or fairly good press freedom in 2016 (Table 6.9). However, in a city like London, which is located in a country that has a fairly good freedom of press, the interviewed experts reported being aware of "oligarchs." Thus, one expert (LO 2) stated that London is a hub of

Table 6.9: Overview of rank, score, and category of nations ranked in the "World Press Freedom Index 2016" (Reporters without Borders, 2016) with reference to the 31 cities investigated.

World Press Freedom Index 2016

Categories	Country	City	Rank	Overall Score 2016
Good	Finland	Helsinki	1	8.59
	Netherlands	Amsterdam	2	8.76
	Sweden	Stockholm	8	12.33
	Austria	Vienna	11	13.18
	Germany	Munich, Frankfurt, Berlin	16	14.8
Fairly good	Canada	Toronto, Montréal, Vancouver	18	15.26
	Australia	Sydney, Melbourne	25	17.84
	Spain	Barcelona	34	19.92
	United Kingdom	London	38	21.70
	United States	New York, Boston, Chicago, San Francisco, Los Angeles	41	22.49
	France	Paris	45	23.83
Problematic	Hong Kong	Hong Kong	69	28.50
	South Korea	Seoul	70	28.58
	Japan	Tokyo	72	28.67
	Italy	Milan	77	28.93
	Brazil	São Paulo	104	32.62
Bad	UAE	Dubai	119	36.73
	Malaysia	Kuala Lumpur	146	46.57
	Singapore	Singapore	154	52.96
Very bad	China	Shanghai, Beijing, Shenzhen	176	80.96

mass media information, publishing information globally, but that, in the end, mass media information is mostly dictated by a few big corporations. Interviewees living in countries with problematic or bad press freedom are also aware of the situation. In Milan, for instance, there is self-censorship in the media because there are things that cannot be said (MI 2). In Japan, the press also seems to be free, but acknowledged to be cooperating in one cartel (TOK 1). In such situations, a small number of mass media "oligarchs" can decide which information will be broadcasted (TOK 6). In some nation, such as Malaysia, the news seem to be deliberately designed to teach the population how to behave by focusing on their fears (KL 1). For example, to end waste in the streets the press is publishing that malaria could break out if the garbage is not collected appropriately (KL 1). The same problem is acknowledged for Hong Kong, but according to one interviewee (HK 3) "Hong Kong is pretty free... Singapore is an example of an informational city that is not free."

Nowadays, a lot of news and information can also or even exclusively be found online. Blogs and social media sites are used by many to consume current news, but suffer from massive differences in quality and trustworthiness. This lead to the problem of "fake news" being circulated throughout a big user group due to dissemination within social networks. To avoid those "fake news," social media platforms like Facebook are asked and expected to implement algorithms that will detect them (Beuth, 2016). However, the users will also have to enhance their information and media literacy to identify those fakes by themselves. Thus, another worry is that "trustworthy" printed material will lose in significance in the future (BO 4). But regardless of the medium (printed or online), independent and trustworthy sources are important. Those can be, for example, independent online blogs that inform and present current news. In Singapore, which has a bad press freedom index score, blogs have begun to supersede printed newspapers (SG 10). But even in regions with a fairly good press freedom index like Canada, many people tend to retrieve information online: "Young people do not read newspapers. They don't trust them, they don't rely on them, they think it is just a kind of garbage getting thrown at them" (TOR 2). Hence, social media and blogs are becoming the main source of information for them.

To investigate the free flow of information through the internet, Freedom House (2016) conducted a survey on the national level based on three main indicators – Obstacles to Access, Limits on Content, and Violations of User Rights (see subchapter 3.2, "Political will") – and comparing 65 countries throughout the world. However, not all the respective nations of the informational world cities investigated are represented. Figure 6.39 shows the freedom of the net and freedom of press for 14 countries in ascending order of average freedom scores. The higher a score reached by a country, the less freedom exists. Thus, countries with a high degree of freedom in both indices are Canada, Germany, and Australia. And those with a low degree are Singapore, UAE, and China. According to both ratings, the freedom of the press as well as the freedom of the internet is declining. Hence, 67% of internet users live in countries where criticism of the government, the military, or the ruling family is censored (Freedom House, 2016). In 2016, authorities from 38 countries arrested people due to posts at social media platforms. Furthermore, secure messaging apps like What's App or Telegram are increasingly being blocked by governments. Overall, merely eight out of the 14 compared countries have a high freedom of the net which implies that in cities like Kuala Lumpur, Singapore, Shanghai, or Dubai a free flow of information does not exist.

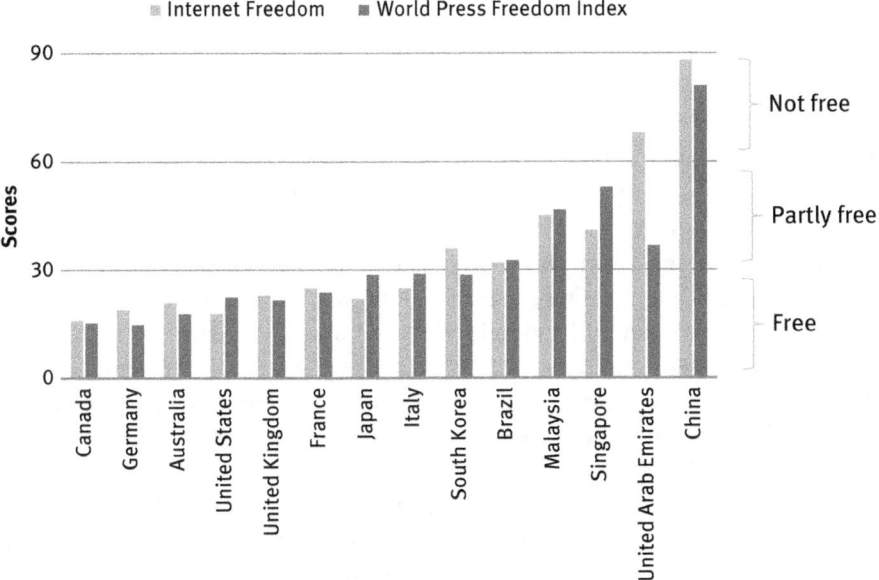

Figure 6.39: Freedom of information based on the internet freedom and the world press freedom index. Nations represent cities investigated in the work at hand and listed in ascending order according to average scores of both indices, with higher scores indicating less freedom. Data source: Freedom House (2016), Reporters without Borders (2016).

Transparency and open data

In addition to mass media, an increasing demand for open data is acknowledged and stated by several experts. Benefits are, for example, improved decisions based on big data (Castelnovo, 2015). "The idea to be data-driven is to understand that you can actually use data to help make better decisions and make things better for citizens. And Toronto does some of that. They never do as much as anyone would like but they are at least doing some of it" (TO 1). In New York, data-driven decision-making started with mayor Bloomberg (NY 1). And in Boston

> City Hall has got a whole bunch of folks together and we created this really unique apps... One of them is called street bomb and it uses the accelerometer in your phone which you have put in your car and you drive around the city. Every time you go over a bomb it takes your GPS, shoots it up to the sky and it goes to the Boston public works department and they collect and aggregate all this data to figure out where all the problems are and resolve them... That's saves money, is creative, and let citizen participate. All the things you want in a city... It's a very interesting city. We have an office in the mayor's department called the office of new urban mechanics... [This] office have done this really creative cutting edge things. (BO 1)

Open data may also help to make information such as data on air quality available. The US Consulate in Shanghai, for instance, has installed a pollution sensor on its building and offers real-time measurements which are accessible online (U.S. Consulate General in Shanghai, n.d.). This globally available and open information put pressure on the city government of Shanghai to improve the air quality (SH 1). It should be noted that an informational or smart city is not the same as a big data city (VI 1). However, opening data to the public may result in an increase in creativity and innovation (MU 1). Startups and the economy may use this data to build new projects with it.

According to the "Open Data Barometer" published by the World Wide Web Foundation and Open Data for Development network, robust data is necessary to promote democracy and drive development, as well as to reach the sustainable development goals: "end poverty, fight inequality and tackle climate change by 2030" (Web Foundation, 2015). Only in very few cases data on public spending, health, education, maps, or census data is available online and free of cost. In the "Open Data Barometer", the development on the national level is listed for 92 countries. The underlying indicators investigate the country's readiness (with regard to open data initiatives and policies), implementation (with regard to the actual realization of governmental commitments), and impact (based on the practical benefits that come with the existing open data).

A further indicator of open data are the right to information laws, which the World Justice Project uses to investigate the development of open government on the national level. However, "countries with relatively weak laws may nonetheless be very open, due to positive implementation efforts, while even relatively strong laws cannot ensure openness if they are not implemented properly" (World Justice Project, n.d.). Comparing the results on the national level based on the cities investigated in the present work, a direct impact of the right to information on the open data barometer cannot be found (Figure 6.40). On the one hand, the United Kingdom as well as the United States reach relatively high scores in both ratings. But on the other hand, China and the UAE both reach high scores with regard to their right to information laws, but very low scores when it comes to the actual implementation and real impact scores as calculated by the Open Data Barometer. Nevertheless, according to one expert from Amsterdam (AM 2), "the city of Amsterdam thinks that they are creating systems to become even more knowledge city than they already are, but Amsterdam is a core example of a city that already was a knowledge city before they started the policy." Accordingly, the Netherlands have reached fairly good scores in both ratings.

According to the interviewed experts, the current state of open data development in informational world cities differs greatly between cities. For instance,

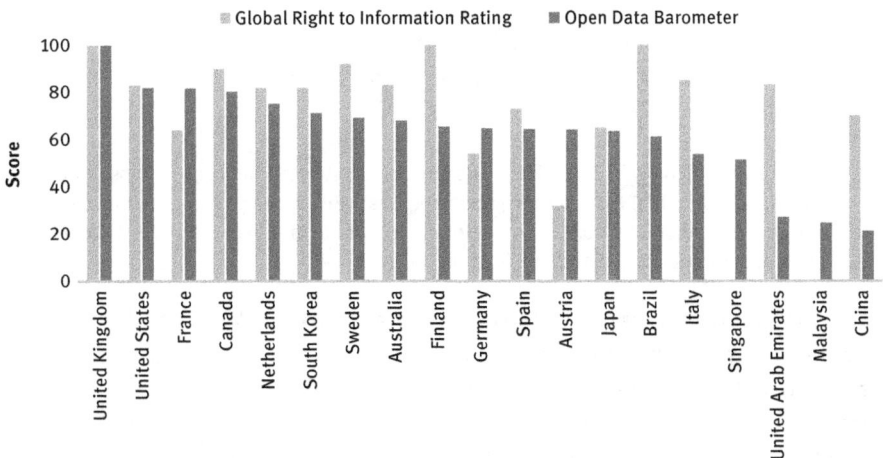

Figure 6.40: Access to information on the national level according to the "Global Right to Information Rating" for the year 2016 and the "Open Data Barometer" for the year 2015. Nations are listed in descending order according to the scores reached in the Open Data Barometer. The higher the score, the more open data/broader right to information is available. Data source: "Global right to information rating. Country data," (n.d.), "Open Data Barometer 2015," (n.d.).

one expert from Toronto (TOR 1) stated that open data and access to information is pretty good in North America. Open data and crowd sourcing happen in Toronto (TOR 1), and, as mentioned by another expert, also in Los Angeles (LA 2), where many records and statistics are online. Even the police department is making their statistics publicly available, and all representatives of the city have an email address to easily contact them. However, in other cities, open data is not as commonplace and rather something that the experts wish for. Thus, one expert from Melbourne (ME 1) stated: "Free access to state information is highly wanted. City council is currently setting up a new digital strategy." With regard to open data and e-government, Munich is also in its infancy and could profit from stronger support (MU 1). Milan is not far developed, as well, so "it's difficult to find information concerning traffic mobility, [or] public transportation" (MI 6). Even Barcelona is another example of low open data access (BA 3).

Finally, one expert argues: "I think we always want more information than we have" (BA 6). Data may also be subject to certain restrictions, such as in Frankfurt, where it is necessary to pay for access (FR 3). Furthermore, in Frankfurt it is frustrating to find information online (FR 5). Thus, information literacy is an absolute necessity when it comes to open data: "The information is pretty open but you need to know how to find the information" (LA 3). For instance, if you like to have access to information, then you need to have the language skills (SP 1). People in China and Russia do not speak English, meaning that a lot of information is not accessible for them. The same problem is acknowledged in Seoul (SE 1).

Ultimately, free access to information enables equal possibilities for everybody in society.

> You look at somebody like Bill Gates. His father was a wealthy man. Steve Jobs didn't have anything. And both of them built gigantic empires... Bill Gates' parents paid for his access to computer that let him do the programming. It is not to diminish even one of them. But it was easier for Bill Gates than it was for Steve. (LA 1)

A provider of this access can be the public library: "We at LA Public Library, we do our best to support that. Again with our Wi-Fi access and free... hardware computers that people also can sign up for two hours per day. That's all free access" (LA 2). Making open data available entails costs, sometimes enormous, for the city, which need to be calculated (MU 1). In addition, many local authorities fear publishing the data, strongly inhibiting the open data development. Therefore, transparency and openness are needed, as described by one expert from Boston (BO 2):

> When you put out the information then take the responsibility. And then take responsibility for correcting it ... like look: here it is. The good the bad and all is here together. Now help us fix it. Help us make it better. Engage with people. I think removing the restrictions on data and information helps with it. Whether it's research from universities and peer reviewed journals. Whether it is open source code and technology stuff. Whether it is government institutions opening up data to the media, press, researchers, advocacy groups, so they can make better decisions and better inform people. I think on every level more information is critical to ... till the whole process. ...information wants to be free" (BO 2).

Free flow of information opens new possibilities for democracy (FR 1). For instance, city government communication can often be more transparent (BA 1). According to one interviewee from Berlin (BE 6), much information is hidden because the local authorities are afraid of raising too many questions, e.g. about contracts with private businesses. Individuals in senior positions both in government and commerce make decisions for society or for the company and often tend to think that the public does not have to know everything (BE 6, own translation). In Chicago, one interviewee also mentioned that "there is knowledge in the government which is not completely transparent, decisions are made through our ... officers, political parties [behind closed doors]" (CH 1).

In a society with open data it is important to still maintain space for privacy. It is important that politicians can discuss things behind closed doors, openly discussing and improving their political strategies etc. before they have to make them public. Too much openness can be used against an organization. This has happened, for instance, to the pirate party in the Berlin Senate: "They had an open SMS conversation channel and everybody was able to see that they fight against each other within this party. This was not very professional and hurt the image of the party" (BE 7, own translation).

In Montréal, for instance, an open contract platform was established, which has made all contracts between the public and private sector visible. This was intended to help overcome corruption issues (MO 3). However, even though Montréal has a "free access information law" (MO 6), it still battles with corruption, recently losing two mayors who were entangled in corruption scandals (MO 4). However, in Vancouver, another Canadian city, one expert has claimed that "corruption wouldn't really work here, because people just work their way around it" (VA 1).

Conclusion: Free flow of information
In summary, the free flow of information can be investigated by looking at the freedom of information granted to journalists and the mass media on the one hand and open data on the other. Both draw upon the human rights, the right of free speech and the right to information, as their foundation. According to the "World Press Freedom Index 2016", not all of the cities investigated are located in nations with a good or fairly good rating. The worst situation is detected in China, followed by Singapore, Malaysia, and the UAE. Thus, six informational world cities can be said to have no or only very limited freedom of the press. Comparing these results with the freedom of the internet, the same conclusion can be made. The most serious problems are laws against publications the government dislikes, which not only has open censorship as a consequence, but also results in self-censorship by journalists. Another threat to the freedom of the press is the growing conglomeration of media outlets, resulting in "oligarchies" which can more or less dictate the news. Laws against social media posts and the blocking or banning of secure chat applications such What's App or Telegram erode the freedom of speech in many countries. Overall, an increase of violations against the right of free speech has been identified both in the press and the internet.

A rather new indicator for the free flow of information is the accessibility of open data. Laws guaranteeing the right to information were first established in the twenty-first and twenty-first century in the Scandinavian countries, but have spread throughout the world since. However, a right-to-information law that looks strong on paper may well be undercut by its actual implementation in the field of open data. This is the case, for example, in China and the UAE. Both have strong right to information laws, but weak open data development and implementations. In comparison, the UK and the US also have strong laws, but with much better actual implementation, demonstrated by their high rating in the Open Data Barometer, which investigates the readiness, implementation, and impact of open data. Finally, open data can be an accelerator of innovation and creativity but it needs to be implemented for the right data. Hence, the World Wide Web Foundation checks for the availability of data on issues such as public spending, health, education, maps, and census data. However, open data does not mean that everything should be public: protection for private data still needs to be guaranteed.

Regarding the hypotheses on political will (H8–H10), the increasing importance of online information has been dominant in each of them, with transparency and web-accessible political plans as important aspects of the political willingness. E-governance can also be used to bridge the institutional levels of the economy, university, and community to foster innovation processes. Still, openness is not easy to handle at all levels. According to the interviewed experts, some discussions and decisions need to be made behind closed doors. Despite the shrinking freedom of information, information overload and fake news are becoming a serious problem, particularly on social media platforms, demonstrating the need for better information literacy and education. In addition, English language skills open up access to global information, while also encouraging an open and welcoming culture, which will be further investigated as part of the cityness factor of world cities in the next subchapter.

6.3 World City

All the investigated cities are world cities. Not necessarily by their physical size and the size of their populations, but by virtue of their importance within the global network of cities. To become a successful player within the global market, the financial sector is of crucial importance. That is why the theory of flows of capital, power, and information is the foundation of informational city construction as described by Manuel Castells (1996). World cities are the hubs within this global network. Accordingly, the world city network can be explained through "central flow theory", where the economy, namely the firms located in a city, are building networks via branches in other cities (Taylor, Hoyler, & Verbruggen, 2010). But cityness also refers to the local process that happens within the space (Sassen, 2001). For Castells, this is the "glocality" of cities. They interact locally within the physical space and their surrounding area, and digitally with the global network. Based on the already discussed definitions in subchapter 3.3, "World city", I will investigate in the following two hypotheses:

H11 An informational world city has to be a financial hub with a large number of banks and insurance companies.

H12   An informational city is supposed to be a global city ("world city").

H11 Financial hub

To be located in a hegemony state has been of advantage for a city to emerge as a global hub (Jacobs, 1969). Hegemony states have always been leading in production, commerce, and finance (Taylor, Hoyler, & Smith, 2012). The financial sector is also acknowledged as part of the advanced producer service firm network that forms the global network of cities in the twenty-first century (Sassen, 2001). Financial markets have been investigated by several researchers and institutions such as the GAWC (Taylor, Derudder, Hoyler, & Witlox, 2012) or the Z/Yen Group with their "Global Financial Centres Index" (GFCI) (Yeandle & Z/Yen Group Limited, 2016). In this section, I will investigate the following hypothesis with the help of the expert interviews and existing indices:

H11   An informational world city has to be a financial hub with a great number of banks and insurance companies.

According to the interviewed experts, many of the investigated cities are financial hubs –not necessarily global hubs, but at least national or regional ones. Milan, for instance, is a financial hub within Italy (MI 4), Vienna is a hub for Central and Eastern Europe (VI 1), and London is a global financial hub (LO 2). In total, 44 experts have stated that being an informational world city does not require being a financial hub. However, 40 of these experts live in a city that they would describe as such a hub at least within a specific region. This is further reflected in the SERVQUAL evaluation in Figure 6.41. On average the quality value is positive (+1.15), but the mean expectation is +4.41. The highest quality values are observed for Kuala Lumpur and Hong Kong. One expert from Hong Kong explains his scoring as follows: "Hong Kong is big in financial things, but weak in other, e.g. culture. Hong Kong is too much about money [and it] is not a balanced city" (HK 3). It is considered important for a city to find the balance between finance and other economic sectors (TOR 1).

Another reason to rate the cities' financial sector rather highly, compared with the expected value of an informational world city, is the historical development. Historically, banks have always played an important role (AM 4), but in the future, this will be less related to a city's development (VI 3). Another expert describes this hypothesis as a "Chicken-Egg-Question. It creates each other. Banks create infrastructure or come there where infrastructure is. Helsinki has this infrastructure and therefore attracts banks and other companies" (HE 4). Some of the interviewed experts mentioned that the financial sector is important

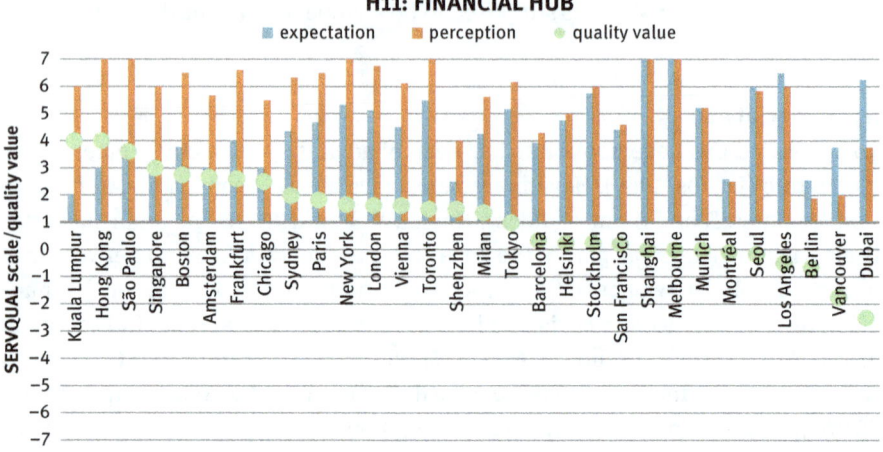

Figure 6.41: Quantitative interview results according to SERVQUAL (quality value = perception − expectation) for H11 (An informational world city has to be a financial hub with a large number of banks and insurance companies).

for the job market (e.g. LO 2). But others do not agree, pointing out that in their respective cities the financial sector has lost many jobs (e.g. AM 6). One expert also stated that "they do business in the entire country. I am not sure the physical location of head offices is as important" (TOR 2). As global players, financial sector firms are part of the producer service firm network, which will be investigated accordingly in section H12 ("Cityness"). However, as a location factor they play a tremendous role with regard to funding (LO 1). Finally, the financial sector is increasingly information-driven, which enhances technology development and the infrastructure in cities where they maintain offices (CH 1). On the flipside, jobs may also be lost due to the increasing digitization. Ideally, the local financial sector can also be a catalyst for venture capital and startups, e.g. in the financial technology sector (FinTech).

With this in mind, I will investigate the correlation of financial hubs and cities of the knowledge society, first with regard to the financial sector as important part of a city's economic success, and second with regard to the future role of financial hubs in providing venture capital and encouraging FinTech companies.

Financial sector as indicator of the knowledge society

That London and New York are financial hubs is not a recent development. Since decades the two cities have played a crucial role within the financial sector (LO 5). "If the informational city is a world city then probably the city has a financial market or can also be a financial hub. But if the informational city is a small or medium size city then the financial sector is not that important" (PA 5, own translation). As described above, many experts do not see a connection between the financial sector and the informational city. Of course, money can be a massive advantage, but it is the city that should be able to invest, not the private banks (BE 7). Hence, an informational city could be a university town without any financial headquarters (AM 2). In fact, four experts emphasized that knowledge institutions and research are of higher importance than financial institutions.

Comparing the financial market ranking and the performance of the best university for each city, the experts' arguments seem to be correct. Figure 6.42 presents the performance of informational world cities in the "Global Financial Center Index" (GFCI) and in the "Center for World University Rankings" (CWUR). For five cities, data was not available. In sum, there is a slight positive correlation of both scores, but overall the numbers are too small to allow any significant statement. The cities that are the biggest financial hubs are London, New York, and Singapore. Within the top ten of the best performing universities only Boston, San Francisco, and New York rank highly in both indices. Thus, New York is a city that is a financial hub as well as a university city. Boston has already been mentioned multiple times as the home of the elite universities Harvard and MIT, but not particularly as a financial hub. Nevertheless, according to one expert (BO 2):

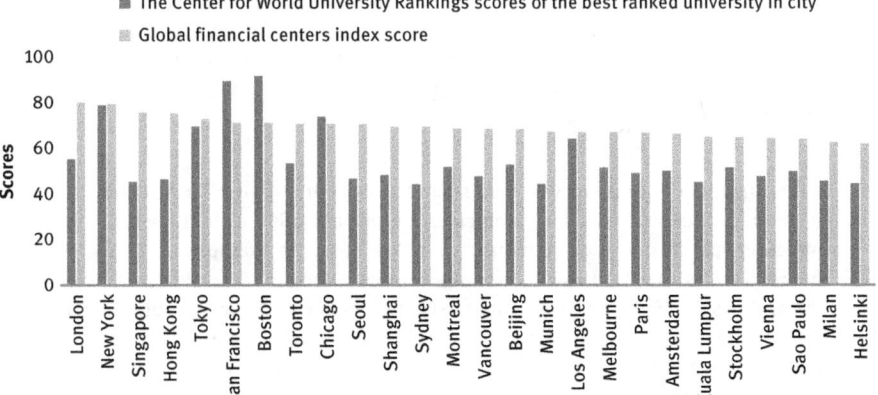

Figure 6.42: Cities' performance within the Global Financial Center Index (GFCI) and the best performing university in each city according to the scores of the Center for World University Rankings. Cities presented in descending order of their GFCI score. GFCI scores are adjusted (n/10). Source: O'Neill (2016), Yeandle & Z/Yen Group Limited (2016).

> Boston has a strong financial stuff here. There are several hedge fund type banks. There is a strong history of finance here, but I think the balance in other data-intensive things such as research, health care, medicine, biopharmaceutical, types of staff [is of importance]. There is all those kinds of things that are at least as critical to pushing that kind of informational city economy.

According to the world city researchers Beaverstock, Smith, and Taylor (1999), global financial centers are categorized as prime, major, and minor global banking service centers. Within the prime category are Frankfurt, Hong Kong, London, Milan, New York, Paris, San Francisco, Singapore, Tokyo, and Zurich. However, not all cities that are prime financial centers also have a university that is ranked within the top of the world, for instance in Frankfurt and Milan. The investigation of Beaverstock et al. (1999) is based on the theory of spaces of flows. Hence, cities or, more explicitly, headquarters in the cities are related to flows of power (Stock, 2011), with headquarters control branches in other cities (Beaverstock et al., 1999). However, there are many financial headquarters within each of the global banking services centers. Finally, no city can control the others, which makes them all financial hubs of the first category. Accordingly, one could argue that universities within the top ten or top twenty within a global ranking could qualify a city as university hub. For this, it should be noted that there are also university branches such the Rochester Institute of Technology, which is headquartered in New York and operates a branch in Dubai. However, the relation of city networks and flows of power will be further discussed in section H12 ("Cityness").

Economic success
"From a historic view banks are important. Banks attract companies to settle down here" (AM 4). But "banks and insurances are not creative and thus do not impact innovation" (FR 4, own translation).

> The economy of a future-focused city shouldn't be over-leveraged in any one industry ... There are leading industries in given cities and it's actually a liability in cities if you are over-leveraged in one specific industry. So New York, for example, is highly leveraged by finance. Los Angeles is highly leveraged in the cinema movies. What is unique about Chicago, it is one of the major cities in the US in which no one sector accounts for more than 30% of our economy. So we are the big economic driver for this region in the United States and one of the things that makes it a much more stable economy is that we have multiple sectors that make up the pie. (CH 3)

Montréal is an example that has found a new role after the financial institutions had left the city. Due to conflicts between Québec and the rest of Canada, many investors and companies moved to Toronto (MO 4). But following their departire, the cost of living increased in Toronto, and Montréal in turn became more attractive for creative people. Today, Montréal is no longer a financial hub, but has become a cultural hub (MO 4). Another way a city can find its strength is demonstrated by Singapore: "Singapore is a phenomenal example of an informational city based on logistics. Because they really run that port and they are really smart about it. A financial hub it is as well. But I think of the balance of requirements" (TOR 1).

Thus, it is of importance to be competitive in various economic sectors. Following Stock (2011), the turnover of the stock exchange market can quantify the city's economic significance and serve an indicator of the flows of capital. In simple terms, the turnover of a stock market is the measure of how often shares have changed hands (The Economist, 2001). It does not concentrate on the headquarters of banks and insurance companies, but rather quantifies the regional economy. Taking a look at the cities that have been identified as primary global banking service centers in the fiscal year 1999 (Beaverstock et al., 1999), some of them are also the cities with the highest turnover according to the local stock exchange in 2015 (Table 6.10). Exceptions are Shanghai, Amsterdam, Shenzhen, and Toronto.

The overall economic activity is also frequently measured by the Gross Domestic Product (GDP) of a region: "The GDP is the monetary value of all goods and services produced in a nation during a given time period" (Brezina, 2011, p. 4).

Table 6.10: The ten biggest stock exchanges according to the highest market capitalization value of shares in 2015. Data source: "Biggest Stock Exchanges in the World" (2016).

Rank	Stock exchange, country	Headquarter city	Market capitalization value of shares in 2015 (USD)
1	New York Stock Exchange, United States	New York*	$19.223 Trillion
2	NASDAQ, United States	New York*	$6.831 Trillion
3	London Stock Exchange Group, United Kingdom, and Italy	London*	$6.187 Trillion
4	Japan Exchange Group, Japan	Tokyo*	$4.485 Trillion
5	Shanghai Stock Exchange, China	Shanghai	$3.986 Trillion
6	Hong Kong Stock Exchange, Hong Kong (SAR China)	Hong Kong*	$3.325 Trillion
7	Euronext, United Kingdom, Belgium, Portugal, France, and the Netherlands	Amsterdam	$3.321 Trillion
8	Shenzhen Stock Exchange, China	Shenzhen	$2.285 Trillion
9	TMX Group, Canada	Toronto	$1.939 Trillion
10	Deutsche Börse AG, Germany	Frankfurt*	$1.762 Trillion

*Cities are primary banking service cities according to Beaverstock, Smith, and Taylor (1999)

The Global Metro Monitor Map investigates metropolitan regions with regard to their GDP, their GDP per capita and their employment ratio (Parilla, Leal, Berube, & Ran, 2015). The change ratio of the GDP for the fiscal years 2013 and 2014 and the GDP per capita in US$ for the informational world cities under investigation are represented in Figure 6.43. As can be seen, the cities with the highest GDP per capita in 2015 were Boston, San Francisco, and New York. Only New York also ranks among the global top ten stock exchanges (Table 6.10). However, the highest growth rates of the GDP per capita were reached by East Asian cities: Shanghai, Shenzhen, Dubai, and Kuala Lumpur.

Looking at the overall development in East Asian metro areas, the growth rates between the years 2009 to 2014 have dropped compared to the rates between 2000 to 2007 (Parilla et al., 2015). However, comparing the GDP per capita with the turnover of the global top ten stock exchanges a positive Pearson correlation can be observed (+0.60). This result is not surprising, since all the cities are successful global hubs. As a rather coarse instrument, the GDP or stock exchange turnover is not able to represent the diversity of industries within an economy, but it can indicate how successful the entire economy of a city is. But as Boston demonstrates, being at the top in the GDP per capita ranking neither depends on being a prime global banking service center nor on ranking among the global top ten stock exchanges. Still, having a strong financial sector can be of help in having a successful economy in general.

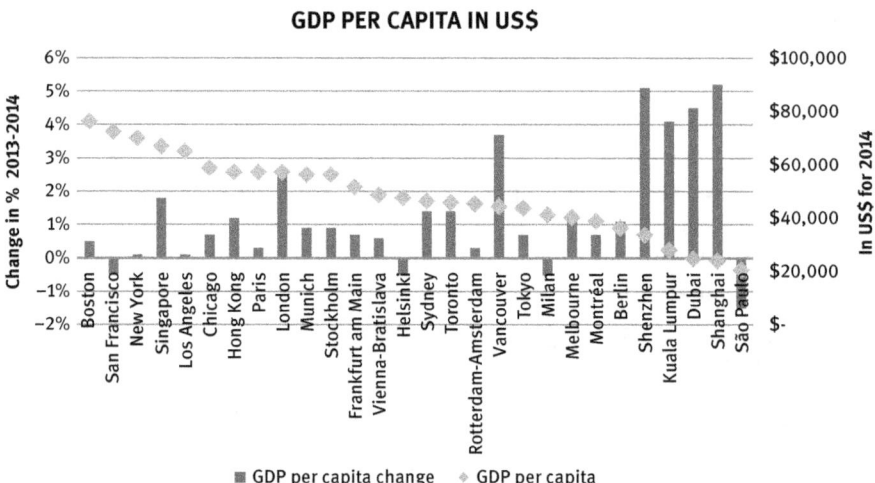

Figure 6.43: Combined change in employment and GDP per capita per metropolitan region for the time period 2013–2014. Metros are listed in descending order according to their GDP per capita in US$ for the fiscal year 2014. Data source: Parilla, Leal, Berube, and Ran (2015).

Future role of finance

Similar to the creative city, physical space for face-to-face meetings and a dense city can also bring together entrepreneurs with potential investors. A dense area where both can meet is of advantage, and then the physical location of a bank is of interest (BE 1). Today, venture capital is becoming increasingly important compared with conservative banks and insurances (BE 1). According to one expert (LO 1), "London is the best place for funding", whereas in Berlin only a few venture capitalists have settled down in the city so far (BE 1). But in order to stimulate the startup scene, financing, business angels, and others who support entrepreneurs in starting their business are needed (MU 1). Following one expert from Munich (MU 1), the advantage of the city is that it is rich, with a lot of small finance institutions and rich private investors that have the ability to provide venture capital. While Munich will likely never become the financial hub that Frankfurt is, it does have a strong financial sector. According to another expert from Vienna (VI 3), the importance of the global financial hub used to be of interest, but will no longer be of interest in the future. When a city cannot provide venture capital, entrepreneurs and startups will have to leave – one expert from Montréal (MO 3) has stated that the entrepreneurs there increasingly head to the US to apply for funding.

According to the global financing trends of venture capital backed companies, the total quarterly investment has grown by about 241% within four years, looking at the third quarter in 2011 in comparison to the third quarter in 2015 (KPMG, 2016). The peak of investment has been in the quarter in 2015 which was about US$ 39.6 billion. Following the report of EY (2016), the total amount raised through venture capital deals was about US$ 72.3 billion for the fiscal year 2015. On a quarterly calculation, a global decline of about -39% can be observed for the time period between the third quarters in 2015 and 2016 (KPMG, 2016). Following the report of KPMG, the highest number of venture capital deals are made for the tech sector. In 2015 and 2016 about 77% to 79% of the deals were signed within the tech sector, around 11% were signed in the health sector, and further 11% cumulated in other sectors per quarter. However, venture capital has become a growing market in recent years. On the one hand it has become part of the financial sector and on the other hand, the subject of funding are startups and young businesses within the tech sector.

Florida and King (2016) investigated the venture capital investments on a metropolitan level based on data from Thomson Reuters for 2012. Accordingly, Table 6.11 represents the metropolitan regions in descending order of their venture capital investments in million US$ (only for the informational world cities under investigation). San Francisco is at the top of this list, followed by three more cities in the United States: Boston, New York, and Los Angeles. London follows in fifth place. However, San Francisco's share of global venture capital investment is

Table 6.11: Overview of metro areas according to their venture capital investment in million US$. Only metro areas that are investigated in the present work as informational world cities and are mentioned within the top 20 venture capital cities within their geographical region are listed. Data source: Florida and King (2016).

Metro Name	Venture Capital Investment (millions US$)
San Francisco	6,471
Boston	3,144
New York	2,106
Los Angeles	1,450
London	842
Beijing	758
Chicago	688
Toronto	628
Shanghai	510
Paris	449
Montréal	267
Amsterdam-Rotterdam	205
Berlin	178
Seoul-Incheon	156
Stockholm	148
Munich	120
Shenzhen	117
Helsinki	99
Tokyo	94
Frankfurt	78
Singapore	57

about 15.9%. Looking at the whole San Francisco Bay Area, adding San Jose, the total share is 25.8%. For San Francisco, this makes up 1.95% of its total economic output. On a national level, the venture capital investments make up 0.17% of the total U.S. GDP as calculated by the OECD (OECD iLibrary, 2013). In comparison, the total information industry makes up 4.9% and the financial industry 17.8% of the total GDP in the United States in 2016 (U.S. Department of Commerce & Bureau of Economic Analysis, 2016). Due to the size of some of the investigated metros, a high amount of venture capital is accumulated there. Thus, for instance, London is the city with the highest concentration of venture capital in European cities, but calculated on a per capita basis the metropolitan region Copenhagen-Malmö has the highest investment rates (Florida & King, 2016). Comparing the most economically powerful cities with the leading venture capital investment cities, some overlaps can be identified (Florida & King, 2016). Many cities like New York, London, Paris, San Francisco, or Boston can be found in the list of the top 20

global cities as well as in the list of the top 20 venture capital cities. Nevertheless, a city like Tokyo which is a leading global city reaches only 54th place among the venture capital cities. According to one interviewee (TOK 2), this is not surprising, because in Tokyo, very few private investments are made.

Following one expert from New York (NY 2) "financial industries, particularly stock markets, and marketplaces... very heavily drive advances in technology in terms of the speed and performance. There is a lot of research like whether developing the fastest network router to go faster on the stock exchange." Hence, "finance is a major type of hub and very information-dependent" (CH 1). Another expert stated: "I think you attract a certain kind of talent having banks... and to some degree, banks and trading options attract network connectivity because you really want fast transactions... even San Francisco is not a really banking hub and yet it is an informational city" (CH 2). Increasingly, however, the finance sector and tech startups are merging into the FinTech industry.

According to several experts, the physical role of banks will decline, with them being relevant mainly for the older population and those who do not feel comfortable using technology (MI 1, TOK 1, ST 1). Following EY (2015b), the adoption of FinTech has large potential to increase and may rise, because they identified that 53.2% of persons interviewed in their survey did not even know that FinTech services exist. Looking at the ranking of the top global providers of financial technology by IDC Financial Insights (2016), the majority of the listed companies have their headquarters in the United States (51 FinTech companies, see Table 6.12). India is represented with eight and France with seven companies in the top 100. On the city level, the highest number of FinTech company headquarters is in New York, followed by London and Paris. However, all three cities were already identified as global banking service centers 17 years ago (Beaverstock

Table 6.12: Top three cities and countries according to their number of headquarters of FinTech companies based on the 100 most successful global providers of financial technology. Data source: IDC Financial Insights (2016).

Headquarters city	Number of companies in the IDC FinTech 100
New York	6
London	4
Paris	4
Headquarters country	**Number of companies in the IDC FinTech 100**
USA	51
India	8
France	7

et al., 1999), and except for Paris are also listed in the top ten of cities with the highest stock exchange turnover ("Biggest Stock Exchanges in the World," 2016).

Conclusion: Financial hub
In summary, there is no evidence that being a financial hub is important in becoming an informational world city. According to the interviewed experts, it could be of advantage, but it is not necessary. Therefore, a total of 44 experts did not agree with this hypothesis, although 40 of them live at least in a regional financial hub. Comparing the rankings of financial hubs with those of university hubs, a slight positive correlation can be observed. However, the rankings cannot be equated. Generally, it is not only the financial sector that is of importance, but an informational world city should rather be competitive in various industries, to guard against one sector not surviving in the future. The economic success can be measured for instance by the flow of capital. Stock (2011) has suggested the turnover of the stock exchanges as an indicator. The results show some overlaps between cities that were acknowledged as financial hubs in 1999 by Beaverstock et al. and those with the highest turnover in 2015. Shanghai, Amsterdam, Shenzhen, and Toronto are cities that have not played that role fifteen years ago, whereas New York, London, Tokyo, Hong Kong, and Frankfurt are at the top in both measurements. A further indicator of the economic success is the GDP. Comparing informational world cities, it becomes clear that it is not necessary to be within the top cities according to the turnover of stock exchanges or within the prime global financial service hubs as represented by Boston. However, Boston is the city with the highest GDP per capita and is one of the top global university hubs.

With reference to the future development of the financial market, venture capital and financial technology (FinTech) firms grow in importance. U.S. metropolises have the highest amounts of venture capital investment. At the top of the list are San Francisco, Boston, and New York. Venture capital is invested heavily in the technology industry. However, the financial sector itself is also highly information-driven, resulting in FinTech firms emerging around the globe. Still, the most successful ones have their headquarters in the United States, as demonstrated by the "Top 100 IDC FinTech" ranking. On the city level, the most companies of the top 100 are headquartered in New York, London, and Paris. Hence, these three cities are also hubs within the world city network, which will be further discussed in the following section.

H12 Cityness
Cityness can be understood in two ways: First, as the urbanity and local process that constitutes the inner-city, and second, as the flows between cities as an inter-city

relation. According to the second definition, cities interact with each other, setting up the world city hierarchy (Friedmann, 1986). Following Castells (2001), informational cities are "glocal" cities. Hence, they act as the physical concentration of the society and economy (locally), and through the spaces of flows (capital, power, and information) they are hubs within a global network. The flows of capital have already been discussed in the previous section (H11, "Financial hub") and the flows of information and knowledge refer to the ICT and cognitive infrastructure (see subchapter 6.1, "Infrastructures"). Finally, the flow of power refers to world cities that are "large, urbanized regions that are defined by dense patterns of interactions rather than by political-administrative boundaries" (Friedmann, 1986). Those interactions primarily refer to the economic market. Producer service firms, as well as multinational corporations, link between cities on hierarchical and horizontal levels (Friedmann, 1986). This constitutes economic hierarchies with regard to headquarters in cities interacting with branches in other cities (see subchapter 3.3, "World city", for a detailed explanation). Accordingly, I will investigate the flows of power of informational cities based on the expert interviews and on secondary research data with the following hypothesis:

H12 An informational city has to be a global city ("world city").

All of the cities investigated in the present work have been identified as world cities in the research of John Friedmann, Peter J. Taylor or Saskia Sassen. However, not all interviewed experts agree that their city is actually a world city. 17 experts stated that they may be living in a regional center or in a sort of world city, but that the only real world cities were New York, London, and Tokyo. All of the interviewed experts from London and New York do agree that their respective city is a world city. Tokyo is an exception in this regard, as two experts did not agree due to what they perceive as a lack of openness (TOK 3, TOK 4). This is also reflected by the negative quality value of -2 for Tokyo in the SERVQUAL evaluation (Figure 6.44). Comparing the results of all cities, 11 have a negative quality value, five a neutral and 14 have a positive value. Thus, no definitive conclusion can be drawn as to whether an informational city is supposed to be a world city or not. In total, 22 experts mentioned explicitly that it is not important and 17 that it is important for an informational city to be a world city. Hence, in London, the interviewed experts on average see it as very beneficial to be a world city as indicated by an expectation value of 6. That they see London as the leading world city is clear through the scaled experience with 7. For the experts in New York, it is a must to be a world city, as they marked their perceptions and expectations both as seven on average. In Tokyo, the experts also rated their expectation very high but not their experience, which points to a lack in cityness.

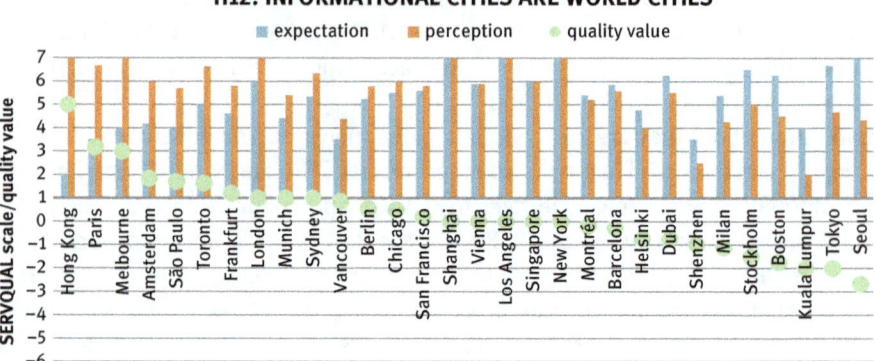

Figure 6.44: Quantitative interview results according to SERVQUAL (quality value = perception − expectation) for H12 (An informational city has to be a global city ('world city').

During the interviews, the question arose whether the city is a world city when it is a global city. For many of the interviewed experts, it is no question that they live in a global city (37 experts agreed), as through import and export trade, a city is integrated in the global market (BA 6, PA 6). Through the internet, information can be retrieved globally (seven experts). Even through a city's own or a nearby airport, the city can be reached within a certain time (BE 1). But to be globally connected, whether digitally or physically, does not automatically lead to being a world city. Thus, nine experts mentioned that they are not living in a world city, respectively the experts are from Barcelona, Berlin, Frankfurt, Milan (three experts), Stockholm, and Tokyo (two experts).

However, several options have been discussed as to what constitutes a world city. "World city, does this mean like we are world class or that we interact with the world and across the world?" (TOR 2). Seven experts mentioned that it is important that the economic market is globally oriented through global trade and an international workforce. If the city is not a world city, then it probably does not have such a multilayered structure. For example, Oxford is a knowledge city and is less economically oriented and Tallinn, for example, is an IT hub but lacks in other aspects (LO 2). Cities that are acknowledged as a regional hub may also convey a cosmopolitan lifestyle. São Paulo, for example, is the most important city in South America (SP 5), and Stockholm is the hub of Scandinavia (ST 4). However, an informational city does not necessarily have to be a world city if the city is able to invest in the right infrastructures of knowledge and information. Examples of informational cities that are no world cities are Oulu in Finland or

Aalborg in Denmark (PA 2), as well as Vanarasi, a knowledge hub in India (TOR 1). For two experts (BO 2, MO 2), to be a hub for research and development is more important than to fulfill the conditions of being a world city.

An informational city can be a top tech city with universities and a high graduate population, but without direct flights (MO 2). However, it could not be a world city due to the lack of physical connectivity. For nine experts, it is of crucial importance to be very well connected with the world. A global airport (nine experts), conferences (two experts), and trade fairs (MI 4) enhance the physical infrastructure. Furthermore, to be open and welcoming has been emphasized as very important. The mix of people (four experts), ethnical and cultural diversity (eight experts), and openness to immigrants (five experts) represent this attitude. In addition, it is of advantage if the population can speak English (seven experts). "The language helps in London to be open and attract people" (LO 6). "Over half of Toronto's population have been born outside Canada and about half of our population are minorities. I think it's probably the most diverse city in the world sometimes... Our diversity is our strength. That is the slogan that Toronto has" (TOR 2). This culture, diversity, and openness is an important aspect in attracting people. Cities that lack in culture have the problem to be not acknowledged as a world city. For instance, Shenzhen is no cultural hub and has also struggled to be accepted as a hub within the global economy (SHE 1). Many cities that are built from scratch have to overcome such problems. Thus, for instance, Songdo, a new planned city near Seoul has to identify its own urbanity (Ilhan, Möhlmann, & Stock, 2015). Accordingly, Saskia Sassen has stated in an interview that those cities are deurbanized places (Meister, 2012). It is possible to build a global economic hub, but the culture is anchored with the people and has to arise from human interaction.

In summary that an informational city is a world city may be a consequence and not the starting point of a development (MI 1). During the interviews diverse aspects have been mentioned that constitute a world city or global city. In the following, I will therefore further investigate the importance, first, of the economy, second, of the physical connectivity and finally, of an open and welcoming culture.

Hubs of power, knowledge, and innovation
Diverse measuring methods have been established to identify flows between cities (see chapter 3, "Measuring cities of the knowledge society"). Economic measures are related to the flows of power between cities in a global network. Accordingly, various rankings, indices, and benchmarks have been published that rank a city's position in comparison to others. Following Clark, Moonen, and

Couturier (2015b), as of 2015 29 publications in total were based on city investigations dealing with business, finance, and investment, and another 16 dealing with the economic growth and performance have been published. In the majority of world city comparisons, six cities consistently rank at the top: New York, London, Paris, Singapore, Hong Kong, and Tokyo. They are referred to as established world cities (Clark, Moonen, & Couturier, 2015a).

However, other cities can be or become global hubs by virtue of specific activities. For example, "Boston doesn't have the same business corporations like London or New York. What you find in Boston is a lot of companies open up their research and development offices here. [It] ... is not about having the CEO ... here. It's about having the real knowledge arms of these corporations located here. ... Microsoft has its base in Seattle, Google in San Francisco... but because of the richness of Boston's information they choose to put engineering [and] research ... here" (BO 2). Frankfurt, for example, is a small city going purely by its size, and only due to its financial market is it a global hub (FR 2). A further niche is served by Shenzhen, namely that of a production hub for the world. It is able to attract talents within China, but not on a global scale (SHE 1). Taylor, Derudder, Faulconbridge, Hoyler, and Ni (2014) emphasize that specific strategic places can constitute global hubs with regard to "one" economic sector. Hence, a global city can find its specific activity within the world economy.

However, global or world city comparisons refer to a city's role within the global economic market and thus to its economic success overall. Hence, to be a capital-intensive city is of advantage, as is the ability to finance the necessary infrastructure (SG 13). And if you are an informational world city, then the information flows are also of crucial importance (MO 3), as, by definition, an informational city is home to the knowledge society (Stock, 2011). Thus, flows of information and knowledge dominate. To identify the control and command centers of the contemporary economy, the number of headquarters in a city is used as an indicator (Taylor & Csomós, 2012). And to measure the knowledge production of a city, the amount of STN publications can be counted (Stock, 2011). Finally, the knowledge that results in innovation is measured by the number of patents granted (Stock, 2011).

According to an expert from Tokyo (TOK 3), Japanese firms have branches around the world. Thus, there is a power flow from the company headquarters to all branches in other cities. But Tokyo as city is not internationally oriented, rather preferring to act within the closed national market. This results in fewer investments, for instance, in R&D. However, this has insulated Japan from the global financial crisis in 2008/2009. Following Taylor and Csomós (2012), Tokyo has developed most of its powerful corporations before the 1990s. This is also true for some corporations in the U.S. However, newcomers like Silicon Valley or Shanghai are expected to shake up the hubs of power in the global economic market. As presented in Figure 6.45, the typical global cities (Tokyo, New York, London, and Paris) are at the top in terms of the number of headquarters in the city, listed in the Fortune 2000 (Taylor & Csomós, 2012).

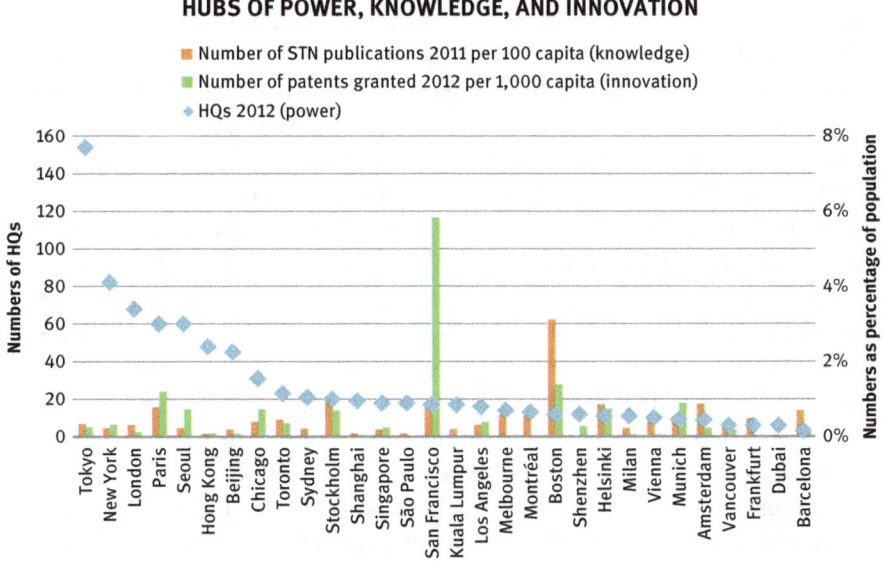

Figure 6.45: Hubs of power, knowledge, and information according to the number of company headquarters, STN publications, and patents granted in a city. Data source: Web of Science (retrieved in 2014), Derwent World Patent Index (retrieved in 2013), Taylor & Csomós (2012).

Looking at the knowledge and innovation output as represented by the number of publications and patents, a slightly different order emerges. According to the total amount of publications, Beijing, Tokyo, London, and Seoul are at the top, and on a per capita basis Boston is topping the list followed by Stockholm, San Francisco, and Amsterdam. Comparing the cities' innovation output, the highest number of patents was granted in Seoul, followed by San Francisco, Shenzhen, and Paris, and on a per capita basis San Francisco leads the field, followed by Boston, Paris, and Munich. In conclusion, being a world or global city based on the economic market is not a guarantee to be similarly competitive with regard to the other flows. Hence, cities like San Francisco, Shenzhen, Munich, Stockholm, and Beijing have already established or are growing in importance as hubs within the global information and knowledge flows that are of crucial according to the definition of informational world cities. Nevertheless, the presented figures just reflect a snapshot of the year 2011/2012. For future research, it would be interesting to investigate the development for a longer time period.

Physical connectivity

To be a global or a world city the city needs to be connected with the rest of the world (nine experts). That all of the 31 cities investigated in the work at hand have an enhanced ICT infrastructure, and some of them are also hubs of the ICT economic market, has already been discussed in section 6.1.1 ("ICT infrastructure"). But next to the digital connectedness, the physical infrastructure is also of importance, connecting a world city not only with its hinterlands but with the rest of the world, as well. Hence, the global digital world needs to be connected to the local physical world (SG 5). In world city research, physical connectivity is measured by the number of flights, especially the number of direct destinations (Budd, 2012), with hubs of international flights referred to as "aerocities." Budd (2012) counted the connectivity of the top five busiest international airports for the year 2008/2009 and ranked them by the number of international passengers (Table 6.13). Comparing the investigated airports, the number of international passengers is not correlated with the number of direct destinations served or of airlines serving the airport. However, it might be of advantage to be a direct neighbor of a world city and to profit from all existing infrastructures (BE 1). Examples are Silicon Valley, which is located within the area of San Francisco, or Cambridge, which is directly affiliated to Boston. Other modes of transport can increase the connectivity further. Thus, for instance, in Singapore, the port has particularly pushed the development towards a global city (SG 4). According an expert from São Paulo (SP 4), the airport

Table 6.13: Connectivity of the top five busiest international airports for the year 2008/2009, *2007. Data source: Budd (2012).

Number of international passengers (in million)	Airport	Direct destinations served	Number of countries served	Number of airlines
62	London – Heathrow	180	90	90
55	Paris – Charles de Gaulle	294	106	66
48	Amsterdam – Schiphol*	267	87	98
47	Frankfurt	304	106	119
46	Hong Kong	180	46	85

of São Paulo is well connected, the local infrastructure is lacking. Hence, the connection from the airport to downtown São Paulo is rather poor. According to the world city researcher Taylor (2012), the main users of the physical infrastructure are multinational corporations. Building this infrastructure, however, is planned on a political level. Thus, political decisions are able to influence the global connectivity, but in the end, a city's government can only invest if the economy provides the necessary capital. Dubai, for example, is growing in importance. Hence, in 2015 Dubai was the city with the most international passengers (77 million), and London was only ranked second (66 million international passengers). Dubai is also aiming to become the world's leading container port by 2030 ("Growing up. The Gulf state's expansion is more sustainable than its previous boom," 2015). Today Dubai is named "gateway city" and it has the potential to reach an equal status as Singapore in the future (Clark et al., 2015a).

World city culture
A world city is not merely made up of all the offices and private residences in a city. It is further constituted by a process of "complexity, incompleteness, and making" (Sassen, 2013, p. 209). Thus, as a world city, it is important to be an open city – open for globalization, open for culture and open for people with diverse backgrounds (BE 1). Xenophobic attitudes and right-wing extremists can severely harm a city's public perception (MU 1). This has been the case at the Max Planck research institute in Dresden, Germany. The institute currently has problems to attract talented people to the city due to widely reported xenophobic incidents. This will, in turn, harm the economy, because talent cannot be attracted from around the world. In contrast, Boston is a city that is attracting these talents to educate them at Harvard or at MIT. The problem is, as one expert (BO 3) stated, that the people come there, graduate, and in many cases leave the city afterwards. Seoul, for instance, is a globally known city, but it is lacking in openness (SE 2). According to one interviewed expert (SE 2), the population is rather closed. They cannot speak English and have to overcome many barriers. Tokyo, as well, is lacking in internationality, openness, and language skills, as stated by one expert (TOK 1). Japanese people tend to stick to themselves. However, openness to immigrants and diversity in culture is not always recognized as positive: "Milan doesn't have the willingness to become a global city because people do not want to lose their soul" (MI 8). Feeling lost due to the quickly growing population and immigrants has also been mentioned as a problem in Dubai. Especially the older generation struggles with the diversity of culture and people they have not grown up with (DU 2).

Nevertheless, to become a global hub, the city has to attract the creative class. An investigation of New York's labor market has identified that the cultural sector is the most dominant in the city, which helps in attracting further talents (Currid, 2006). However, the salaries are not the highest in comparison with other cities. "New York's great attraction may be that it is an integrated production and distribution system that provides access to high volumes of knowledge, ideas, information, skill sets, and greater possibilities for individual success within one's field" (Currid, 2006, p. 343). Thus, the attraction of talent is an important factor of a world city.

Whether a city is open to immigrants can be determined by the share of foreign-born population. Figure 6.46 represents these shares for the informational world cities investigated in the present work. Due to missing data, São Paulo had to be excluded. Overall, Dubai is the city with the highest share of foreign-born population (83%). This is not surprising, as Dubai has undergone a growth boom ("Growing up. The Gulf state's expansion is more sustainable than its previous boom," 2015). In the 2000's its economy was mainly based on the oil reserves. But economic free zones have attracted many foreign corporations to open subsidiaries in Dubai. In some of these economic zones, businesses were even allowed to open a branch without any local partner. Thus, the labor market was open to non-local firms as well as non-local talents. According to an interviewed expert (DU 3), local men prefer to work for the state, as the jobs are very well paid and the living costs are very high. Dubai also has a large number of immigrants work

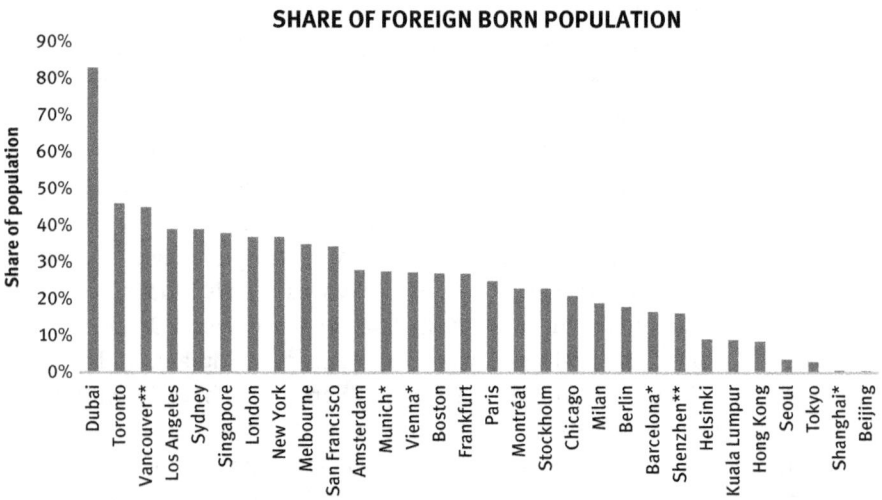

Figure 6.46: Share of foreign-born population of informational world cities (Year of statistics **2011, 2015, *2016). Data based on diverse resources see Appendix VII, "Share of foreign born population."

at the construction sites to build the high-rise buildings as well as talents that are responsible for the economic market growth (Kosior et al., 2015). The second and third highest share of foreign-born population can be found in Toronto and Vancouver. The interviewed experts from both cities have mentioned that their city is very open to immigrants and that this is of importance for the local labor market (TO 2, VA 3). Montréal, the third Canadian city in this investigation, is less open, due to the language barriers (MO 3) and less representation of international organizations (MO 6). The established world cities London and New York both have a share of 37% of foreign-born population, whereas Tokyo is the big exception with only 3%. Further low rates can also be identified for other Asian cities like Beijing and Shanghai, which both have rates below one percent.

The "Global Talent Competitiveness index" (GTCi) investigates the ability of countries and businesses to attract a talented workforce. This index is produced jointly by INSEAD, Adecco, and Singapore's Human Capital Leadership Institute, and was so far published for the years 2013, 2014, and 2015/16. The index focuses on the openness of nations towards a global workforce. According to the "World Migration Report 2015," migrants tend to concentrate in global cities and not spread in the nation (IOM, 2015). Thus, for example, Australia has a share of 28% of foreign-born population, but the majority is concentrated in the world cities Sydney and Melbourne. Therefore, it is not hard to see why Singapore, a city-state, reached the top of the talent competitiveness ranking in 2015 (see Figure 6.47).

Figure 6.47: Global Talent Competitiveness index 2015 on national data in comparison to the share of foreign born population based on city data. Data source INSEAD, Adecco, and HCLI, (2015) and Appendix VII, "Share of foreign-born population."

Nevertheless, the GTCi aims to be neutral, globally respected and to contribute to the global debate on talent attraction and economic success, but as cities are more agile in attracting talents, not much advice is given how governments can contribute to this development on a national level (INSEAD, Adecco, & HCLI, 2015). The calculation of the index is based on two sub-indices. First the competitiveness sub-index, which measures, for instance, the climate of the regulatory market, the business environment, and further what a country is doing to attract talent, to grow the talent (e.g. by education), and to retain it (e.g. through lifestyle and sustainability). The second sub-index is based on the talent competitiveness output, which measures the actual situation on the labor market as an indicator of the success of current regulation and talent attraction.

Comparing the talent competitiveness on a national level and the share of foreign-born population on city level, a slight positive Pearson correlation of +0.5 can be identified. Comparing the U.S. cities, which have slightly lower scores than Singapore on a national level, the share of the foreign population varies between 27% and 39% on the city level. Nevertheless, these are very high scores compared to the cities located in Asia. These cities have a share of between 0.5% and 16.2%, and in consequence China ranks lowest in the ability to attract foreign talents.

Finally, the findings of all countries investigated in the global talent competitiveness ranking show a correlation of a nation's income with its ability to attract talent (INSEAD et al., 2015). Further, in a world of talent circulation, cities become "talent hubs." Cities can best contribute to attracting talent by offering "(1) high-quality infrastructure, (2) competitive market conditions and business environment (including clusters), (3) an existing critical mass of talents, with excellent networking and cooperation possibilities, and (4) superior living conditions (including factors as diverse as climatic conditions, cultural environment, safety and easy access to key services such as health or education)" (INSEAD et al., 2015, p. 37).

Conclusion on cityness

In summary, the investigation has not proven that it is important for an informational city to be a world city. Inter-city relations constitute the global city, as cities are physically and digitally connected with the rest of the world, especially those located in nations with a high level of development. These cities are then able to become hubs within the global economy. However, world cities are measured primarily according to their status as a hub of multinational corporations or producer service firms. To be home to the headquarters of these firms is considered to be the place of command and control, controlling subsidiaries in other cities. This can be referred to as the flows of power that build inter-city connectivity. Comparing the flows of power with those of knowledge and information, no direct relation can be found. Hence, the cities that are recognized as knowledge hubs, like Boston, or innovation hubs, like San Francisco, are not the leading cities of power flows. An economic definition of the term world city in the twenty-first century will result in Tokyo, New York, and London as primary cities within the hierarchy of power. Economic success can therefore provide the necessary infrastructure, but it is no guarantee to be a leading world city of knowledge or innovation.

A further indicator of a world city is to be physically connected with the rest of the world. This can be measured by the number of international passengers, direct destinations, and airlines per airport. The amount of cargo handled at a container port is also used as a secondary measure. However, infrastructure development is always based on political decisions, requiring businesses and the government to work together to accomplish a successful development. One city that is profiting from its enhanced physical infrastructure is Dubai. The city has the busiest airport and aims to become the largest container port in the world in the future.

Finally, the inner-city process is acknowledged as an important aspect of being a world city. Urbanity constitutes real cities, but is based on human interaction and not on planned dense areas of offices and residences. In world cities, the global labor force is welcomed, something that can be detected through the number of foreign-born residents. This number reflects openness to cultural diversity, but should not be equated with the ability to attract a talented workforce. Multinational corporations and producer service firms are needed to provide an adequate labor market. In this, business and cities can mutually benefit from each other: The city offers the superior infrastructure and metropolitan lifestyle, and the firms create the necessary capital to sustain it. A slight positive correlation can be identified between the share of foreign-born residents and the ability to attract foreign talent. Nevertheless, a city like Tokyo is home to the highest number of multinational corporations, but is not able to attract international talents. Other Asian cities also have a very limited international population. According to the experts, the language barriers could be a problem in attracting talents. Thus, cities in the United States or Singapore, with its bilingual population, score very high in the GTCi, in contrast to China or Malaysia. Among the Canadian cities, the Anglophone cities Toronto and Vancouver have a much higher percentage of foreign-born population than the Francophone Montréal.

In summary, both hypotheses (H11 and H12) investigated with respect to being a world city do not deliver a clear result. A city such as New York is an established world city, and has been for decades. It has a high GDP, has the highest capital flow at the stock exchanges and is also home to six of the 100 most successful FinTech companies. But New York is not the city with the highest knowledge or innovation output. Nevertheless, it is able to attract talents and multinational corporations, both being world city indicators. However, this does not prevent smaller cities from being successful, as the case of Boston demonstrates. In such cities, however, the economic market will be more specialized. As identified by Taylor et al. (2014), they can constitute global hubs with regard to "one" economic sector. And under the assumption that information, innovation and talent may flow globally, it will eventually suffice to become a regional hub.

References

Academic Ranking of World Universities. (n.d.). University and college rankings list. Retrieved from http://www.shanghairanking.com/resources.html
Ács, Z. J. Á., Szerb, L., & Autio, E. (2016). *Global entrepreneurship index.* Washington, DC. Retrieved from https://thegedi.org/global-entrepreneurship-and-development-index/
Ajuntament de Barcelona. (n.d.). Barcelona WiFi. Retrieved from http://www.bcn.cat/barcelonawifi/en/
Albino, V., Berardi, U., & Dangelico, R. M. (2015). Smart cities: Definitions, dimensions, performance, and initiatives. *Journal of Urban Technology, 22*(1), 3–21. https://doi.org/10.1080/10630732.2014.942092
Alderson, A. S., & Beckfield, J. (2012). Corporate networsk of world cities. In B. Derudder, M. Hoyler, P. J. Taylor, & F. Witlox (Eds.), *Handbook of globalization and world cities* (pp. 126–134). Cheltenham, UK; Northampton, MA: Edward Elgar Publishing.
Arcadis. (2015). *Sustainable cities index 2015.* Retrieved from www.arcadis.com/sustainablecities
Beaverstock, J. V., Smith, R. G., & Taylor, P. J. (1999). A roster of world cities. *Cities, 16*(6), 445–458. https://doi.org/10.1016/S0264-2751(99)00042-6
Bertelsmann Stiftung. (2016). *Policy performance and governance capacities in the OECD. Sustainable governance indicators.* Gütersloh, DE.
Beutelspacher, L. (2014). Assessing information literacy: Creating generic indicators and target group-specific questionnaires. In *Information literacy. lifelong learning and digital citizenship in the 21st century* (pp. 521–530). Springer International Publishing. https://doi.org/10.1007/978-3-319-14136-7_55
Beuth, P. (2016, December 15). Fake News: Facebook will Falschmeldungenn kennzeichnen [Fake News: Facebook will flag false messages]. *Zeit Online.* Retrieved from www.zeit.de/digital/internet/2016-12/fake-news-facebook-massnahmen-details-news-feed
Beyers, W. B., & Lindahl, D. P. (1996). Lone eagles and high fliers in rural producer services. *Rural Development Perspectives, 11,* 2–10.
Biggest Stock Exchanges in the World. (2016, July 7). *WorldAtlas.com.* Retrieved from http://www.worldatlas.com/articles/biggest-stock-exchanges-in-the-world.html
BMWi. (2016). Mehr Rechtssicherheit bei WLAN [More security with WLAN]. Retrieved from http://www.bmwi.de/DE/Themen/Digitale-Welt/Netzpolitik/rechtssicherheit-wlan,did=695728.html

Board Of Governors of the Federal Reserve System. (2016). *Consumers and mobile financial services 2016. Federal reserve report. Federal Reserve Report*. Washington, DC. Retrieved from http://www.federalreserve.gov/econresdata/mobile-devices/files/mobile-device-report-201203.pdf

Bond Hill, C. (2015). Income inequality and higher education. *American Council on Education*. Retrieved from https://www.acenet.edu/the-presidency/columns-and-features/Pages/Income-Inequality-and-Higher-Education.aspx

BOP Consulting. (2015). World cities culture report 2012, 77. https://doi.org/10.1017/CBO9781107415324.004

Borgman, C. L. (1999). What are libraries? Competing visions. *Information Processing & Management*, *35*(3), 227–243. https://doi.org/10.1016/S0306-4573(98)00059-4

Born, C. (2015). *Die Anpassung öffentlicher Bibliotheken an die Bedürfnisse der Nutzer im 21. Jahrhundert. Eine empirische Untersuchung von prototypischen Städten der Wissensgesellschaft [The adaptation of public libraries to the needs of users in the 21st century...]*. Unpublished master's thesis, Heinrich-Heine University Düsseldorf, DE.

Bratzel, S. (1995). *Extreme der Mobilität [Extreme mobility]* (Vol. 51). Wiesbaden, DE: VS Verlag für Sozialwissenschaften. https://doi.org/10.1007/978-3-322-92225-0

Brezina, C. (2011). *Understanding the gross domestic product and the gross national product*. New York, NY: The Rosen Publishing Group.

British Council. (2012). *Going global 2012. The shape of things to come: Higher education global trends and emerging opportunities to 2020*. Retrieved from http://hdl.voced.edu.au/10707/252974%5Cnwww.britishcouncil.org/higher-education

Bruno, G., Esposito, E., Genovese, A., & Gwebu, K. L. (2011). A critical analysis of current indexes for digital divide measurement. *The Information Society*, *27*, 16–28. https://doi.org/10.1080/01972243.2010.534364

Budd, L. C. S. (2012). Airports: From flying fields to twenty-first century aerocities. In B. Derudder, M. Hoyler, P. J. Taylor, & F. Witlox (Eds.), *International handbook of globalization and world cities* (pp. 151–161). Cheltenham, UK; Northampton, MA: Edward Elgar Publishing.

Busher, V., Doody, L., Webb, M., & Aoun, C. (2014). *Urban mobility in the smart city age. Smart cities cornerstone series*. London, UK; New York, NY.

Cairns, S., Atkins, S., & Goodwin, P. (2002). Disappearing traffic? The story so far. *Proceedings of the ICE – Municipal Engineer*, *151*(1), 13–22. https://doi.org/10.1680/muen.2002.151.1.13

Camagni, R. P. (1995). The concept of innovative milieu and its relevance for public policies in European lagging regions. *Papers in Regional Science*, *74*, 317–340. https://doi.org/10.1111/j.1435-5597.1995.tb00644.x

Canada's Open Data Exchange. (n.d.). 2016 Open cities index - top 20 results. Retrieved from https://publicsectordigest.com/2016-open-cities-index-top-20-results

Capdevila, I. (2014). Coworking spaces and the localized dynamics of innovation. The case of Barcelona. *XXIVe Conférence Internationale de Management Stratégique*, 1–26. https://doi.org/10.2139/ssrn.2502813

Carrillo, F. J., Yigitcanlar, T., García, B., & Lönnqvist, A. (2014). *Knowledge and the city: Concepts, applications and trends of knowledge-based urban development*. New York, NY: Routledge.

Castells, M. (1989). *The informational city: Information technology, economic restructuring, and the urban-regional process*. Oxford, UK: Blackwell.

Castells, M. (1996). *The rise of the network society*. Malden, MA: Blackwell.
Castells, M. (2001). *The internet galaxy. Reflections on the internet, business, and society*. Oxford, UK: Oxford University Press.
Castelnovo, W. (2015). Citizens as sensors/information providers in the co-production of smart city services. In *Proceedings of the 12th Conference of the Italian Chapter of AIS*. Rome, IT.
Chamberlain, F., & Riggs, W. (2016). Shifting the tide: Transit-oriented development and active transportation planning in Los Angeles. *Focus: The Journal of Planning Practice and Education / Journal of the City and Regional Planning Department / Cal Poly, San Luis Obispo, 12*(1), 52–59. Retrieved from http://digitalcommons.calpoly.edu/focus/vol12/iss1/1
Chandra, A., & Kaiser, U. (2015). *Newspapers and Magazines* (Forthcoming Handbook of Media Economics No. 2479359). Retrieved from https://ssrn.com/abstract=2479359
Chellapandi, S., Wun Han, C., & Chiew Boon, T. (2010). The National Library of Singapore experience: Harnessing technology to deliver content and broaden access. *Interlending & Document Supply, 38*(1), 40–48. https://doi.org/10.1108/02641611011025361
Chesbrough, H. (2003). *Open innovation. The new imperative for creating and profiting from technology*. Boston, MA: Harvard Business School.
Choo, C. W. (1995). National computer policy management in Singapore. Planning an intelligent island. In T. Kinney (Ed.), *Proceedings of the 58th Annual Meeting of the American Society for Information Science* (pp. 152–156). Chicago, IL.
Chow, G. C. (2004). Economic reform and growth in China. *Annals of Economics and Finance., 152*, 127–152. Retrieved from http://down.aefweb.net/AefArticles/aef050107.pdf
City of Melbourne. (2013). Melbourne the most admired knowledge city. Retrieved from http://www.melbourne.vic.gov.au/news-and-media/Pages/Melbournethemostadmiredknowledgecity.aspx
Clark, G., Moonen, T., & Couturier, J. (2015a). *Globalisation and competition: The new world of cities*.
Clark, G., Moonen, T., & Couturier, J. (2015b). *The business of cities 2015*. Retrieved from http://www.jll.com/Research/jll-business-of-cities-report.pdf
Cohendet, P., Grandadam, D., & Simon, L. (2011). Rethinking urban creativity: Lessons from Barcelona and Montréal. *City, Culture and Society, 2*(3), 151–158. https://doi.org/10.1016/j.ccs.2011.06.001
Collis, C., Felton, E., & Graham, P. (2010). Beyond the inner city: Real and imagined places in creative place policy and practice. *The Information Society, 26*(2), 104–112. https://doi.org/10.1080/01972240903562738
Comunian, R. (2011). Rethinking the creative city: The role of complexity, networks and interactions in the urban creative economy. *Urban Studies, 48*(6), 1157–1179.
Corrêa, A. S., Corrêa, P. L. P., & da Silva, F. S. C. (2014). Transparency portals versus open government data. In *Proceedings of the 15th Annual International Conference on Digital Government Research – dg.o '14* (pp. 178–185). New York, New York, USA: ACM Press. https://doi.org/10.1145/2612733.2612760
Cullen, R. (2010). Defining the transformation of government. E-government or e-governance paradigm? In H. J. Scholl (Ed.), *E-Government: Information, Technology, and Transformation* (pp. 52–71). London, UK; New York, NY: Routledge.
Currid, E. (2006). New York as a global creative hub: A competitive analysis of four theories on world cities. *Economic Development Quarterly, 20*(4), 330–350. https://doi.org/10.1177/0891242406292708

Dapp, T. F., & Ehmer, P. (2011). *Cultural and creative industries*. Frankuft am Main, DE. Retrieved from https://www.dbresearch.com/PROD/DBR_INTERNET_EN-PROD/PROD0000000000272899/Cultural_and_creative_industries:_Growth_potential.pdf

Dargay, J., & Gately, D. (1999). Income's effect on car and vehicle ownership, worldwide: 1960–2015. *Transportation Research Part A: Policy and Practice, 33*(2), 101–138. https://doi.org/10.1016/S0965-8564(98)00026-3

Davies, A. (2014). Strava's cycling app is helping cities build better bike lanes. *Wired*. Retrieved from https://www.wired.com/2014/06/strava-sells-cycling-data/

Dehua, J., & Beijun, S. (2012). Internet of knowledge. The soft infrastructure for smart cities [in Chinese]. In *In Smart City and Library Service. The Proceedings of the Sixth Shanghai International Library Forum* (pp. 147–154). Shanghai, CHN.

Dembosky, A. (2013, June 9). Silicon Valley rooted in backing from US military. *US Politics & Policy*. Retrieved from http://www.ft.com/cms/s/0/8c0152d2-d0f2-11e2-be7b-00144feab7de.html#axzz4K3Dk1wAr

Die Umweltplakette/Feinstaubplakette. (n.d.). Retrieved from http://www.umweltplakette.org/

Dubai private education landscape. (2014). Dubai, UAE. Retrieved from https://www.khda.gov.ae/CMS/WebParts/TextEditor/Documents/LandscapePEEnglish.pdf

Dubai Statistics Center. (2015). Education. Retrieved from https://www.dsc.gov.ae/en-us/Themes/Pages/Education.aspx?Theme=37

Duncan, D. T., Aldstadt, J., Whalen, J., Melly, S. J., & Gortmaker, S. L. (2011). Validation of walk score for estimating neighborhood walkability: An analysis of four US metropolitan areas. *International Journal of Environmental Research and Public Health, 8*(12), 4160–4179. https://doi.org/10.3390/ijerph8114160

Ergazakis, E., Ergazakis, K., Metaxiotis, K., & Charalabidis, Y. (2009). Rethinking the development of successful knowledge cities: An advanced framework. *Journal of Knowledge Management, 13*(5), 214–227. https://doi.org/10.1108/13673270910988060

Ericsson. (2014). *Networked society city index 2014*. Stockholm, SWE. Retrieved from http://www.ericsson.com/res/docs/2014/networked-society-city-index-2014.pdf

Ericsson. (2016). *Networked society city index 2016*. Stockholm, SW. Retrieved from https://www.ericsson.com/assets/local/networked-society/reports/city-index/2016-networked-society-city-index.pdf

EWContributor. (2015, September). 8 of the world's best bike sharing programs. *EcoWatch*. Retrieved from http://www.ecowatch.com/8-of-the-worlds-best-bike-sharing-programs-1882105476.html

EY. (2015a). *Cultural times. The first global map of cultural and creative industries*. Retrieved from http://www.unesco.org/new/fileadmin/MULTIMEDIA/HQ/ERI/pdf/EY-Cultural-Times2015_Low-res.pdf

EY. (2015b). *EY fintech adoption index. Key findings*. Retrieved from http://www.ey.com/gl/en/industries/financial-services/ey-fintech-adoption-index

EY. (2016). *Back to reality. EY global venture capital trends 2015*. Retrieved from http://www.ey.com/Publication/vwLUAssets/ey-global-venture-capital-trends-2015/$FILE/ey-global-venture-capital-trends-2015.pdf

Fietkiewicz, K. J., Mainka, A., & Stock, W. G. (2016). EGovernment in cities of the knowledge society. An empirical investigation of smart cities' governmental websites. *Government Information Quarterly*. https://doi.org/10.1016/j.giq.2016.08.003

Florida, R. L. (2003). Cities and the creative class. *City and Community, 2*(1), 3–19. https://doi.org/10.1111/1540-6040.00034

Florida, R. L. (2012). *The rise of the creative class, revisited* (2nd ed.). New York, NY: Basic Books.
Florida, R. L. (2014). The creative class and economic development. *Economic Development Quarterly, 28*(3), 196–205. https://doi.org/10.1177/0891242414541693
Florida, R. L., & Gates, G. (2001). Technology and tolerance: To high-technology growth. The Brookings Institution. Retrieved from http://www.urban.org/sites/default/files/alfresco/publication-pdfs/1000492-Technology-and-Tolerance.PDF
Florida, R. L., & King, K. M. (2016). *Rise of the global startup city*. Toronto, CA. Retrieved from http://martinprosperity.org/content/rise-of-the-global-startup-city/
Florida, R. L., Mellander, C., & King, K. (2015). *The global creativity index 2015*. Toronto, CA. Retrieved from http://martinprosperity.org/media/Global-Creativity-Index-2015.pdf
Forman, C., Goldfarb, A., & Greenstein, S. (2016). Agglomeration of invention in the Bay Area: Not just ICT. *American Economic Review, 106*(5), 146–151. https://doi.org/10.1257/aer.p20161018
Freedom House. (2016). About freedom on the net. Retrieved from https://freedomhouse.org/
Freifunk – Berlin. (n.d.). Retrieved from https://berlin.freifunk.net/
Friedmann, J. (1986). The world city hypothesis. *Development and Change, 17*(1), 69–83. https://doi.org/10.1111/j.1467-7660.1986.tb00231.x
Gibson, J., Robinson, M., & Cain, S. (2015). *CITIE: City initiatives for technology, innovation and entrepreneurship. A resource for city initiatives*. Retrieved from http://citie.org/wp-content/uploads/2015/04/CITIE_Report_2015.pdf
Global right to information rating. Country data. (n.d.). Retrieved from http://www.rti-rating.org/country-data/
Goethe Universität Frankfurt am Main. (n.d.). Studierendenstatistik. Retrieved from http://www.uni-frankfurt.de/52565158
Google Cultural Institute. (n.d.). The lab at the cultural institute, a crossroads of ideas, art and technology. Retrieved from https://www.google.com/culturalinstitute/thelab/
Gottschalk, P. (2009). Maturity levels for interoperability in digital government. *Government Information Quarterly, 26*(1), 75–81. https://doi.org/10.1016/j.giq.2008.03.003
Growing up. The Gulf state's expansion is more sustainable than its previous boom. (2015, June 6). *The Economist*. Retrieved from http://www.economist.com/news/middle-east-and-africa/21653621-gulf-states-expansion-more-sustainable-its-previous-boom-growing-up
Hall, P. (2000). Creative cities and economic development. *Urban Studies, 37*(4), 639–649. https://doi.org/10.1080/00420980050003946
Haller, M., Hadler, M., & Kaup, G. (2013). Leisure time in modern societies: A new source of boredom and stress? *Social Indicators Research, 111*(2), 403–434. https://doi.org/10.1007/s11205-012-0023-y
Hansen, H. K., Asheim, B., & Vang, J. (2009). The European creative class and regional development: How relevant is Florida's theory for Europe? In L. Kong & J. O´Connor (Eds.), *Creative economies, creative cities. Asian-European perspectives* (pp. 99–120). London, UK; New York, NY; a.o.: Springer. https://doi.org/10.1007/978-1-4020-9949-6_7
Harlan, C. (2012, March 24). In South Korean classrooms, digital textbook revolution meets some resistance. *The Wahington Post*. Retrieved from https://www.washingtonpost.com/world/asia_pacific/in-south-korean-classrooms-digital-textbook-revolution-meets-some-resistance/2012/03/21/gIQAxiNGYS_story.html
Harrison, T. M., Burke, G. B., Cook, M., Cresswell, A., & Hrdinová, J. (2011). Open government and e-government: Democratic challenges from a public value perspective. In *12th Annual International Conference on Digital Government Research* (pp. 245–253). College Park, MD: ACM.

Hartley, J., Potts, J., & MacDonald, T. (2012). The CCI creative city index 2012. *Cultural Science Journal, 5*(1), 138.

Hartmann, S., Mainka, A., & Stock, W. G. (2017). Citizen relationship management in local governments: The potential of 311 for public service delivery. In A. A. Paulin, L. G. Anthopoulos, & C. G. Reddick (Eds.), *Beyond bureaucracy: Towards sustainable government informatisation [in press]*. Springer.

Hawkins, A. J. (2016, January 8). The entire New York City subway system will have Wi-Fi by the end of 2016. *The Verge*. Retrieved from http://www.theverge.com/2016/1/8/10737408/nyc-mta-subway-wifi-mobile-payment-cuomo

Hedrick-Wong, D. Y., & Choong, D. (2014). *MasterCard global destination cities index. MasterCard Worldwide Insights*. Retrieved from http://newsroom.mastercard.com/wp-content/uploads/2014/07/Mastercard_GDCI_2014_Letter_Final_70814.pdf

Herrmann, B. L., Gauthier, J., Holtschke, D., Bermann, R. D., & Marmer, M. (2015). *The global startup ecosystem report 2015. The Startup Ecosystem Report Series*. Retrieved from https://startup-ecosystem.compass.co/ser2015/

Herrmann, B. L., Marmer, M., Dogrultan, E., & Holtschke, D. (2012). *The Global Startup Ecosystem Report 2012*. https://doi.org/10.4324/9780203165829_PART_ONE

Hiller, J. S., & Bélanger, F. (2001). *Privacy strategies for electronic government* (E-government series). *Center for Global Electronic Commerce*. Arlington, VA, USA. Retrieved from http://www.businessofgovernment.org/sites/default/files/PrivacyStrategies.pdf

Hitters, E., & Richards, G. (2002). The creation and management of cultural clusters. *Creativity and Innovation Management, 11*(4), 234–247. https://doi.org/10.1111/1467-8691.00255

Holmes, R. (2016). Looks like THE isn't big in Japan anymore. *University Ranking Watch*. 2016. Retrieved from http://rankingwatch.blogspot.de/2016/07/looks-like-isnt-big-in-japan-anymore.html

Hornidge, A., & Kurfürst, S. (2010). *Envisioning the future, conceptualising public space. Hanoi and Singapore negotiating spaces for negotiation* (ZEF Working Paper Series No. 58). Retrieved from www.zef.de/fileadmin/webfiles/downloads/zef_wp/wp58.pdf

IDA. (n.d.-a). iN2015 Masterplan. Retrieved from www.ida.gov.sg/Tech-Scene-News/iN2015-Masterplan

IDA. (n.d.-b). Programmes & partnership – Infocomm Development Authority of Singapore. Retrieved from https://www.ida.gov.sg/Programmes-Partnership

IDC Financial Insights. (2016). *IDC financial isights fintech rankings top 100*. Retrieved from http://www.idc.com/download/finalrankings2016.pdf

Ilhan, A., Möhlmann, R., & Stock, W. G. (2015). Citizens' acceptance of u-life services in the ubiquitous city Songdo. In M. Foth, M. Brynskov, & T. Ojala (Eds.), *Citizen's right to the digital city: Urban interfaces, activism, and placemaking* (pp. 215–229). Singapore, SG: Springer.

INSEAD, Adecco, & HCLI. (2015). *The global talent competitiveness index 2015*. (B. Lanvin & P. Evans, Eds.). Fontainbleau: INSEAD.

IOM. (2015). *World migration report 2015. Migrants and cities: New partnerships to manage mobility*. Geneva, CHE: International Organization for Mirgation (IOM). Retrieved from http://publications.iom.int/system/files/wmr2015_en.pdf

ISO/IEC JTC 1. InformationTechnology. (2015). *Smart cities. Preliminary Report 2014*. Geneva, CHE. Retrieved from http://www.iso.org/iso/smart_cities_report-jtc1.pdf

Jacobs, J. (1969). *The economy of cities*. New York, NY: Random House.

Jamrisko, M., & Lu, W. (2016, January 19). These are the world's most innovative economies. *Bloomberg Markets*. Retrieved from https://www.bloomberg.com/news/articles/2016-01-19/these-are-the-world-s-most-innovative-economies

Janssen, M., & Helbig, N. (2016). Innovating and changing the policy-cycle: Policy-makers be prepared! *Government Information Quarterly [in Press]*. https://doi.org/10.1016/j.giq.2015.11.009

Jin, S., & Cho, C. M. (2015). Is ICT a new essential for national economic growth in an information society? *Government Information Quarterly*, *32*(3), 253–260. https://doi.org/10.1016/j.giq.2015.04.007

Kamel Boulos, M. N., & Al-Shorbaji, N. M. (2014). On the internet of things, smart cities and the WHO healthy cities. *International Journal of Health Geographics*, *13*(1), 10. https://doi.org/10.1186/1476-072X-13-10

Kane, T. (2010). *The importance of startups in job creation and job destruction* (Kauffman Foundation Research Series: Firm Formation and Economic Growth The). Kansas City, MI.

Kazis, N. (2011, June 16). From London to D.C., bike-sharing is safer than riding your own bike [Blog post]. *Streetsblog New York City*. Retrieved from http://www.streetsblog.org/2011/06/16/from-london-to-d-c-bike-sharing-is-safer-than-riding-your-own-bike/

Ke, W., & Wei, K. K. (2004). Successful e-government in Singapore. *Communications of the ACM*, *47*(6), 95–99. https://doi.org/10.1145/990680.990687

Kelly, B. C., Carpiano, R. M., Easterbrook, A., & Parsons, J. T. (2014). Exploring the gay community question: Neighborhood and network influences on the experience of community among urban gay men. *The Sociological Quarterly*, *55*(1), 23–48. https://doi.org/10.1111/tsq.12041

Kenworthy, J. R. (2006). The eco-city: Ten key transport and planning dimensions for sustainable city development. *Environment and Urbanization*, *18*(1), 67–85. https://doi.org/10.1177/0956247806063947

Khveshchanka, S., Mainka, A., & Peters, I. (2011). Singapur: Prototyp einer informationellen Stadt [Singapore: Prototype of an informational city]. *Information – Wissenschaft & Praxis*, *62*(2–3), 111–121.

Kim, J. H.-Y., & Jung, H.-Y. (2010). South Korean digital textbook project. *Computers in the Schools*, *27*(3–4), 247–265. https://doi.org/10.1080/07380569.2010.523887

Kosior, A., Barth, J., Gremm, J., Mainka, A., & Stock, W. G. (2015). Imported expertise in world-class knowledge infrastructures: The problematic development of knowledge cities in the Gulf Region. *Journal of Information Science Theory and Practice*, *3*(3), 17–44.

KPMG. (2016). *Venture pulse Q3 2016. Global analysis of funding*. Retrieved from https://assets.kpmg.com/content/dam/kpmg/xx/pdf/2016/10/venture-pulse-q3-2016-report.pdf

Kyttä, M., Broberg, A., Tzoulas, T., & Snabb, K. (2013). Towards contextually sensitive urban densification: Location-based softGIS knowledge revealing perceived residential environmental quality. *Landscape and Urban Planning*, *113*, 30–46. https://doi.org/10.1016/j.landurbplan.2013.01.008

La Rue, F. (2011). *Report of the Special Rapporteur on the promotion and protection of the right to freedom of opinion and expression*. New York, NY. Retrieved from http://www2.ohchr.org/english/bodies/hrcouncil/docs/17session/A.HRC.17.27_en.pdf

Landry, C. (2008). *The creative city: A toolkit for urban innovators* (2nd ed.). New York, NY: Earthscan Publications.

Landry, C. (2011a). *Creativity, culture & the city: A question of interconnection.* Edden, DE. Retrieved from http://www.e-c-c-e.de/fileadmin/content_bilder/Aktivitaeten/Forum_dAvignon_Ruhr/Downloadbereich/CREATIVITY__ CULTURE___THE_CITY.pdf

Landry, C. (2011b). The creativity city index. *City, Culture and Society, 2*(3), 173–176. https://doi.org/10.1016/j.ccs.2011.09.003

Levy, D. (2000). Digital libraries and the problem of purpose. *D-Lib Magazine, 6*(1). https://doi.org/10.1045/january2000-levy

LMU. (n.d.). Geschichte der LMU – LMU München [History of the LMU – LMU Munich]. Retrieved from http://www.uni-muenchen.de/ueber_die_lmu/portraet/geschichte/index.html

Lor, P. J., & Britz, J. J. (2007). Challenges of the approaching knowledge society: Major international issues facing LIS professionals. *Libri, 57*(3), 111–122. https://doi.org/10.1515/LIBR.2007.111

Madanipour, A. (2013). The identity of the city. In S. Serreli (Ed.), *City project and public space* (pp. 49–63). Dordrecht: Springer. https://doi.org/10.1007/978-94-007-6037-0_3

Mainka, A. (2011). *Singapur: Prototyp einer informationellen Stadt [Singapore: Prototype of an informational city].* Heinrich-Heine University, Düsseldorf, DE.

Mainka, A., Castelnovo, W., Miettinen, V., Bech-Petersen, S., Hartmann, S., & Stock, W. G. (2016). Open innovation in smart cities: Participation and co-creation of public services. In A. Grove, D. H. Sonnenwald, L. Harrison, C. Blake, C. Schlögl, I. Peters, ... Y.-L. Theng (Eds.), *Proceedings of the 79th ASIS&T Annual Meeting (Vol. 53). Creating Knowledge, Enhancing Lives through Information & Technology.* Copenhagen, DNK: Richard B. Hill. Retrieved from https://www.asist.org/files/meetings/am16/proceedings/openpage16.html

Mainka, A., Fietkiewicz, K. J., Kosior, A., Pyka, S., & Stock, W. G. (2013). Maturity and usability of e-government in informational world cities. In W. Castelnovo & E. Ferrari (Eds.), *Proceedings of the 13th European Conference on e-Government University* (pp. 292–300). Como, IT: Academic Conferences and Publishing International Limited Reading.

Mainka, A., Hartmann, S., Meschede, C., & Stock, W. G. (2015a). Mobile application services based upon open urban government data. In Proceedings of the iConference 2015: Create, Collaborate, Celebrate. Newport Beach, CA

Mainka, A., Hartmann, S., Meschede, C., & Stock, W. G. (2015b). Open government: Transforming data into value-added city services. In M. Foth, M. Brynskov, & T. Ojala (Eds.), *Citizen's right to the digital city: Urban interfaces, activism, and placemaking* (pp. 199–214). Singapore, SG: Springer.

Mainka, A., Hartmann, S., Orszullok, L., Peters, I., Stallmann, A., & Stock, W. G. (2013). Public libraries in the knowledge society: Core services of libraries in informational world cities. *Libri, 63*(4), 295–319. https://doi.org/10.1515/libri-2013-0024

Mainka, A., Hartmann, S., Stock, W. G., & Peters, I. (2015). Looking for friends and followers: A global investigation of governmental social media use. *Transforming Government: People, Process and Policy, 9*(2), 237–254. https://doi.org/10.1108/TG-09-2014-0041

Malecki, E. J. (2012). Internet networks of world cities: Agglomeration and dispersion. In B. Derudder, M. Hoyler, P. J. Taylor, & F. Witlox (Eds.), *International handbook of globalization and world cities* (pp. 117–125). Cheltenham, UK; Northampton, MA: Edward Elgar Publishing.

Meister, F. (2012). Die Global City ist ein brutaler Ort [The global city is a brutal place]. *Woz,* (25), 15–17.

Mergel, I. (2013). Social media adoption and resulting tactics in the U.S. federal government. *Government Information Quarterly, 30*(2), 123–130. https://doi.org/10.1016/j.giq.2012.12.004

Mesmer, P. (2014). Seoul demolishes its urban expressways as city planners opt for greener schemes. *The Guardian*. Retrieved from https://www.theguardian.com/world/2014/mar/13/seoul-south-korea-expressway-demolished

Miller, C. (2015). The top lgbt-friendly cities in the U.S. Retrieved from https://www.nerdwallet.com/blog/mortgages/top-lgbt-friendly-cities-2015/

Minguillo, D., Tijssen, R., & Thelwall, M. (2015). Do science parks promote research and technology? A scientometric analysis of the UK. *Scientometrics*, *102*(1), 701–725. https://doi.org/10.1007/s11192-014-1435-z

Mobile World Congress. (n.d.). Retrieved from https://www.mobileworldcongress.com/about/mobile-world-capital/

Morandi, C., Rolando, A., & Di Vita, S. (2016). *From smart city to smart region. SpringerBriefs in Applied Sciences and Technology* (Vol. 30). Cham, DE: Springer International Publishing. https://doi.org/10.1007/978-3-319-17338-2

Murugadas, D., Vieten, S., Nikolic, J., Fietkiewicz, K. J., & Stock, W. G. (2015). Creativity and entrepreneurship in informational metropolitan regions. *Journal of Economic and Social Development*, *2*(1), 14–24. Retrieved from http://hdl.handle.net/10125/41496

NACTO. (2015). *Walkable station spacking is key to successful, equitable bike share*. Retrieved from https://www.bicyclenetwork.com.au/media/vanilla_content/files/NACTO_Walkable-Station-Spacing-Is-Key-For-Bike-Share.pdf

Nesta. (2015). The European digital city index. Retrieved from http://www.nesta.org.uk/blog/launching-european-digital-city-index

Newzoo. (2016, April 21). Global games market revenues 2016. Retrieved from https://newzoo.com/insights/articles/global-games-market-reaches-99-6-billion-2016-mobile-generating-37/

Nokia. (2015). HERE and the Netherlands evaluating 4G/LTE-based C-ITS road messaging system. Retrieved from http://company.nokia.com/en/news/press-releases/2015/11/25/here-and-the-netherlands-evaluating-4glte-based-c-its-road-messaging-system

Nonaka, I., & Takeuchi, H. (1995). *The Knowledge-creating Company*. New York, NY: Oxford University Press.

O'Neill, M. (2016). World university rankings. Retrieved from https://www.kaggle.com/mylesoneill/world-university-rankings

OECD. (2014). *Measuring the digital economy*. Paris, FR: OECD Publishing. https://doi.org/10.1787/9789264221796-en

OECD iLibrary. (2013). Statistics. Entrepreneurship at a glance. Retrieved from http://www.oecd-ilibrary.org/sites/entrepreneur_aag-2013-en/06/03/index.html?itemId=/content/chapter/entrepreneur_aag-2013-27-en

Open Data Barometer 2015. (n.d.). Retrieved from http://opendatabarometer.org/data-explorer/?_year=2015&indicator=ODB&lang=en

Palvia, S. C. J., & Sharma, S. S. (2007). E-government and e-governance: Definitions/domain framework and status around the world. In *International Conference on E-governance* (pp. 1–12). https://doi.org/10.3991/ijac.v5i1.1887

Pancholi, S., Yigitcanlar, T., & Guaralda, M. (2015). Place making facilitators of knowledge and innovation spaces: Insights from European best practices. *International Journal of Knowledge-Based Development*, *6*(3), 215. https://doi.org/10.1504/IJKBD.2015.072823

Parilla, J., Leal, T. J., Berube, A., & Ran, T. (2015). *Global metro monitor 2014: An uncertain recovery. The Brookings Institution*. Retrieved from http://www.brookings.edu/~/media/Research/Files/Reports/2015/01/22 global metro monitor/bmpp_GMM_final.pdf

Paris – Vélib'. (n.d.). Retrieved from http://en.velib.paris.fr/How-it-works/Stations
Paterson, I. (2015, July 27). Sydney science park at Luddenham to generate thousands of new jobs for western Sydney. *Penrith Press*. Penrith, AUS. Retrieved from http://www.dailytelegraph.com.au/newslocal/news/sydney-science-park-at-luddenham-to-generate-thousands-of-new-jobs-for-western-sydney/news-story/9ed8e525b6bd00d09aff1bdc82849fbb
Peña-López, I. (2006, March 30). Networked readiness index vs. human development index [Blog post]. *ICTology*. Retrieved from http://ictlogy.net/review/?p=378
Peters, A. (2015, January 13). 7 cities that are starting to go car-free [Blog post]. *CO.Exist*. Retrieved from https://www.fastcoexist.com/3040634/7-cities-that-are-starting-to-go-car-free
Pike, A., & Tomaney, J. (2010). *Handbook of local and regional development*. New York, NY: Taylor & Francis.
Post, R. C. (2007). *Urban mass transit: The life story of a technology*. Westport, CT; Greenwood Press.
Qiang, C. Z., Rosotto, C., & Kimura, K. (2009). Economic impact of broadband. In *Information and communications for development 2009. Extending reach and increasing impact* (pp. 35–50). Washington, DC: World Bank.
Quacquarelli Symonds. (n.d.-a). QS Best Student Cities ranking: 1st. Retrieved from http://www.topuniversities.com/university-rankings-articles/qs-best-student-cities/paris
Quacquarelli Symonds. (n.d.-b). Top universities. Retrieved from http://www.topuniversities.com/
Rantisi, N. M., & Leslie, D. (2010). Materiality and creative production: the case of the Mile End neighborhood in Montréal. *Environment and Planning A*, *42*(12), 2824–2841. https://doi.org/10.1068/a4310
Regio IT. (n.d.). Smart mobility. Retrieved from http://www.regioit.de/produkte-leistungen/mobilitaet/smart-mobility.html
Relly, J. E. E., & Sabharwal, M. (2009). Perceptions of transparency of government policymaking: A cross-national study. *Government Information Quarterly*, *26*(1), 148–157. https://doi.org/10.1016/j.giq.2008.04.002
Reporters without Borders. (2016). World press freedom index 2016. Retrieved from https://rsf.org/en/
Roberts, E. B., & Eesley, C. (2009). *Entrepreneurial Impact: The Role of MIT*. Kansas City, MI. Retrieved from http://www.kauffman.org/~/media/kauffman_org/researchreportsandcovers/2009/02/mit_impact_full_report.pdf
Robinson, R. (2016, February 1). Why smart cities still aren't working for us after 20 years. And how we can fix them [Blog post]. *The Urban Technologist. People. Place. Technology*. Retrieved from https://theurbantechnologist.com/2016/02/01/why-smart-cities-still-arent-working-for-us-after-20-years-and-how-we-can-fix-them/
Sæbø, Ø., Rose, J., & Molka-Danielsen, J. (2010). eParticipation: Designing and managing political discussion forums. *Social Science Computer Review*, *28*(4), 403–426.
Sassen, S. (2001). *The global city: New York, London, Tokyo* (2nd Ed.). Princeton, NJ: Princeton University Press.
Sassen, S. (2013). Does the city have speech? *Public Culture*, *25*(2 70), 209–221. https://doi.org/10.1215/08992363-2020557
Schröder, B., Kuta, K. D., & Haidar, R. (2010). *Marktstudie Arabische Emirate für den Export beruflicher Aus- und Weiterbildung [Market study Arab Emirates for export vocational education and training]*. Bonn, DE. Retrieved from http://www1.imove-germany.de/cps/rde/xbcr/imove_projekt_de/p_iMOVE-Marktstudie_VAE_Vereinigte-Arabische-Emirate_2010.pdf

Schumann, L., & Stock, W. G. (2015). Acceptance and use of ubiquitous cities' information services. *Information Services & Use, 35*(3), 191–206. https://doi.org/10.3233/ISU-140759

Sharma, R. S., Lim, S., & Boon, C. Y. (2009). A vision for a knowledge society and learning nation: The role of a national library system. *The ICFAI University Journal of Knowledge Management, 7*(5/6), 91–113.

Sharma, R. S., Ng, E. W. J., Dharmawirya, M., & Keong Lee, C. (2008). Beyond the digital divide: A conceptual framework for analyzing knowledge societies. *Journal of Knowledge Management, 12*(5), 151–164. https://doi.org/10.1108/13673270810903000

Shaziman, S., Usman, I. M. S., & Tahir, M. (2010). Waterfront as public space case study; Klang River between Masjid Jamek and Central Market, Kuala Lumpur. *Selected Topics in Energy, Environment, Sustainable Development and Landscaping Waterfront*, 344–349. Retrieved from http://www.wseas.us/e-library/conferences/2010/TimisoaraP/EELA/EELA-56.pdf

Siemens. (2012). *The green city index*. Munich. Retrieved from http://www.siemens.com/entry/cc/features/greencityindex_international/all/de/pdf/gci_report_summary.pdf

Skot-Hansen, D., Hvenegaard, C., Jochumsen, H., Skot-Hansen, D., Hvenegaard Rasmussen, C., & Jochumsen, H. (2013). The role of public libraries in culture-led urban regeneration. *New Library World, 114*(1/2), 7–19. https://doi.org/10.1108/03074801311291929

Sohn, D.-W., & Kenney, M. (2007). Universities, clusters, and innovation systems: The case of Seoul, Korea. *World Development, 35*(6), 991–1004. https://doi.org/10.1016/j.worlddev.2006.05.008

Srivastava, S. C., Chandra, S., & Theng, Y.-L. (2010). Evaluating the role of trust in consumer adoption of mobile payment systems: An empirical analysis. *Communications of the Association for Information Systems, 27*, 561–588.

Stanford Research Park. (n.d.). About "past, presence & future." Retrieved from http://stanford-researchpark.com/about

Statista.com. (n.d.). China: Population of Shenzhen from 1995 to 2030 (in millions). Retrieved from http://www.statista.com/statistics/466986/china-population-of-shenzhen/

Stehr, N. (2003). The social and political control of knowledge in modern societies. *International Social Science Journal, 55*(4), 643–655. https://doi.org/10.1111/j.0020-8701.2003.05504014.x

Stock, W. G. (2011). Informational cities: Analysis and construction of cities in the knowledge society. *Journal of the American Society for Information Science and Technology, 62*(5), 963–986. https://doi.org/10.1002/asi

Susha, I., & Grönlund, Å. (2012). EParticipation research: Systematizing the field. *Government Information Quarterly, 29*(3), 373–382. https://doi.org/10.1016/j.giq.2011.11.005

Taylor, P. J. (2012). On city cooperation and city competition. In B. Derudder, M. Hoyler, P. J. Taylor, & F. Witlox (Eds.), *International handbook of globalization and world cities* (pp. 64–72). Cheltenham, UK; Northampton, MA: Edward Elgar Publishing.

Taylor, P. J., & Csomós, G. (2012). Cities as control and command centres: Analysis and interpretation. *Cities, 29*(6), 408–411. https://doi.org/10.1016/j.cities.2011.09.005

Taylor, P. J., Derudder, B., Faulconbridge, J., Hoyler, M., & Ni, P. (2014). Advanced producer service firms as strategic networks, global cities as strategic places. *Economic Geography, 90*(3), 267–291. https://doi.org/10.1111/ecge.12040

Taylor, P. J., Derudder, B., Hoyler, M., & Witlox, F. (2012). Advanced producer servicing networks of world cities. In P. J. Taylor, B. Derudder, M. Hoyler, & F. Witlox (Eds.), *International handbook of globalization and world cities* (pp. 135–145). Cheltenham, UK; Northampton, MA: Edward Elgar Publishing.

Taylor, P. J., Hoyler, M., & Smith, D. (2012). Cities in the making of world hegemonies. In B. Derudder, M. Hoyler, P. J. Taylor, & F. Witlox (Eds.), *International handbook of globalization and world cities* (pp. 22–30). Cheltenham, UK; Northampton, MA: Edward Elgar Publishing.

Taylor, P. J., Hoyler, M., & Verbruggen, R. (2010). External urban relational process: Introducing central flow theory to complement central place theory. *Urban Studies*, *47*(13), 2803–2818. https://doi.org/10.1177/0042098010377367

TGH. (2016). Technology goes home. Retrieved from http://www.techgoeshome.org/#!about/czk3

The Daily Beast. (2011, January 17). 20 most tolerant states [Blog post]. Retrieved from http://www.thedailybeast.com/articles/2011/01/16/ranking-the-most-tolerant-and-least-tolerant-states.html

The Economist. (2001, August 9). Stockmarket turnover. Retrieved from http://www.economist.com/node/731687

Thoening, J.-C. (2015, May 27). Why France is building a mega-university at Paris-Saclay to rival Silicon Valley [Blog post]. *The Conversation*. Retrieved from http://theconversation.com/why-france-is-building-a-mega-university-at-paris-saclay-to-rival-silicon-valley-41786

TUM. (n.d.). Geschichte – TUM [Hystory – TUM]. Retrieved from http://www.tum.de/die-tum/die-universitaet/geschichte/

U.S. Consulate General in Shanghai. (n.d.). U.S. Consulate Shanghai air quality monitor. Retrieved from http://shanghai.usembassy-china.org.cn/airmonitor.html

UAE Embassy in Ottawa. (2016). Business and trade. Free zones & special economic zones. Retrieved from http://www.uae-embassy.ae/Embassies/ca/Content/1151

UITP. (2015). *Mobility in cities database. Synthesis report. UITP*. Retrieved from http://search.ebscohost.com/login.aspx?direct=true&db=bth&AN=18984081&site=ehost-live

UNDP. (2014). *Human development report 2014*. New York, NY. Retrieved from http://hdr.undp.org/en/content/human-development-report-2014

United Nations. (2016). *United Nation e-governmnet survey 2016. E-government in support of sustainable development*. New York. Retrieved from http://workspace.unpan.org/sites/Internet/Documents/UNPAN96407.pdf

U.S. Department of Commerce, & Bureau of Economic Analysis. (2016). Gross output by industry. Retrieved from https://www.bea.gov/iTable/iTable.cfm?ReqID=51&step=1#reqid=51&step=51&isuri=1&5114=a&5102=15

van de Stadt, I., & Thorsteinsdóttir, S. (2007). Going e-only: All Icelandic citizens are hooked. *Library Connect*, *5*(1), 2.

Venkatraman, V. (2014). Mobility beyond transport in smart cities [Video file]. *TEDxCopenhagenSalon*. Retrieved from https://www.youtube.com/watch?v=z5O4Yl6ZB4k

Von Zedtwitz, M., & Heimann, P. (2006). Innovation in clusters and the liability of foreignness of international R&D. In E. G. Carayannis & D. F. J. Campbell (Eds.), *Knowledge creation, diffusion, and use in innovation networks and knowledge clusters*. Westport, CT; London, UK: Greenwood Publishing Group.

Wall, R., & van der Knaap, B. (2012). Centrality, hierarchy and heterarchy of worldwide corporate networks. In B. Derudder, M. Hoyler, P. J. Taylor, & F. Witlox (Eds.), *International handbook of globalization and world cities* (pp. 209–229). Cheltenham, UK; Northampton, MA, UK; Northampton, MA: Edward Elgar Publishing.

Wang, J. (2013). The economic impact of special economic zones: Evidence from Chinese municipalities. *Journal of Development Economics*, *101*(1), 133–147. https://doi.org/10.1016/j.jdeveco.2012.10.009

Web Foundation. (2015). *ODB global report*. Retrieved from http://opendatabarometer.org/
Wilson, S. G., Plane, D. a, Mackun, P. J., Fischetti, T. R., Goworowska, J., Cohen, D. T., ... Hatchard, G. W. (2012). Patterns of metropolitan and micropolitan population change: 2000 to 2010. *2010 Census Special Reports*, (September), 1–102. https://doi.org/C2010SR-01
WIRED. (2016). Shenzhen: The Silicon Valley of hardware (Part 1) [Video file]. Retrieved from https://www.youtube.com/watch?v=hp6F_ApUq-c
World Justice Project. (n.d.). Global right to information rating. Methodology. Retrieved from http://www.rti-rating.org/methodology/
World Justice Project. (2015). *Open government index 2015 report*. Washington, DC, Seattle, WA. Retrieved from http://worldjusticeproject.org/sites/default/files/ogi_2015.pdf
Wright, G. (2016, January 12). The best cities to live in the world 2015. *Global Finance*. Retrieved from www.gfmag.com/global-data/non-economic-data/best-cities-to-live
Yeandle, M., & Z/Yen Group Limited. (2016). *Global financial centres index 19*. Retrieved from http://www.longfinance.net/global-financial-centre-index-19.html
Yigitcanlar, T. (2010). Informational city. In R. Hutchison (Ed.), *Encyclopedia of urban studies* (pp. 392–395). New York, NY: Sage.
Yigitcanlar, T., O'Connor, K., & Westerman, C. (2008). The making of knowledge cities: Melbourne's knowledge-based urban development experience. *Cities*, *25*(2), 63–72. https://doi.org/10.1016/j.cities.2008.01.001

7 Conclusion

The present work was a first attempt to investigate the influence of political willingness, infrastructures, and the status as a world city on the state and development of prototypical cities of the knowledge society. The underlying theories on cities of the 21st century have been developed throughout the last decades, but a global comparison of real cities is missing. Thus, for example, the emergence of the creative class is primarily demonstrated in the United States and so far has not been found for cities in other countries (Murugadas, Vieten, Nikolic, Fietkiewicz, & Stock, 2015). Other case studies, such as for instance the correlation of the "Human Development Index" and the "ICT Development Index", have shown that ICT is growing in importance globally (Stock, 2011). Even though this data was gathered on the national level and not on the city level, it demonstrates that a higher human development positively affects the maturity of the ICT infrastructure and vice versa. But what does this mean for our cities in a global economy? As there are different definitions of prototypical cities of the knowledge society, and as indices on the city level are not available on a global scale, the approach in this work has been to identify the "new field" of real world examples of informational world cities. Hence, the attempt was not to only focus on the increase of ICT, as many previous case study projects of smart cities have, but also on the recent development of cities with regard to governmental, human, and economic interaction.

In this chapter, I will summarize the main findings of each hypothesis and reveal whether they were supported by the data or whether they have to be reconsidered based on the expert interviews. Finally, the identified characteristics in a typical informational world city will be summarized.

7.1 Infrastructures

The main infrastructures of an informational world city are digital and cognitive. The digital refers to investigations concerning the information and communication technology (ICT). Its use and access are highly correlated with a high level of human development on the national level (Stock, 2011). That is why all of the investigated cities are located in a high or very high human development environment and it can be assumed that the underlying ICT infrastructure is advanced. Furthermore, as a global hub, an informational world city adds value to the economic market of the ICT sector. The cognitive infrastructure is used by researchers to refer to the knowledge and creative city (Ergazakis, Metaxiotis, &

Psarras, 2004; Florida, 2002; Landry, 2008). As described in the literature review, it is difficult to separate both city types, as they have many overlapping indicators. Thus, the infrastructure was investigated according to knowledge and creative institutions. In total, seven hypotheses were investigated as part of the digital, knowledge, or creative city. While the hypotheses could not be shown as valid for each of the 31 cities, the results were able to provide some best practice examples.

Looking at the first hypothesis (H1), "Informational world cities are hubs for companies with information market activities, e.g. telecommunication companies," it can be agreed that information market activities play an important role in the majority of the investigated cities. However, the market is changing from being a hub of telecommunication and hardware production, as Helsinki has been in the past due to the former success of Nokia, towards increasing numbers of technology-driven corporations and software products. Technology hubs are for example the San Francisco Bay Area, the Taipei Region, and the Seoul Region. Telecommunication hubs are Tokyo, Beijing, and Dallas. Furthermore, cultural and other institutions located in ICT hubs may profit from synergy effects. In addition, universities are needed to educate the knowledge workforce. Interestingly, most experts associate the ICT market with entrepreneurship and tech startups. However, a job-growth impact has so far only been demonstrated for North American cities and not on a global level (Murugadas et al., 2015). Cities ranked as the most entrepreneurship-friendly are Silicon Valley, New York, London, Los Angeles, and Helsinki, all of which can be interpreted as entrepreneur ICT hubs. Finally, due to the increasing possibilities for how information and technology can be used it can be assumed that an informational world city is an ICT-related hub at least with reference to tech startups.

Hence, the second hypothesis (H2), "The ICT infrastructure in an informational world city is more important than automotive traffic infrastructure," was very provocative since the physical space in a city is still of high importance. Nevertheless, on the city level, an independent measuring of the ICT maturity is missing. Some indicators are only useful in dense areas, such as access to, quality of, and use of Wi-Fi. And it should be emphasized that the shrinking importance of cars in the city is a natural result of population density and not directly caused or amplified by advanced ICT. But as people are going to use other modes of mobility than a car, ICT can make them more comfortable, e.g. through mobile payment. Finally, fewer cars result in a more sustainable city. Because of that, the hypothesis has to be reformulated as follows: "The maturity of ICT and sustainability are highly correlated in an informational world city."

The third hypothesis (H3) "Science parks or university clusters that cooperate with knowledge-intensive companies are important in an informational world city," is acknowledged as being the heart of the knowledge society. Hence, all the best practice examples of ICT hubs as well as most innovative cities are home to "elite universities." Science parks and clusters need collaboration with top quality universities. Best practice examples are Cambridge in the Boston Area and Silicon Valley in the San Francisco Bay Area. Both are agglomerations of universities and knowledge-intensive companies, demonstrating the success of physical concentration. Furthermore, both cities have a flourishing startup scene and are highly innovative. In conclusion, cities with highly ranked universities are more likely to be a knowledge and/or innovation hub.

With regard to the fourth hypothesis (H4), "An informational world city needs to be a creative city," a mixture of the traditional and modern definitions was discussed. Hence, the term creativity was widened from culture and arts to include technology and innovation. Creative institutions (museums, galleries, etc.) are not good indicators to identify a creative city, whereas the innovation output can be measured by the number of patents granted. By the pure number of patents, Seoul is the leading city, while per capita, San Francisco and Boston have the highest numbers in 2012. Furthermore, the creative economy affects the digital economy, for example by online sales of cultural goods. A creative city also allows a free flow of people. This is measured by the city's tolerance. Most tolerant in the US are San Francisco and Los Angeles (among the 5 US cities investigated in the work at hand). Creativity also needs space: "Creative milieus" for artists, and "milieus of innovation" for the technology talents and entrepreneurs. Thus, being a creative city is at the very least an advantage, with San Francisco again being one of the leading best practice examples. Finally, out of all creative city aspects, the most important characteristic is space for and flow of the creative class.

The role of physical space was investigated in the fifth hypothesis (H5), "Physical space for face-to-face interaction is important for an informational world city." Two main aspects stood out: First, the city should have an architecture that encourages face-to-face meeting, e.g. by a "coffee house culture" as it is known in Vienna or Paris. Second, coworking spaces are growing in importance in an increasingly digitized world. Entrepreneurs and startups profit from tacit knowledge exchange and an open innovation process, just like other businesses and corporation. However, most cities offer only commercial places such as cafés, bars or restaurants, and only little free community space. To positively impact creativity, it is important to assemble a diverse mix of people and work across different disciplines. Libraries often offer this space and the necessary infrastructure through Wi-Fi hotspots. On a per capita basis Amsterdam, San Francisco, and Boston have the most coworking space available – and thus spaces for people to meet face-to-face.

The public library serves a special function in an informational world city. This is expressed in the sixth hypothesis (H6), "A fully developed content infrastructure, e.g. supported by digital libraries, is a characteristic feature of an informational world city." Access to specialized databases needs to be adjusted to what is important to the general public. Relevant are, for example, instructions on how to use digital content, assistance in information retrieval, and digital equipment and Wi-Fi connectivity at all branches. As a digital content provider, the library can help the municipality to open and digitize data. Based on an investigation of the offered digital services, the cities New York, San Francisco, and Toronto are best practice examples. Finally, a fully developed content infrastructure is not of great importance. The hypothesis needs to be reformulated into the following: "The knowledge society needs advanced information literacy as well as local information providers (e.g. open data portals)."

As the final part of the investigation of infrastructures, the libraries' importance as physical space was explored with H7, "Libraries are important in an informational world city as a physical place for face-to-face communication and interaction." That physical space for face-to-face communication is needed in an informational world city was already revealed by H5, but it is not a common standard in informational cities that libraries are those places. However, library space must adjust to the community's demands. In Beijing the working spaces and in Shenzhen the bookstores are crowded. In Stockholm space for a variety of activities is given instead. Interestingly, maker spaces add a new dimension to the existing spaces in libraries. Best practice examples of physical library spaces can be found in Montréal, Singapore, and Shenzhen. However, interesting maker spaces have also been introduced in Chicago and Toronto, allowing citizens to create and not merely consume. In some cases, they can also be invited to the public library to be part of the libraries' program as well as to participate in the decision-making regarding future developments. Indeed, the public library is important as face-to-face space for the community, which is why the community should be able to interact and participate in the future of this space.

7.2 Political will

The political will of an informational world city refers to the governmental plans for and energy in pushing forward the knowledge-based urban development. With respect to the development of the ICT infrastructure, e-services can also be made available on the municipal level. Hence, through e-participation and e-democracy, the city can reach the level of e-governance or even become an open government, if it is able to take charge in implementing open innovation processes. Finally, transparency was identified as key to a trustworthy government. However, on a global scale, the free flow of information is shrinking.

Concerning H8, "Political willingness is important to establish an informational world city especially according to knowledge economy activities", the investigation indicated that it is not a necessary requirement. What government can do is to boost the informational city development through funding and incentives. Particularly, incentives to attract the information and knowledge economy are of advantage, e.g. Twitter in San Francisco, Ubisoft in Montréal, or the special economic zones in Dubai and several Chinese cities. Political action in an informational world city can both help and hurt. In Berlin, for example, the startup evolution was a bottom-up development. This is comparable to the creative milieus that have emerged in free spaces. Today, milieus of innovation follow this trend. For instance, Shenzhen's evolution into a Silicon Valley of Hardware was only possible because the government did not interfere. Cities that are acknowledged as having a strong political will towards a knowledge-based urban development are Barcelona, Boston, Melbourne, Montréal, Seoul, Shenzhen, Singapore, Tokyo, Vancouver and Vienna.

A further part of the political will is the integration of e-governance, as investigated by H9, "An informational world city is characterized by e-governance (including e-government, e-participation, e-democracy)." However, most of the cities do not perform as the interviewed experts would like to see. There are still many fears that need to be faced before e-governance can be realized on the municipal level, such as high costs and the security and reliability of the data. However, on the national level, many governments have introduced online identification that can be used to verify government-to-citizen or government-to-business transactions, e.g. SingPass in Singapore. This is a major step forward in the realm of e-services, but still a long way from true open government. Participatory approaches in decision-making processes need engaged citizens as well as an open-minded administration. Ultimately, e-governance could become a characteristic feature of an informational world city, but is currently not yet fully available.

Regarding the political will and e-services, a hypothesis on freedom of information was evaluated, namely H10, "A free flow of all kinds of information (inclusive mass media information) is an important characteristic of an informational world city." Freedom of information can be investigated on the one hand by the political will as evidenced by the laws guaranteeing freedom of information and freedom of expression, and on the other hand by political actions represented through the implementation of open data portals. However, a freedom of information law by itself does not prevent corruption, as the example of Montréal demonstrates. Similarly, mere access to mass media must not be confused with a free flow of information, since self-censorship is growing. Independent publishers are also becoming increasingly rare, even though they are vital within a knowledge society, for instance to foster critical thinking. In conclusion, the investigated cities located in China, Singapore, Malaysia, or the UAE have no or very little freedom of information. Informational cities in "Western" countries tend to have a higher degree of freedom of information than others, but there is a global decline in the free flow of information that should alarm the knowledge society.

7.3 World city

Research on world cities as well as on informational cities deals with the flows of information, capital, and power. Capital and power are related to financial hubs and financial flows, as measured by multinational financial institutions, stock exchanges, and the GDP (Taylor, 2004). Furthermore, the cityness as an aspect of urbanity and inter-city flows was investigated in the case study of the 31 informational world cities.

Looking at H11, "An informational world city has to be a financial hub with a lot of banks and insurance companies," the interviewed experts did not agree. The overall economic success is much more important than the financial sector (measured by the stock exchange and GDP). Many cities like New York, London, Paris, San Francisco, or Boston can be found in the list of the top 20 global cities as well as in the list of the top 20 venture capital cities. Finance and insurance are not creative and therefore do not add value to the informational world city, whereas the FinTech sector may close the gap between the world city and informational city. Global hubs combining both city types are New York, London, and Paris. Interestingly, Boston is no global hub of finance and is not among the top stock exchanges, but it is the city with the highest GDP per capita and one of the top global university hubs. Being successful therefore does not seem to depend on being a global financial hub, at least in the case of Boston.

Finally, concerning H12, "An informational city is supposed to be a global city ('world city')," it could not be determined whether small or medium size cities are unable to be informational cities. For some experts, it is more important to be a hub for research and development. However, those cities generally do not have the kind of infrastructure which attracts multinational corporations, which in turn eventually attract the talents. But as revealed by Boston, this does not seem to be a universal rule, as the city is the hub for R&D of many multinational corporations, but not their headquarters. Therefore, these institutions as well as the elite universities attract talents to graduate and to work there. In a world of talent circulation, cities become "talent hubs" independent of their size but characterized by their openness. Thus, the hypotheses referring to world city research need to be adjusted to questioning the openness of informational cities towards talents and innovation. However, this is just true for cities in "Western" countries, as, surprisingly, cities in Asia are growing in their GDP per year but not in their openness for talents or information. Hence, based on the measurement of power, knowledge, and innovation flows, they are able to compete or even lead the top cities, as Seoul does when it comes to the total number of patents granted per year.

7.4 The typical informational world city

Merging the ideas described above, we are able to draw the prototypical informational world city as follows:

An informational world city is by definition ubiquitously connected and enables a sustainable lifestyle. ICT-related jobs are easy to find there. These jobs will not be IT hardware manufacturing jobs, but rather may be related to culture, health, finance or any other discipline that can enhance its efficiency and possibilities through ICT. Furthermore, it is easy to start one's own tech-related business, as it is supported by the government and will enjoy many benefits and incentives. And if that does not work out, the multinational corporations in the special economic zones are also on the lookout for creative talents.

Whenever needed, highly educated people can be found for a coffee break in any of the countless restaurants or at your favorite coworking space. To share their knowledge with others, entrepreneurs join the library community for a hack event or demonstrate their 3D design skills in an open seminar. The library promotes information literacy skills and keeps users up to date on social and environmental topics. Every now and then space can be found that were redesigned for different kind of activities, such as art, performing or coding workshops.

In terms of the world city status, it becomes less important to be a hub of the financial sector. To make investments and the ability to apply for venture capital will replace traditional finance institutions. However, both are located where they are able to access an enhanced ICT infrastructure and additionally may profit from the synergies with entrepreneurs of the FinTech sector.

Everyone is connected everywhere and anytime. The taxes and bus tickets are paid via mobile application and the news users read consists of pre-filtered pro-government information. Access is not restricted, but limited. In addition, language barriers make it difficult to retrieve the worldwide knowledge. Talent inflow is only observed in predominately English-speaking regions. For reliable full content data, access to the university library is needed, which is often limited to students and faculty. Citizen engagement in political decision processes to fight to realize the freedom of information for all increases, but tweets to the administration often go unanswered – the local government is too busy due to the digitization processes in each department. Ultimately, being one of the most successful cities e.g. in finance, innovation, or education is by definition reserved for a select few. The majority will remain regional hubs and not become command and control destinations.

These findings are based on the investigated case studies. To assemble more evidence about the city development in the 21st century, interdisciplinary approaches as in the present work and further empirical and statistical data are needed. Finally, the identified characteristics based on the literature review and modified according to the expert interviews will contribute to an increased understanding of cities' development, growth, and success in the 21st century. Hence, we are now able to benchmark informational cities on a global scale with regard to their political willingness, infrastructures, and cityness. In the future, research projects at the Department of Information Science of Heinrich-Heine University Düsseldorf will continue to investigate the selected 31 informational world cities, especially with respect to the indicators of the labor market, corporate structure, and weak location factors. Considering the increase of the Internet of Things, open innovation, and sharing industry, it will be interesting to see how the cities will develop in the following years.

References

Ergazakis, K., Metaxiotis, K., & Psarras, J. (2004). Towards knowledge cities: Conceptual analysis and success stories. *Journal of Knowledge Management, 8*(5), 5–15. https://doi.org/10.1108/13673270410558747

Florida, R. L. (2002). *The rise of the creative class: And how it's transforming work, leisure, community and everyday life*. New York, NY: Basic Books.

Landry, C. (2008). *The creative city: A toolkit for urban innovators* (2nd ed.). New York, NY: Earthscan Publications.

Murugadas, D., Vieten, S., Nikolic, J., Fietkiewicz, K. J., & Stock, W. G. (2015). Creativity and entrepreneurship in informational metropolitan regions. *Journal of Economic and Social Development, 2*(1), 14–24. Retrieved from http://hdl.handle.net/10125/41496

Stock, W. G. (2011). Informational cities: Analysis and construction of cities in the knowledge society. *Journal of the American Society for Information Science and Technology, 62*(5), 963–986. https://doi.org/10.1002/asi

Taylor, P. J. (2004). *World city network: A global urban analysis*. London, UK: Routledge. https://doi.org/10.4324/9780203634059

Appendix I: List of all interview partners

Amsterdam; Netherlands:
AM 1, personal communication, January 21, 2014
AM 2, personal communication, January 21, 2014
AM 3, personal communication, January 21, 2014
AM 4, personal communication, January 22, 2014
AM 5, personal communication, January 22, 2014
AM 6, personal communication, January 23, 2014
AM 7, personal communication, January 21, 2014

Barcelona, Spain:
BA 1, personal communication, December 4, 2013
BA 2, personal communication, December 4, 2013
BA 3, personal communication, December 4, 2013
BA 4, personal communication, December 4, 2013
BA 5, personal communication, December 4, 2013
BA 6, personal communication, December 5, 2013
BA 7, personal communication, December 5, 2013
BA 8, personal communication, December 5, 2013
BA 9, personal communication, December 5, 2013

Berlin, Germany:
BE 1, personal communication, November 20, 2013
BE 2, personal communication, November 20, 2013
BE 3, personal communication, November 20, 2013
BE 4, personal communication, November 20, 2013
BE 5, personal communication, November 20, 2013
BE 6, personal communication, November 20, 2013
BE 7, personal communication, November 19, 2013
BE 8, personal communication, November 20, 2013
BE 9, personal communication, November 20, 2013

Boston, USA:
BO 1, personal communication, August 29, 2013
BO 2, personal communication, August 30, 2013
BO 3, personal communication, August 30, 2013
BO 4, personal communication, August 30, 2013

Chicago, USA:
CH 1, personal communication, September 4, 2013
CH 2, personal communication, September 5, 2013
CH 3, personal communication, September 4, 2013

Dubai, VAE:
DU 1, personal communication, February 20, 2013
DU 2, personal communication, February 20, 2013
DU 3, personal communication, February 20, 2013
DU 4, personal communication, February 25, 2013

Frankfurt, Germany:
FR 1, personal communication, May 12, 2014 (phone call)
FR 2, personal communication, November 13, 2013
FR 3, personal communication, November 12, 2013
FR 4, personal communication, November 12, 2013
FR 5, personal communication, November 12, 2013

Helsinki, Finland:
HE 1, personal communication, August 6, 2013
HE 2, personal communication, August 6, 2013
HE 3, personal communication, August 6, 2013
HE 4, personal communication, August 5, 2013

Honk Kong, China:
HK 1, personal communication, July 15, 2011
HK 2, personal communication, July 15, 2011
HK 3, personal communication, February 13, 2014
HK 4, personal communication, February 14, 2014

Kuala Lumpur, Malaysia:
KL 1, personal communication, February 19, 2014

London, UK:
LO 1, personal communication, June 27, 2013
LO 2, personal communication, June 27, 2013
LO 3, personal communication, June 27, 2013
LO 4, personal communication, June 27, 2013
LO 5, personal communication, June 27, 2013

LO 6, personal communication, June 27, 2013
LO 7, personal communication, June 27, 2013
LO 8, personal communication, June 27, 2013

Los Angeles, USA:
LA 1, personal communication, September 9, 2013
LA 2, personal communication, September 9, 2013
LA 3, personal communication, September 12, 2013

Melbourne, Australia:
ME 1, personal communication, February 28, 2012
ME 2, personal communication, March 1, 2012
ME 2, personal communication, March 1, 2012

Milan, Italy:
MI 1, personal communication, November 6, 2013
MI 2, personal communication, November 6, 2013
MI 3, personal communication, November 6, 2013
MI 4, personal communication, November 6, 2013
MI 5, personal communication, November 7, 2013
MI 6, personal communication, November 7, 2013
MI 7, personal communication, November 6, 2013
MI 8, personal communication, November 6, 2013

Montreal, Canada:
MO 1, personal communication, March 17, 2014
MO 2, personal communication, March 14, 2014
MO 3, personal communication, March 14, 2014
MO 4, personal communication, March 15, 2014
MO 5, personal communication, March 14, 2014
MO 6, personal communication, June 3, 2014 (phone call)

Munich, Germany:
MU 1, personal communication, April 29, 2014
MU 2, personal communication, April 28, 2014
MU 3, personal communication, April 28, 2014
MU 4, personal communication, April 28, 2014
MU 5, personal communication, April 28, 2014
MU 6, personal communication, April 30, 2014

Appendix I: List of all interview partners

New York, USA:
NY 1, personal communication, August 27, 2013
NY 2, personal communication, August 21, 2013 (phone call)
NY 3, personal communication, August 21, 2013 (phone call)
NY 4, personal communication, August 21, 2013 (phone call)
NY 5, personal communication, August 28, 2013

Paris, France:
PA 1, personal communication, December 6, 2013
PA 2, personal communication, December 10, 2013
PA 3, personal communication, December 11, 2013
PA 4, personal communication, December 11, 2013
PA 5, personal communication, December 12, 2013
PA 6, personal communication, December 12, 2013

San Francisco; USA:
SF 1, personal communication, September 17, 2013
SF 2, personal communication, September 17, 2013
SF 3, personal communication, September 17, 2013
SF 4, personal communication, September 17, 2013
SF 5, personal communication, September 17, 2013
SF 6, personal communication, September 16, 2013

Sao Paulo, Brasil:
SP 1, personal communication, February 24, 2014
SP 2, personal communication, February 24, 2014
SP 3, personal communication, February 25, 2014
SP 4, personal communication, February 26, 2014
SP 5, personal communication, April 16, 2014 (phone call)

Seoul, South Korea:
SE 1, personal communication, July 23, 2012
SE 2, personal communication, July 23, 2012
SE 3, personal communication, July 23, 2012

Shanghai, China:
SH 1, personal communication, July 13, 2012

Shenzhen, China:
SHE 1, personal communication, February 12, 2014
SHE 2, personal communication, February 12, 2014

Singapore, Singapore:
SG 1, personal communication, February 12, 2014
SG 2, personal communication, June 24, 2010
SG 3, personal communication, June 29, 2010
SG 4, personal communication, June 29, 2010
SG 5, personal communication, June 28, 2010
SG 6, personal communication, June 24, 2010
SG 7, personal communication, June 24, 2010
SG 8, personal communication, June 24, 2010
SG 9, personal communication, June 25, 2010
SG 10, personal communication, June 25, 2010
SG 11, personal communication, June 29, 2010
SG 12, personal communication, June 30, 2010
SG 13, personal communication, June 30, 2010
SG 14, personal communication, June 28, 2010

Stockholm, Sweden:
ST 1, personal communication, August 9, 2013
ST 1, personal communication, August 8, 2013
ST 1, personal communication, August 8, 2013
ST 1, personal communication, August 8, 2013

Sydney, Australia:
SY 1, personal communication, July 7, 2011
SY 2, personal communication, July 9, 2011
SY 3, personal communication, July 8, 2011

Tokyo, Japan:
TO 1, personal communication, July 26, 2012
TO 2, personal communication, April 27, 2013
TO 3, personal communication, April 28, 2013
TO 4, personal communication, April 28, 2013
TO 5, personal communication, April 27, 2013
TO 6, personal communication, July 26, 2012
TO 7, personal communication, April 28, 2013

Appendix I: List of all interview partners

Toronto, Canada:
TOR 1, personal communication, March 11, 2014
TOR 2, personal communication, March 11, 2014
TOR 3, personal communication, March 11, 2014
TOR 4, personal communication, March 10, 2014

Vancouver, Canada:
VA 1, personal communication, March 19, 2014
VA 2, personal communication, March 19, 2014
VA 3, personal communication, March 20, 2014
VA 4, personal communication, April 9, 2014 (phone call)
VA 5, personal communication, April 28, 2014 (phone call)
VA 6, personal communication, March 20, 2014

Vienna, Austria:
VI 1, personal communication, January 29, 2014
VI 2, personal communication, January 29, 2014
VI 3, personal communication, January 30, 2014
VI 4, personal communication, January 30, 2014
VI 5, personal communication, January 31, 2014
VI 6, personal communication, January 31, 2014
VI 7, personal communication, January 31, 2014
VI 8, personal communication, January 31, 2014

Names of interview partners:
Apolloni, Bruno; Montanelli, Stefano; Genta, Lorenzo; De Cindio, Fiorella; Sorrentino, Maddalena; Castelnovo, Walter; Viscusi, Gianluigi; Maurino, Andrea; Nijkamp, Peter; Bontje, Marco; Sleutjes, Bart; Koeman, Willem; Duinker, Candy; Shulmeister, Mike; Lam, Tat; Bunt, Travis; Forrest, Ray; Fauth, Tim; Stiller-Kern, Gabrielle; Malina, Simone; Lucon, Oswaldo; Misue Sato, Celly; Gegner, Martin; Madreiter, Thomas; Feigl, Markus; Weidinger, Norbert; Skerlan-Schuhböck, Thomas; Himpele, Klemens; Wieser, Peter; Ohler, Sabine; Adam, Ursula; Stolarick, Kevin; Berry, Rob; Turner, Lorne S.; Glass, Elisabeth; Archambault, Eric; Lariviere, Vincent; Tomalty, Ray; Stoffl, Manfred; Poulin, Marie-Pierre; Wixted, Brian; Holbrook, Adam; Flaherty, Shelagh; Smith, Richard; Abadal, Ernest; Batlle-Montserrat, Joan; Carrasco Bonet, Marta; Sanz Marco, Lluis; Saller, Raymond; Engel, Sophie; Buettner, Anke; Becker, Peter; Thierstein, Alain; Booth, Jan; Sauer, Peter; Bawden, Peter; Smith, John; Godel, Moritz; Bikakis, Antonis; Moszynski, Michael; Glossob, Catherine; Peterek, Michael; Schröder, Thomas; Zabel, Ralf; Chung, Heejin; Socia, Deb; Savoie, Curt; Lewis, Harry; Bush, Margaret; Erdmann, Nora; Miwa, Makiko; Kleske, Johannes; Schwarzmann, Igor; Petras, Vivien; Trikulja, Violeta; Dröge, Evelyn; Bordfeld, Fred; Wimmer, Ulla; Kobalz Nägele, Rüdiger; Homilius, Sabina; Kenworthy, Jeffrey; Totok, Andreas; Botte, Alexander;

Appendix I: List of all interview partners

Halpern, Joann; Bezman, Steven; Torvi, Lara; Nicklin, Andrew; Sassen, Saskia; Winkle, Curtis R.; Tovla, John; Bannon, Brian; Strauss, Karen; Danelo, Cathy; Sanabria, Alvaro; Anderson, Megan; Hall, Mark; Herrera, Luis; Fleming, Thomas B.; Rudell-Betts, Linda; Pincetl, Stephanie; Engardt, Jonas; Hedenström, Maria; Holmenmark, Agneta; Haas, Tigran; Lempiäinen, Nina; Valinkangas Loytty, Taru; Manninen, Rikhard; Lönnqvist, Henrik; Suleiman, Ali Amour; Kayed, Nasif; Young, Sybille; Roeder, Maya; Eilrich, Claus; Dieth, Regine; Finken, Holger; Schöningh, Ingo; Shipp, John; Pysik, Dagmar; Gnodtke, Hans; Winter, Stephan; Johnson, Paul; Pagell, Ruth; Munoo, Rajen; Pin Pin, Yeo; Khoo Soo Guan, Christopher; Chua, Luke Eng Koon; Lim Yan Hong; Nandivada, Meenakshi; Chan Wai Ling, Belinda; Periasamy, Pelly; Goh, Cindy; Herzog, Klaus; Dresel, Robin; Tan, Margaret; Coulman, Florian; Gomera Cruz, Aldo; Terrin, Jean-Jacques; Krzatala-Jaworska, Ewa; Ritte, Léonard.

Appendix II: Literature Review

City Name	World/Global City Author	Year	Knowledge City Author	Year	Creative City Author	Year	Digital City Author	Year	Smart City Author	Year
Amsterdam	Taylor, P. J.	2000	Musterd, S.	2004	Hitters, E., & Hospers, G.-J.	2002	Ishida, T.	2000	Hoolands, R. G.	2008
	Hall, P.	2005	Van den Berg, L.	2005	Musterd, S.	2003b	van de Besselaar, P. Riemens, P., & Lovink,	2001	Abdoullaev, A.	2011
			Musterd, S., & Deurloo,	2006	Ultermark, J.	2004	Couclelis, H.	2002	Komninos, N.	2011
			Matthiessen, C. W.,	2006	Landry, C.	2004	Schuler, D.	2004	Walravens, N.	2015
			Van Winden. W., van	2007	Heywood, P.	2008	Horne, M., Thompson,	2005		
			Gilderbloom, J.I.,	2009	Evans, G.	2008	Shin, D., Nah, Y., Lee, I.-	2007		
			Musterd, S., Gritsai,	2009	Bontje, M., &	2009		2008		
			Pethe, H., Hafner, S., &	2010	Romein, A., &	2009				
			Pareja-Eastaway, M.,	2010		2012				
			Pareja-Eastaway, M.,	2013						
			Streit, A., & Lange, B.	2013						
Barcelona	Taylor, P. J.	2000	Ergazakis, K.,	2004	Foord, J.	2008	Gdaniec, C.	2000	Bakici, T., Almirall, E.,	2013
	Hall, P.	2005	Musterd, S.	2004	Landry, C.	2008	Schuler, D.	2005		
			Walliser, A.	2004	Costa, P., &	2009	Evans, G.	2009		
			Dvir, R., & Pasher, E.	2004	Bontje, M., &	2009				
			Edvinsson, L.	2006	Cohendet, P.,	2011				
			Ergazakis, E., Ergazakis,	2009	Marti-Costa, M.,	2011				
			Musterd, S., & Gritsai,	2009	Landry, C.	2012				
			Yigitcanlar, T.	2009	Hospers, G.-J.	2003a				
			Pethe, H., Hafner, S., &	2010	Hospers, G.-J.	2003b				
			Pawlowsky, P.	2011						
			Mataxoists, K., &	2012						
			Wesselman, S.,	2012						
			Pareja-Eastaway, M.,	2013						

City Name	World/Global City Author	Year	Knowledge City Author	Year	Creative City Author	Year	Digital City Author	Year	Smart City Author	Year
Beijing	Taylor, P. J. Hall, P. Bassens, D.	2000 2005 2012	Ergazakis, K., Matthiessen, C. W., Zhaa, P.	2004 2006 2010	Keane, M. Kong, L., &	2009 2009	Keane, M. Song, J., Zhang, J., & Anthopoulos, L., &	2009 2009 2010	Zhu, Z.	2011
Berlin	Taylor, P. J. Eckardt, F. Hall, P.	2000 2005 2005	Franz, P. Streit, A., & Lange, B.	2009 2013	Hospers, G.-J. Hall, P. Ebert, R., & Foord, J. Lange, B., Landry, C. Evans, G. Brake, k. Streit, A., &	2003b 2004 2007 2008 2008 2008 2009 2012 2013	Horne, M., Thompson,	2007	Walravens, N.	2015
Boston	Taylor, P. J. Hall, P.	2000 2005	Kotkin, J., & DeVol, R. Edvinsson, L. Matthiessen, C. W., Ergazakis, E., Ergazakis, Reffat, R. M. Yigitcanlar. T. Carrillo, F. J.,	2001 2006 2006 2009 2010 2012 2014	Hospers, G.-J. Wu, W. Evans, G.	2003b 2005 2009	Kotkin, J., & DeVol, R. Horne, M., Thompson,	2003b 2007	Glaser, E. L. and Berry,	2001 2006
Chicago	Friedmann, J. Taylor, P. J. Hall, P.	1986 2000 2005	Kotkin, J., & DeVol, R. Reffat, R. M.	2001 2010	Lloyd, R. Landry, C.	2002 2008	Deren, L., Qing, Z., & Kotkin, J., & DeVol, R.	2000 2001	Walravens, N.	2015

City Name	World/Global City Author	Year	Knowledge City Author	Year	Creative City Author	Year	Digital City Author	Year	Smart City Author	Year
Dubai	Bassens, D., Derudder,	2010	Edvinsson, L.	2006	Durmaz, B.,	2008	Horne, M., Thompson,	2007	Al-Harder, M., & Rodzi,	2009
	Taylor, P. J.	2000	Alraouf, A. A.	2008			Shin, D., Nah, y., Lee, I.-	2008		
	Bassens, D.	2012	Pawlowsky, P.	2011						
			Carrillo, F. J.,	2014						
Frankfurt	Friedmann, J.	1986	Szogs, G. M.	2011	Landry, C.	2008	Horne, M., Thompson,	2007		
	Taylor, P. J.	2000								
	Hall, P.	2005								
Helsinki	Taylor, P. J.	2000	Van den Berg, L.	2005	Landry, C.	2008	Ishida, T.	2000	Landry, C.	2008
			Van Winden, W., van	2007	Bontje, M., &	2009	Couclelis, H.	2004		
			Yigitcanlar, T.	2009			Roper, S., & Grimes, S.	2005		
			Inkinen, T., &	2010			Horne, M., Thompson,	2007		
			Pethe, H., Hafner, S., &	2010			Yigitcanlar, T.	2009		
			Stachowiak, K.,	2013						
			Carrillo, F. J.,	2014						
Hong kong	Friedmann, J.	1986	Edvinsson, L.	2006	Foord, J.	2008	Blythe, S.E	2005	Nam, T., & Pardo, T. A.	2011
	Taylor, P. J.	2000	Pawlowsky, P.	2011	Heywood, P. Kong, L., &	2008	Horne, M., Thompson,	2007	Komninos, N.	2011
	Hall, P.	2005					Shin, D., Nah, Y, Lee, I.-	2008		
	Bassens, D.	2012								
Kuala Lumpur	Taylor, P.J.	2000	Edvinsson, L.	2006	Yusuf, S., &	2005	Yigitcanlar, T., &	2010		
	Hall, P.	2005	Yigitcanlar, T., &	2010						
			Carrillo, F. J.,	2014						

City Name	World/Global City Author	Year	Knowledge City Author	Year	Creative City Author	Year	Digital City Author	Year	Smart City Author	Year
London	Friedmann, J.	1986	Musterd, S.	2004	Hospers, G.-J.	2003b	Aurigi, A.	2000	Angelidou, M.	2015
	Taylor, P. J.	2000	Matthiessen, C. W.,	2006	Foord, J.	2008	Horne, M., Thompson,	2007	Ben Letaifa, S.	2015
	Sassen, S.	2001			Evans, G.	2009			Walravens, N.	2015
	Hall, P.	2005			Maitland, R.	2010				
Los Angeles	Friedmann, J.	1986	Kotkin, J., & DeVol, R.	2001	Hospers, G.-J.	2003b	Ishida, T.	2000		
	Taylor, P. J.	2000	Winkel Schwarz, A., &	2006	Foord, J.	2006	Kotkin, J., & DeVol, R.	2001		
	Hall, P.	2005					E. M., & Podevyn, M.	2007		
Melbourne	Taylor, P. J.	2000	Ergazakis, K.,	2000	Landry, C., &	2004	Deren, L, Qing, Z., &	1995	Nam, T., & Pardo, T. A.	2011
	Hall, P.	2005	Dvir, R., & Pasher, E.	2005	Baum, S.,	2008				
			Edvinsson, L.	2004	Yigitcanlar, T.	2009				
			Yigitcanlar, T.,	2006						
			Yigitcanlar, T., &	2008						
			Yigitcanlar, T.	2010						
			Metaxiotis, K., &	2012						
			Carrillo, F. J.,	2012						
				2014						
Milan	Friedmann, J.	1986	Musterd, S.	2004	Landry, C., &	2004	Hospers, G.-j.	1995		
	Taylor, P. J.	2000	Musterd, S., & Gritsai,	2009	Hospers, G.-J.	2003b	Benini, M., De Cindio,	2003b		
	Hall, P.	2005	Pareja-Eastaway, M.,	2010	Foord, J.	2008	Schuler, D.	2005		
					Evans, G.	2009	De Cindio, F.	2005		
							Evans, G.	2009		
Montereal	Hall, P.	2005	Ergazakis, E., Ergazakis,	2009	Ley, D.	2003	Hampton, K.N., &	2008	Hollands, R. G.	2008
			Yigitcanlar, T.	2009	Gertler, M. S.	2004			Leydesdorff, L., &	2011
			Mataxiotis, K., &	2012	Cohendet, P.,	2010				
					Cohendet, p.,	2011				

City Name	World/Global City Author	Year	Knowledge City Author	Year	Creative City Author	Year	Digital City Author	Year	Smart City Author	Year
Munich	Taylor, P.J.	2000	Van den Berg, L.	2005	Landry, C., Bontje, M., &	2008	Horne, M., Thompson,	2007		
	Hall, P.	2005	Van Winden, W., van Ergazakis, E., Ergazakis, Musterd, S., & Gritsai, Yigitcanlar, T. Pethe, H., Hafner, S., & Metaxiotis, K., &	2007 2009 2009 2009 2010 2012		2009				
New York	Friedmann, J.	1986	Kotkin, J., & DeVol, R.	2001	Hospers, G.-J.	2001	Kotkin, J., & DeVol, R.	2001	Angelidou, M.	2015
	Taylor, P.J.	2000	Bugliarello, G.	2004	Currid, E.	2006	Horne, M., Thompson,	2007		
	Sassen, S.	2001	Edvinsson, L.	2006	Foord, J.	2008	Hampton, K. N., &	2008		
	Hall, P.	2005	Matthiessen, C. W.,	2006	Heywood, P. Evans, G. Zukin, S., &	2008 2009 2011				
Paris	Friedmann, J.	1986	Matthiessen, C. W.,	2006	Hospers, G.-J. Evans, G. Vivant, E.	2003b 2009 2010	Hospers, G.-J. Horne, M., Thompson,	2003b 2007		
	Hall, P.	2005								
	Taylor, P.J.	2000								
San Francisco	Friedmann, J.	1986	Kotkin, J., & DeVol, R.	2001	Heywood, P.	2008	Kotkin, J., & DeVol, R.	2001	Glaeser, E. L. and Berry,	2006
	Taylor, P.J.	2000	Matthiessen, C. W.,	2006	Evans, G.	2009	Horne, M., Thompson,	2007	Hollands, R. G.	2008
	Hall, P.	2005	Reffat, R. M.	2010					Nam, T., & Pardo, T. A.	2011
									Lee, J. H., Hancock, M.	2014
Sao Paulo	Friedmann, J.	1986	Ergazakis, Metaxiotis,	2004	Costa, P., &	2009				
	Taylor, P.J.	2000	Rocco, R.	2012						
	Hall, P.	2005								

City Name	World/Global City Author	Year	Knowledge City Author	Year	Creative City Author	Year	Digital City Author	Year	Smart City Author	Year
Shenzhen	Cartier, C.	2002	Wang, D., Wu, Z., Li, Y.,	2012			Wang. C. C.	2012	Wang, D., Wu, Z., Li, Y.,	2012
			Carrillo, F. J.,	2014					de jong, M., Yu, C.,	2013
									Kang, Y., Lei, Z., Ca, C.,	2014
Seoul	Friedmann, J.	1986	Reffat, R. M.	2010	Yusuf, S., &	2005	Lee, S.-H., Yigitcanlar,	2008	Nam, T., & Pardo, T. A.	2011
	Taylor, P. J.	2000			Lee, Y.-S., &	2012	Shin, D., Nah, Y., Lee, I.-	2008	Lee, J. H., Hancock, M.	2014
	Hall, P.	2005					Evans, G.	2009		
							Choi, J. H., &	2009		
Shanghai	Taylor, P. J.	2000	Sigurdson, J.	2005	Yusuf, S., &	2005	Ding, P., Lin, D., &	2005	Sigurdson, J.	2005
	Hall, P.	2005	Reffat, R. M.	2010	Kong, L., &	2009	Song, J., Zhang, J., &	2009	Nam, T., & Pardo, T. A.	2011
	Bassens, D.	2012			Wei, L. W., & Chen, Y.	2009	Lagerkvist, A.	2010	Zhu, Z.	2011
						2012				
Singapore	Friedmann, J.	1986	Heng, T. M., & Low, L.	1993	Landry, C., &	1995	Ishida, T.	2005	Al-Hader, M., & Rodzi,	2009
	Taylor, P. J.	2000	Edvinsson, L.	2006	Yusuf, S., &	2005	Horne, M., Thompson,	2007	Nam, T., & Pardo, T. A.	2011
	Hall, P.	2005	Hornidge, A-K.	2007	Wong, C. Y. L., Evans, G.	2006	Baum, S., Yigitcanlar,	2008	Zhu, Z.	2011
			Wong, C.	2008		2008	Shin, D., Nah, Y., Lee, I.-	2009		
			Ergazakis, E., Ergazakis,	2009	Kong, L., &	2009				
			Yigitcanlar, T.	2009						
			Reffat, R. M.	2010						
			Edvinsson, L.	2011						
			Carrillo, F. J.,	2014						

City Name	World/Global City Author	Year	Knowledge City Author	Year	Creative City Author	Year	Digital City Author	Year	Smart City Author	Year
Stockholm	Taylor, P.J.	2000	Edvinsson, L.	2006	Landry, C., &	1995			Abdoullaev, A.	2011
	Hall, P.	2005	Ergazakis, E., Ergazakis,	2009	Hospers, G.-J.	2003b			Nam, T., & Pardo, T. A.	2011
			Yigitcanlar, T.	2009					Shahrokini, H.,	2015
			Metaxiotis, K., &	2012						
Sydney	Friedmann, J.	1986	Yigitcanlar, T., &	2010	Baum, S.,	2008	Horne, M., Thompson,	2007	Abdoullaev, A.	2011
	Taylor, P.J.	2000	Hu, R.	2012						
	Hall, P.	2005								
Tokyo	Friedmann, J.	1986	Matthiessen, C. W.,	2006			Horne, M., Thompson,	2007		
	Taylor, P.J.	2000								
	Sassen, S.	2001								
	Hall, P.	2005								
Toronto	Friedmann, J.	1986	Evans, G.	2009	Ley, D.	2003	Horne, M., Thompson,	2007	Nam, T., & Pardo, T. A	2011
	Taylor, P.J.	2000			Gertler, M. S.	2004	Hampton, K. N., &	2008		
	Hall, P.	2005			Heywood, p.	2008				
					Catungal, J. P.	2009				
					Evans, G.	2009				
					Henry, R.	2010				
Vancouver	Hall, P.	2005	Yigitcanlar, T.	2012	Ley, D.	2003	Horne, M., Thompson,	2007	Hollands, R. G	2008
					Gertler, M. S.	2004			Nam, T., & Pardo, T. A.	2011
					Heywood, P.	2008				
Vienna	Friedmann, J.	1986	Trippl, M.	2005	Hospers, G.-J.	2012	Ishida, T.	2003b	Hofstetter, K., & Vogl,	2011
					Hall, P.	2004	Horne, M., Thompson,	2007		

Appendix III: Bike sharing

City	URL	Comments
Amsterdam	no	Bike renting: http://www.iamsterdam.com/en/visiting/plan-your-trip/getting-around/rental/bike-hire
Barcelona	https://www.bicing.cat/	
Beijing	yes - but no official website found	http://www.cbsnews.com/news/china-beijing-bike-share-more-lanes-aim-to-reduce-congestion-pollution/
Berlin	http://www.nextbike.de/de/berlin/	
Boston	https://www.thehubway.com/	
Chicago	http://www.divvybikes.com/	
Dubai	http://www.bykystations.com/en/dubai/	Available around tourist destinations (Pal Jumeirah and Burj Khalifa)
Frankfurt	http://www.nextbike.de/de/frankfurt	
Helsinki	https://www.hsl.fi/en/citybikes	
Hong Kong	no	Just as touristic attraction and freetime activity within a park http://www.westkowloon.hk/en/smartbike
Kuala Lumpur	no	First Bikelines are introduced https://www.theguardian.com/cities/2015/sep/18/how-crowd-sourced-map-kuala-lumpurs-ideas-cycling
London	https://tfl.gov.uk/modes/cycling/santander-cycles	
Los Angeles	https://bikeshare.metro.net	
Melbourne	http://www.melbournebikeshare.com.au	
Milan	https://www.bikemi.com	
Montreal	https://montreal.bixi.com/	Project: renting bikes with OPUS card which is the usual public transportation card which can be preloaded and used as payment
Munich	http://www.nextbike.de/en/muenchen/	
New York	https://www.citibikenyc.com/	
Paris	http://en.velib.paris.fr/	
San Francisco	http://www.bayareabikeshare.com/	
Sao Paulo	http://www.mobilicidade.com.br/bikesampa.asp	
Shenzhen	http://www.qfggzxc.com/index.php	
Seoul	https://www.bikeseoul.com:447/main.do?lang=en	

City	URL	Comments
Shanghai	http://www.chinarmb.com/	
Singapore	no	Prototype http://www.zaibike.com/
Stockholm	http://www.citybikes.se/home	
Sydney	no	
Tokyo	http://docomo-cycle.jp/minato/en/	
Toronto	https://toronto.bixi.com	
Vancouver	https://www.mobibikes.ca/	
Vienna	https://www.citybikewien.at/de/	

Appendix IV: Best ranked university in city

City	Institution	Academic Ranking of World Universities (Shanghai Ranking Consultancy)	QS World University Rankings	Center for World University Rankings (CWUR)
Boston	Massachusetts Institute of Technology	3	1	3
San Francisco	Stanford University	2	3	2
London	University College London	18	7	27
Chicago	University of Chicago	9	10	8
Singapore	National University of Singapore	101	12	65
New York	Columbia University	8	22	6
Paris	ENS - Paris	72	23	37
Montreal	McGill University	64	24	42
Beijing	Tsinghua University	101	25	78
Los Angeles	University of California, Los Angeles	12	27	15
Hong Kong	Hong Kong University of Science and Technology	201	28	319
Toronto	University of Toronto	25	34	32
Seoul	Seoul National University	101	36	24
Tokyo	University of Tokyo	21	39	13
Melbourne	University of Melbourne	44	42	93
Sydney	University of Sydney	101	45	88
Vancouver	University of British Columbia	40	50	62
Shanghai	Fudan University	151	51	195
Amsterdam	University of Amsterdam	101	55	114
Munich	Technical University of Munich	51	60	101
Stockholm	Royal Institute of Technology	201	92	126
Helsinki	University of Helsinki	67	96	111
Berlin	Free University of Berlin	301	119	147
Sao Paulo	University of Sao Paulo	101	143	132
Kuala Lumpur	University of Malaya	301	146	498
Vienna	University of Vienna	151	153	223
Barcelona	University of Barcelona	151	166	116
Milan	Polytechnic University of Milan	201	187	397

Source:
Academic Ranking of World Universities. (n.d.). University and college rankings list. Retrieved from http://www.shanghairanking.com/resources.html
Quacquarelli Symonds. (n.d.). Top universities. Retrieved from http://www.topuniversities.com/
O'Neill, M. (2016). World university rankings. Retrieved from https://www.kaggle.com/mylesoneill/world-university-rankings

Appendix V: Patents granted 2000–2012

Publication Year Basic	2000	2001	2002	2003	2004	2005	2006	2007	2008	2009	2010	2011	2012	Total 2000–2012
Dubai	n.v.	5	7	7	12	18	22	32	39	33	29	44	24	272
Frankfurt	8	24	33	43	19	53	52	47	54	36	38	50	19	476
Kuala Lumpur	2	12	18	15	25	37	33	47	69	65	91	86	39	539
Melbourne	37	50	51	55	56	57	80	80	78	59	44	46	9	702
Sydney	68	72	115	92	93	115	107	101	111	117	105	111	64	1271
Sao Paulo	24	54	76	95	106	123	140	160	151	180	148	139	70	1466
Vienna	58	127	177	215	249	178	223	220	212	196	207	164	100	2326
Montreal	384	351	185	164	168	171	165	215	200	167	171	148	30	2519
Barcelona	42	115	187	236	244	281	339	346	330	385	340	273	169	3287
Vancouver	110	214	360	433	410	455	535	524	605	558	500	489	374	5567
Amsterdam	106	207	327	351	471	731	820	677	605	563	494	344	233	5929
Milan	221	398	475	555	573	646	687	639	582	527	520	445	227	6495
Hong Kong	26	101	246	383	468	574	704	707	830	736	691	738	668	6872
Helsinki	312	444	518	627	570	670	658	670	729	648	605	563	437	7451
Berlin	270	479	678	713	636	731	831	845	878	938	862	762	365	8988
Shanghai	27	54	106	206	348	481	652	977	1057	1196	1458	1437	1031	9030
Stockholm	519	726	825	660	553	658	701	765	975	1086	1093	1113	607	10281
Toronto	215	385	602	747	759	849	1051	1083	1127	1044	950	1067	924	10803
Beijing	56	114	216	280	401	626	838	991	1397	1732	1903	1994	1535	12083
Boston	414	590	925	932	964	1165	1113	1149	1242	1121	1092	1051	853	12611
Singapore	161	298	704	987	964	1394	1531	1503	1527	1593	1625	1530	1318	15135
Los Angeles	278	532	975	1152	1218	1373	1487	1434	1557	1493	1488	1444	1475	15906
London	1321	1484	1573	1562	1480	1516	1460	1623	1665	1535	1537	1444	1048	19248
Shenzhen	13	28	80	205	367	549	1307	1966	2303	2901	4099	3887	2958	20663
Chicago	432	572	1082	1642	1770	1991	2111	2208	2280	2203	2060	2012	1989	22352

Publication Year	2000	2001	2002	2003	2004	2005	2006	2007	2008	2009	2010	2011	2012	Total 2000–2012
Basic														
Paris	891	1087	1681	2065	2103	2124	2269	2365	2453	2491	2609	2638	2706	27482
New York	826	1317	2201	2346	2426	2808	3036	3006	3197	2886	2797	2591	2699	32136
Munich	1257	2034	2834	3152	3076	3375	3507	3364	3199	3155	2490	2086	1225	34754
San Francisco	446	808	1584	2093	1806	2566	2834	3244	3775	3996	3932	4175	4702	35961
Tokyo	1545	2009	2417	2525	3025	3485	3021	3112	3445	4183	3687	4656	2323	39433
Seoul	827	1784	2892	3439	4894	6751	8622	9647	9405	8568	8555	8645	7163	81192

Appendix VI: Coworking spaces

City Name	Number of spaces	Source
Amsterdam	60	http://www.launchdesk.nl/en/desk-amsterdam?flexdesk=no&days=1&price=25-450&persons=1
Barcelona	31	http://barcelonanavigator.com/barcelona-co-working-spaces/
Beijing	7	http://www.labsterx.com/blog/beijing-coworking-spaces/
Berlin	93	http://www.berlin.de/projektzukunft/standortinformationen/coworking-spaces-in-berlin/
Boston	34	https://bostonstartupsguide.com/guide/boston-coworking-spaces-roundup/
Chicago	152	https://www.desktimeapp.com/
Dubai	10	https://www.wamda.com/2013/07/10-popular-coworking-spaces-in-the-arab-world https://www.quora.com/What-are-the-best-co-working-spaces-in-Dubai own search on google
Frankfurt	15	https://www.designoffices.de/standorte/frankfurt-westend/coworking-spaces/ http://www.the-office-frankfurt.de/ https://www.coworking.de own search on google https://www.yelp.de/search?find_loc=Frankfurt+am+Main,+Hessen&start=10&cflt=sharedofficespaces
Helsinki	15	http://www.helsinkibusinesshub.fi/article/find-your-coworking-space-in-helsinki/
Hong Kong	24	http://www.9amconsulting.com/coworking-spaces-hong-kong.html
Kuala Lumpur	19	http://www.startupblink.com/blog/kuala-lumpur-startup-ecosystem/ http://www.lifestyleasia.com/450062/thrive-10-co-working-spaces-around-kuala-lumpur-to-know/ http://www.nextupasia.com/9-awesome-coworking-space-around-kuala-lumpur-malaysia/ http://zafigo.com/stories/zafigo-stories/your-essential-guide-to-12-co-working-spaces-in-kuala-lumpur/
London	167	http://www.coworkinglondon.com/co-working-spaces/
Los Angeles	65	https://www.desktimeapp.com
Melbourne	38	http://www.creativespaces.net.au/find-a-space#!/?sort=newest&page=1&usage=567&municipal_ids%5B%5D=342
Milan	11	https://www.sharedesk.net/search/list/Milan-Italy http://www.coworking-news.de/european-coworking-directory/#italy

Appendix VI: Coworking spaces

City Name	Number of spaces	Source
Montreal	21	https://www.quora.com/Where-are-the-coworking-spaces-in-Montreal
Munich	13	http://www.coworking-news.de/coworking-verzeichnis/#muc http://www.t3hero.com/blog/die-besten-co-working-spaces-in-hamburg-muenchen-und-berlin.html https://www.sharedesk.net/search/list/Munich,+Germany
New York	50	https://blog.getkisi.com/top-coworking-spaces-in-nyc/
Paris	19	http://www.eu-startups.com/2016/01/overview-of-the-best-coworking-spaces-in-paris/ http://frenchtechhub.com/blog/2015/10/the-10-best-coworking-sites-in-the-paris-region/
San Francisco	48	http://wiki.coworking.org/w/page/16583935/SanFranciscoCoworking
Sao Paulo	109	https://coworkingbrasil.org/espacos/?city=S%C3%A3o+Paulo
Shenzhen	14	http://startuplivingchina.com/top10-best-coworking-spaces-in-china/ https://www.globalfromasia.com/shenzhen-startup-overview/ own google search
Seoul	6	https://www.coworker.com/south-korea/seoul
Shanghai	11	http://startuplivingchina.com/top10-best-coworking-spaces-in-china/ https://www.coworker.com/search/shanghai
Singapore	19	http://thehoneycombers.com/singapore/co-working-spaces-in-singapore-shared-offices-for-freelancers-budding-entrepreneurs-and-start-up-companies/#WkQqozlZ4VhL75O8.97
Stockholm	17	http://www.swedishwire.com/business/19627-top-co-working-spaces-in-stockholm- http://www.yourlivingcity.com/stockholm/community/networking-community/co-working-spaces/ https://www.routesnorth.com/stockholm/six-of-the-best-co-working-spaces-in-stockholm/ own search on google http://lukeryan.me/the-best-co-working-spaces-in-stockholm/
Sydney	38	https://www.coworker.com/search/sydney/australia http://wiki.coworking.org/w/page/16583720/CoworkingSydney
Tokyo	25	https://tokyocheapo.com/business/drop-in-coworking-spaces-tokyo/
Toronto	43	http://startupheretoronto.com/support-category/coworking-spaces/

Appendix VI: Coworking spaces

City Name	Number of spaces	Source
Vancouver	15	http://wiki.coworking.org/w/page/16583740/CoworkingVancouver own search on google http://www.bcliving.ca/5-best-coworking-spaces-in-metro-vancouver https://www.yelp.com/search?cflt=sharedofficespaces&find_loc=Vancouver%2C+WA
Vienna	26	https://goodnight.at/magazin/freizeit/216-co-working-spaces-wien own search on google http://wiki.coworking.org/w/page/16583742/CoworkingVienna

www.ingramcontent.com/pod-product-compliance
Lightning Source LLC
Chambersburg PA
CBHW062002220426
43662CB00010B/1203